SEMI-CLASSICAL ANALYSIS FOR NONLINEAR SCHRÖDINGER EQUATIONS

Rémi Carles

*CNRS & University of Montpellier 2,
France*

SEMI-CLASSICAL ANALYSIS FOR NONLINEAR SCHRÖDINGER EQUATIONS

World Scientific

NEW JERSEY · LONDON · SINGAPORE · BEIJING · SHANGHAI · HONG KONG · TAIPEI · CHENNAI

Published by

World Scientific Publishing Co. Pte. Ltd.

5 Toh Tuck Link, Singapore 596224

USA office: 27 Warren Street, Suite 401-402, Hackensack, NJ 07601

UK office: 57 Shelton Street, Covent Garden, London WC2H 9HE

British Library Cataloguing-in-Publication Data
A catalogue record for this book is available from the British Library.

ISBN-13 978-981-279-312-6
ISBN-10 981-279-312-7

Printed in Singapore.

Preface

These pages describe the semi-classical limit for nonlinear Schrödinger equations in the presence of an external potential. The motivation of this study is two-fold. First, it is expected to provide interesting models for physics. For instance, the nonlinear Schrödinger equation is a common model for Bose–Einstein condensation. To describe the physical phenomenon, qualitative properties of the solutions of these equations may be helpful. According to the various régimes considered, different asymptotic behaviors associated to the equations may be interesting. One of them is the semi-classical limit, where the behavior of the wave function as the (rescaled) Planck constant goes to zero is studied. On the other hand, this study also has purely analytical motivations. It is well-known that the semi-classical limit (also called geometrical optics) yields useful information for problems related to functional analysis, even when there is no Planck constant in the initial problem. For instance, such methods have proven efficient in the construction of parametrices or in studying the propagation of singularities. In these notes, we emphasize some applications of WKB analysis for the study of qualitative properties of super-critical nonlinear Schrödinger equations.

The book consists of two parts. The first one is dedicated to the WKB methods and the semi-classical limit before the formation of caustics. The second part treats the semi-classical limit in the presence of caustics in the special geometric case where the caustic is reduced to a point (or to several isolated points). Both parts are essentially independent. The first part may be viewed simply as a motivation for the second part. Most of the content of the second part does not refer to WKB analysis. Also, the technical aspects are fairly independent in the two parts. The first part relies greatly on techniques coming from the study of hyperbolic partial differential equations, and especially, from quasi-linear ones. The second part is more typical of nonlinear Schrödinger equations in a way, since it involves scattering theory, as well as Strichartz estimates related to Schrödinger equations.

These lecture notes correspond to an extended version of a course given in Vienna at the Wolfgang Pauli Institute during a workshop organized by Jean-Claude Saut in May 2007, and in Beijing in October 2007, during a special semester organized by Ping Zhang at the Morningside Center of Mathematics of the Chinese Academy of Sciences. The goal of this course was to give an overview of the current knowledge and techniques in the study of the semi-classical limit for nonlinear Schrödinger equations. These notes are a good opportunity to fix (some of) the typographical errors that have remained in the papers written by the author, and to present several results in a unified way. We have also tried to point out some interesting open questions, especially when they appear natural in the course of the text. Since the course was addressed to researchers and to graduate students, the present text is essentially self-contained, and only for technical details and further developments have we chosen to direct the reader to the original references.

I wish to thank warmly Thomas Alazard for his careful reading of the initial version of the manuscript.

General Notations

Functions

By default, the functions that we consider are complex-valued.

The space variable, denoted by x, belongs to $\mathbb{R}^n$. The time variable is denoted by t.

The partial derivatives with respect to the time variable and to the j-th space variable are denoted by ∂_t and ∂_j, respectively.

We denote by Λ the Fourier multiplier $(\mathrm{Id} - \Delta)^{1/2}$, where Δ stands for the Laplacian

$$\Delta = \sum_{j=1}^{n} \partial_j^2.$$

Function spaces

We denote by $L^p(\mathbb{R}^n)$, or simply L^p, the usual Lebesgue spaces on $\mathbb{R}^n$. The inner product of $L^2(\mathbb{R}^n)$ is defined as

$$\langle f, g \rangle = \int_{\mathbb{R}^n} f(x) \overline{g}(x) dx.$$

Consider $f = f(t, x)$ a function from $I \times \mathbb{R}^n$ to $\mathbb{C}$, where I is a time interval. If $f \in C(I; L^p(\mathbb{R}^n))$, we write

$$\|f\|_{L^\infty(I;L^p)} = \sup_{t \in I} \|f(t)\|_{L^p(\mathbb{R}^n)}.$$

The Schwartz class of smooth functions $\mathbb{R}^n \to \mathbb{C}$ which decay rapidly as well as all their derivatives is denoted by $\mathcal{S}(\mathbb{R}^n)$.

For $f \in \mathcal{S}(\mathbb{R}^n)$, we define its Fourier transform by

$$\widehat{f}(\xi) = \mathcal{F}f(\xi) = \frac{1}{(2\pi)^{n/2}} \int_{\mathbb{R}^n} e^{-ix\cdot\xi} f(x) dx,$$

so that the inverse Fourier transform is given by

$$\mathcal{F}^{-1} f(x) = \frac{1}{(2\pi)^{n/2}} \int_{\mathbb{R}^n} e^{ix\cdot\xi} f(\xi) d\xi.$$

For $s \geqslant 0$, we define the Sobolev space $H^s(\mathbb{R}^n) = H^s$ as

$$H^s(\mathbb{R}^n) = \left\{ f \in \mathcal{S}'(\mathbb{R}^n) \; ; \; \xi \mapsto \langle\xi\rangle^s \widehat{f}(\xi) \in L^2(\mathbb{R}^n) \right\},$$

where we have denoted $\langle\xi\rangle = (1 + |\xi|^2)^{1/2}$. Note that if $s \in \mathbb{N}$, then

$$H^s(\mathbb{R}^n) = \left\{ f \in L^2(\mathbb{R}^n) \; ; \; \partial^\alpha f \in L^2(\mathbb{R}^n), \; \forall \alpha \in \mathbb{N}^n, \; |\alpha| \leqslant s \right\}.$$

Recall that if $s > n/2$, then $H^s(\mathbb{R}^n)$ is an algebra, and $H^s(\mathbb{R}^n) \subset L^\infty(\mathbb{R}^n)$.

The set $H^\infty(\mathbb{R}^n)$, or simply H^∞, is the intersection of all these spaces:

$$H^\infty = \cap_{s \geqslant 0} H^s(\mathbb{R}^n).$$

This is a Fréchet space, equipped with the distance

$$d(f,g) = \sum_{s \in \mathbb{N}} 2^{-s} \frac{\|f - g\|_{H^s}}{1 + \|f - g\|_{H^s}}.$$

Semi-classical limit

The dependence of functions upon the semi-classical parameter ε is denoted by a superscript. For instance, the wave function is denoted by u^ε.

All the irrelevant constants are denoted by C. In particular, C stands for a constant which is independent of ε, the semi-classical parameter.

Let $(\alpha^h)_{0 < h \leqslant 1}$ and $(\beta^h)_{0 < h \leqslant 1}$ be two families of positive real numbers.

- We write $\alpha^h \ll \beta^h$, or $\alpha^h = o(\beta^h)$, if $\limsup_{h \to 0} \alpha^h / \beta^h = 0$.
- We write $\alpha^h \lesssim \beta^h$, or $\alpha^h = \mathcal{O}(\beta^h)$, if $\limsup_{h \to 0} \alpha^h / \beta^h < \infty$.
- We write $\alpha^h \approx \beta^h$ if $\alpha^h \lesssim \beta^h$ and $\beta^h \lesssim \alpha^h$.

If u^h and v^h are functions, we write $u^h \approx v^h$ if $\|u^h - v^h\| \ll \|v^h\|$, for some norm to be precised (or not, when computations are purely formal).

Contents

Caustic Crossing: The Case of Focal Points   109

PART I
WKB Analysis

Chapter 1

Preliminary Analysis

We consider nonlinear Schrödinger equations in the presence of a parameter $\varepsilon \in]0, 1]$,

$$i\varepsilon \partial_t u^\varepsilon + \frac{\varepsilon^2}{2}\Delta u^\varepsilon = V u^\varepsilon + f\left(|u^\varepsilon|^2\right) u^\varepsilon, \qquad (1.1)$$

where $u^\varepsilon = u^\varepsilon(t, x)$ is complex-valued. Throughout this book, the space variable, denoted by x, lies in the whole Euclidean space $\mathbb{R}^n$, $n \geqslant 1$. Many of the results presented in this first part can easily be adapted to the case of the torus $\mathbb{T}^n$. The external potential $V = V(t, x)$ and the (local) nonlinearity f are supposed to be smooth, *real-valued*, and independent of ε. The aim of these notes is to describe some results about the asymptotic behavior of the solution u^ε as the parameter ε goes to zero. We shall be more precise about the initial data that we consider below. The nonlinearity f is *local* (e.g. power-like nonlinearity): in particular, we choose not to mention results related to nonlocal nonlinearities, such as the Schrödinger–Poisson system

$$i\varepsilon \partial_t u^\varepsilon + \frac{\varepsilon^2}{2}\Delta u^\varepsilon = V u^\varepsilon + V_p u^\varepsilon \quad ; \quad \Delta V_p = \lambda\left(|u^\varepsilon|^2 - c\right),$$

or the Hartree equation

$$i\varepsilon \partial_t u^\varepsilon + \frac{\varepsilon^2}{2}\Delta u^\varepsilon = V u^\varepsilon + \lambda\left(\frac{1}{|x|^\gamma} * |u^\varepsilon|^2\right) u^\varepsilon.$$

We do not consider ε-dependent potential either, an issue for which the main model we have in mind is that of a lattice periodic potential, whose period is of order ε:

$$i\varepsilon \partial_t u^\varepsilon + \frac{\varepsilon^2}{2}\Delta u^\varepsilon = V u^\varepsilon + V_\Gamma\left(\frac{x}{\varepsilon}\right) u^\varepsilon + f\left(|u^\varepsilon|^2\right) u^\varepsilon,$$

where the potential V_Γ is periodic with respect to some regular lattice $\Gamma \simeq \mathbb{Z}^n$. See for instance [Bensoussan *et al.* (1978); Robert (1998); Teufel

(2003)] for an introduction to the asymptotic study in the linear case of the above equation, and [Carles *et al.* (2004)] for an example of asymptotic behavior in a nonlinear régime. Our choice is to focus on (1.1), and to describe as precisely as possible the variety of known phenomena in the limit $\varepsilon \to 0$.

There are several reasons to study the asymptotic behavior of u^ε in the semi-classical limit $\varepsilon \to 0$. Let us mention two. First, (1.1) with $f(|u|^2)u = |u|^4 u$ (quintic nonlinearity) is sometimes used as a model for one-dimensional Bose–Einstein condensation in space dimension $n = 1$ ([Kolomeisky *et al.* (2000)]). When $n = 2$ or 3, a cubic nonlinearity, $f(|u|^2)u = |u|^2 u$, is usually considered. The external potential V can be an harmonic potential (isotropic or anisotropic), or a lattice periodic potential (see e.g. [Dalfovo *et al.* (1999); Pitaevskii and Stringari (2003)]). According to the different physical parameters at stake, the asymptotic behavior of u^ε as $\varepsilon \to 0$ may provide relevant informations to describe u^ε itself. This approach is similar to the theory of geometric optics, developed initially to describe the propagation of electro-magnetic waves, such as light. In that context, the propagation of the wave is also described by partial differential equations, and ε usually corresponds to a wavelength, which is small compared to the other parameters. For Maxwell's equations, ε corresponds to the inverse of the speed of light. We invite the reader to consult [Rauch and Keel (1999)] for an overview of this theory, mainly in the context of hyperbolic equations. We shall not develop further on the physical motivations, but rather focus our attention on the mathematical aspects. The term "geometric optics" means that it is expected that the propagation of light is accurately described by rays. For Schrödinger equations, the analogue of this notion is usually called "classical trajectories". These notions are identical, and follow from the notion of bicharacteristic curves. As a consequence, the limit $\varepsilon \to 0$ relates classical and quantum wave equations. In particular, the semi-classical limit $\varepsilon \to 0$ for u^ε is expected to be described by the laws of hydrodynamics. We will come back to this aspect more precisely later.

Another motivation lies in the study the Cauchy problem for nonlinear Schrödinger equations with no small parameter ($V \equiv 0$ and $\varepsilon = 1$ in (1.1), typically). One can prove ill-posedness results for energy-supercritical equations by reducing the problem to semi-classical analysis for (1.1). This aspect is discussed in details in Sec. 5.1 and Sec. 5.2. Note that the application of the theory of geometric optics to functional analysis has a long history.

In [Lax (1957)], it was used to construct parametrices. It has also been used to study the propagation of singularities (see e.g. [Taylor (1981)]), or of quasi-singularities [Cheverry (2005)]. In the case of Schrödinger equations, semi-classical analysis has proven useful for instance in control theory [Lebeau (1992)], in the proof of Strichartz estimates [Burq *et al.* (2004)], and in the propagation of singularities for the nonlinear equation [Szeftel (2005)].

We underscore the fact that the WKB analysis for (nonlinear) Schrödinger equations is rather specific to this equation. An important feature is the fact that for gauge invariant nonlinearities, it is possible to describe the solution with one phase and one harmonic only, provided that the initial data are of this form: $u^\varepsilon \approx a e^{i\phi/\varepsilon}$. For several other equations (e.g. Maxwell equations), the analysis is rather different, even on the algebraic level. We invite the reader to consult for instance [Joly *et al.* (1996b); Métivier (2004b); Rauch and Keel (1999); Whitham (1999)], and references therein, to have an idea of the important results for equations different from the Schrödinger equation. However, the general framework presented in §1.1 (derivation of the equation, and steps toward a justification) is not specific to the equation: the main specificity of gauge invariant nonlinear Schrödinger equations (as in Eq. (1.1)) is that the equations derived at the formal step look simpler than for other equations, due to the fact that we work with only one phase (and one harmonic).

Before introducing the approach developed in this first part, we present two basic results, which will be used throughout these notes.

Lemma 1.1 (Gronwall lemma and a continuity argument).
(1) *Let* $u, a, b \in C([0,T]; \mathbb{R}_+)$ *be such that*

$$u(t) \leqslant u(0) + \int_0^t a(\tau)u(\tau)d\tau + \int_0^t b(\tau)d\tau, \quad \forall t \in [0,T].$$

Denote $A(t) = \int_0^t a(\tau)d\tau$. *Then*

$$u(t) \leqslant u(0)e^{A(t)} + \int_0^t b(s)e^{A(t)-A(s)}ds, \quad \forall t \in [0,T].$$

(2) *Let* $u, b \in C([0,T]; \mathbb{R}_+)$ *and* $f \in C(\mathbb{R}_+; \mathbb{R}_+)$ *such that*

$$u(t) \leqslant u(0) + \int_0^t f(u(\tau))u(\tau)d\tau + \int_0^t b(\tau)d\tau, \quad \forall t \in [0,T].$$

Let $M = \sup\{f(v); \ v \in [0, 2u(0)]\}$. *There exists* $\underline{t} \in]0,T]$ *such that*

$$u(t) \leqslant u(0)e^{Mt} + \int_0^t b(s)e^{M(t-s)}ds, \quad \forall t \in [0, \underline{t}].$$

Proof.   (1) Denote

$$w(t) = u(0) + \int_0^t a(\tau)u(\tau)d\tau + \int_0^t b(\tau)d\tau.$$

By assumption, $w \in C^1([0,T])$ and $w'(t) = a(t)u(t) + b(t) \leqslant a(t)w(t) + b(t)$. Therefore,

$$\left(w(t)e^{-A(t)}\right)' \leqslant b(t)e^{-A(t)},$$

and the first point follows by integrating this inequality, since $u(t) \leqslant w(t)$. (2) Suppose that there exists $t \in]0,T]$ such that $u(t) > 2u(0)$. Since u is continuous, we can define

$$\underline{t} = \min\{\tau \in [0,T]; \ u(\tau) = 2u(0)\} > 0.$$

The assumption implies

$$u(t) \leqslant u(0) + M \int_0^t u(\tau)d\tau + \int_0^t b(\tau)d\tau, \quad \forall t \in [0,\underline{t}].$$

Gronwall lemma then yields

$$u(t) \leqslant u(0)e^{Mt} + \int_0^t b(s)e^{M(t-s)}ds, \quad \forall t \in [0,\underline{t}].$$

The right hand side is continuous, and is equal to $u(0)$ for $t = 0$. Up to decreasing $\underline{t}$, this right hand side does not exceed $2u(0)$ for $t \in [0,\underline{t}]$, hence the conclusion of the lemma.

If $u(t) \leqslant 2u(0)$ for all $t \in [0,T]$, then we can trivially take $\underline{t} = T$.   □

Lemma 1.2 (Basic energy estimate). *For $\varepsilon > 0$, consider $\mathbf{u}^\varepsilon$ solving*

$$i\varepsilon\partial_t\mathbf{u}^\varepsilon + \frac{\varepsilon^2}{2}\Delta\mathbf{u}^\varepsilon = F^\varepsilon\mathbf{u}^\varepsilon + R^\varepsilon \quad ; \quad \mathbf{u}^\varepsilon_{|t=0} = \mathbf{u}^\varepsilon_0. \tag{1.2}$$

Assume that $F^\varepsilon = F^\varepsilon(t,x)$ is real-valued. Let I be a time interval such that $0 \in I$. Then we have, at least formally:

$$\sup_{t \in I}\|\mathbf{u}^\varepsilon(t)\|_{L^2} \leqslant \|\mathbf{u}^\varepsilon_0\|_{L^2} + \frac{1}{\varepsilon}\int_I\|R^\varepsilon(\tau)\|_{L^2}\,d\tau.$$

Proof.   Since the statement is formal, so is the proof. Multiply (1.2) by $\overline{\mathbf{u}}^\varepsilon$, and integrate over $\mathbb{R}^n$:

$$i\varepsilon\int_{\mathbb{R}^n}\overline{\mathbf{u}}^\varepsilon\partial_t\mathbf{u}^\varepsilon dx + \frac{\varepsilon^2}{2}\int_{\mathbb{R}^n}\overline{\mathbf{u}}^\varepsilon\Delta\mathbf{u}^\varepsilon dx = \int_{\mathbb{R}^n}F^\varepsilon|\mathbf{u}^\varepsilon|^2 dx + \int_{\mathbb{R}^n}\overline{\mathbf{u}}^\varepsilon R^\varepsilon dx.$$

Taking the imaginary part, the second term of the left hand side vanishes, since Δ is self-adjoint. Similarly, since F^ε is real-valued, the first term of the right hand side disappears, and we have:

$$\varepsilon\frac{d}{dt}\int_{\mathbb{R}^n}|\mathbf{u}^\varepsilon|^2 = \varepsilon\int_{\mathbb{R}^n}\partial_t|\mathbf{u}^\varepsilon|^2 = 2\operatorname{Im}\int_{\mathbb{R}^n}\overline{\mathbf{u}}^\varepsilon R^\varepsilon.$$

Cauchy–Schwarz inequality yields

$$\varepsilon\frac{d}{dt}\|\mathbf{u}^\varepsilon\|_{L^2}^2 \leqslant 2\|\mathbf{u}^\varepsilon\|_{L^2}\|R^\varepsilon\|_{L^2}.$$

Let $\delta > 0$. We infer from the above inequality:

$$\varepsilon\frac{d}{dt}\left(\|\mathbf{u}^\varepsilon\|_{L^2}^2 + \delta\right) \leqslant 2\left(\|\mathbf{u}^\varepsilon\|_{L^2}^2 + \delta\right)^{1/2}\|R^\varepsilon\|_{L^2}.$$

Since $\|\mathbf{u}^\varepsilon\|_{L^2}^2 + \delta \geqslant \delta > 0$, we can simplify:

$$\varepsilon\frac{d}{dt}\left(\|\mathbf{u}^\varepsilon\|_{L^2}^2 + \delta\right)^{1/2} \leqslant \|R^\varepsilon\|_{L^2}.$$

Integration with respect to time yields, for $t \in I$:

$$\varepsilon\left(\|\mathbf{u}^\varepsilon(t)\|_{L^2}^2 + \delta\right)^{1/2} \leqslant \varepsilon\left(\|\mathbf{u}_0^\varepsilon\|_{L^2}^2 + \delta\right)^{1/2} + \int_I\|R^\varepsilon(t)\|_{L^2}dt.$$

The lemma follows by letting $\delta \to 0$. $\qquad\qquad\qquad\qquad\qquad\square$

1.1 General presentation

The general approach of WKB expansions (after three papers by Wentzel, Kramers and Brillouin respectively, in 1926) consists of mainly three steps. The first step, which is described in more details in this section, consists in seeking a function v^ε that solves (1.1) up to a small error term:

$$i\varepsilon\partial_t v^\varepsilon + \frac{\varepsilon^2}{2}\Delta v^\varepsilon = Vv^\varepsilon + f\left(|v^\varepsilon|^2\right)v^\varepsilon + r^\varepsilon,$$

where r^ε should be thought of as a "small" (as $\varepsilon \to 0$) source term. Typically, we require

$$\|r^\varepsilon\|_{L^\infty([-T,T];L^2)} = \mathcal{O}\left(\varepsilon^N\right)$$

for some $T > 0$ independent of ε, and $N > 0$ as large as possible. In this first step, we derive equations that define v^ε, which are hopefully simpler than (1.1). The second step consists in showing that such a v^ε actually exists, that is, in solving the equations derived in the first step. The last step is the study of *stability* (or *consistency*): even if r^ε is small, it is not clear that $u^\varepsilon - v^\varepsilon$ is small too. Typically, we try to prove an error estimate of the form

$$\|u^\varepsilon - v^\varepsilon\|_{L^\infty([-T,T];L^2)} = \mathcal{O}\left(\varepsilon^K\right)$$

for some $K > 0$ (possibly smaller than N). Note also that for the nonlinear problem (1.1), it is not even clear from the beginning that an L^2 solution can be constructed on a time interval independent of $\varepsilon \in]0,1]$.

The initial data that we consider for WKB analysis are of the form

$$u^\varepsilon(0,x) = \varepsilon^\kappa a_0^\varepsilon(x) e^{i\phi_0(x)/\varepsilon}. \tag{1.3}$$

The phase ϕ_0 is independent of ε and real-valued. The initial amplitude a_0^ε is complex-valued, and may have an asymptotic expansion as $\varepsilon \to 0$,

$$a_0^\varepsilon(x) \underset{\varepsilon \to 0}{\sim} a_0(x) + \varepsilon a_1(x) + \varepsilon^2 a_2(x) + \dots, \tag{1.4}$$

in the sense of formal asymptotic expansions, where the profiles a_j are independent of ε. Note that ε^κ then measures the size of $u^\varepsilon(0,x)$ in $L^\infty(\mathbb{R}^n)$. We shall always consider cases where $\kappa \geqslant 0$. When the nonlinearity is non-trivial, $f \neq 0$, the asymptotic behavior of u^ε as $\varepsilon \to 0$ strongly depends on the value of κ, as is discussed below. An important feature of Schrödinger equations with gauge invariant nonlinearities like in (1.1) is that if the initial data are of the form (1.3), then for small time at least (before caustics), the solution u^ε is expected to keep the same form, at least approximately:

$$u^\varepsilon(t,x) \underset{\varepsilon \to 0}{\sim} \varepsilon^\kappa a^\varepsilon(t,x) e^{i\phi(t,x)/\varepsilon}, \tag{1.5}$$

where a^ε is expected to have an asymptotic expansion as well. This is in sharp contrast with the analogous problems for hyperbolic equations (e.g. Maxwell, wave, Euler): typically, because the solutions of the wave equations are real-valued, the factor $e^{i\phi_0/\varepsilon}$ is replaced, say, by $2\cos(\phi_0/\varepsilon) = e^{i\phi_0/\varepsilon} + e^{-i\phi_0/\varepsilon}$. By nonlinear interaction, other phases are expected to appear, like $e^{ik\phi/\varepsilon}$, $k \in \mathbb{Z}$, for instance. This can be guessed by looking at the first iterates of a Picard's scheme. Moreover, phases different from ϕ might be involved in the description of u^ε, by nonlinear mechanisms too. We will see that unlike for these models, such a phenomenon is ruled out for nonlinear Schrödinger equations, provided that only one phase is considered initially, see (1.3). This is an important geometric feature in this study. On the other hand, studying the asymptotic behavior of u^ε whose initial data are *sums* of initial data as in (1.3) is an interesting open question so far.

To describe the expected influence of the parameter κ on the asymptotic behavior of u^ε, assume that the nonlinearity f is homogeneous:

$$f\left(|u^\varepsilon|^2\right)u^\varepsilon = \lambda|u^\varepsilon|^{2\sigma}u^\varepsilon, \quad \lambda \in \mathbb{R}, \ \sigma > 0.$$

The case $\sigma \in \mathbb{N} \setminus \{0\}$ corresponds to a smooth nonlinearity. Even though the parameter κ may be viewed as a measurement of the size of the (initial) wave function, we shall rather consider data of order $\mathcal{O}(1)$, by introducing $\widetilde{u}^\varepsilon = \varepsilon^{-\kappa}u^\varepsilon$. Dropping the tildes, we therefore consider

$$i\varepsilon\partial_t u^\varepsilon + \frac{\varepsilon^2}{2}\Delta u^\varepsilon = Vu^\varepsilon + \lambda\varepsilon^\alpha|u^\varepsilon|^{2\sigma}u^\varepsilon \quad ; \quad u^\varepsilon(0,x) = a_0^\varepsilon(x)e^{i\phi_0(x)/\varepsilon}, \quad (1.6)$$

where $\alpha = 2\sigma\kappa \geqslant 0$.

1.2 Formal derivation of the equations

Assuming that the initial data have an asymptotic expansion of the form (1.4), we seek $u^\varepsilon(t,x) \sim a^\varepsilon(t,x)e^{i\phi(t,x)/\varepsilon}$, with

$$a^\varepsilon(t,x) \underset{\varepsilon\to 0}{\sim} a(t,x) + \varepsilon a^{(1)}(t,x) + \varepsilon^2 a^{(2)}(t,x) + \dots$$

We use the convention $a^{(0)} = a$. On a formal level at least, the general idea consists in plugging this asymptotic expansion into (1.6), and then ordering in powers of ε. The lowest powers are the ones we really want to cancel, and if we are left with some extra terms, we want to be able to consider them as small source terms in the limit $\varepsilon \to 0$ (by a perturbative analysis for instance). To summarize, we first find $b^{(0)}, b^{(1)}, \dots$, such that

$$i\varepsilon\partial_t u^\varepsilon + \frac{\varepsilon^2}{2}\Delta u^\varepsilon - Vu^\varepsilon - \lambda\varepsilon^\alpha|u^\varepsilon|^{2\sigma}u^\varepsilon \underset{\varepsilon\to 0}{\sim} \left(b^{(0)} + \varepsilon b^{(1)} + \varepsilon^2 b^{(2)} + \dots\right)e^{i\phi/\varepsilon}.$$

Then we consider the equations $b^{(0)} = 0$, $b^{(1)} = 0$, etc. Note that this makes sense provided that $\alpha \in \mathbb{N}$, for otherwise, non-integer powers of ε appear in the above right hand side.

Denoting by ∂ a differentiation with respect to the time variable, or any space variable, we compute formally:

$$\partial u^\varepsilon \underset{\varepsilon\to 0}{\sim} \left(i\varepsilon^{-1}\left(a + \varepsilon a^{(1)} + \varepsilon^2 a^{(2)} + \dots\right)\partial\phi\right.$$
$$\left. + \partial a + \varepsilon\partial a^{(1)} + \varepsilon^2\partial a^{(2)} + \dots\right)e^{i\phi/\varepsilon}.$$

Similarly, for $1 \leqslant j \leqslant n$,

$$\partial_j^2 u^\varepsilon \underset{\varepsilon \to 0}{\sim} \Big(-\varepsilon^{-2} \left(a + \varepsilon a^{(1)} + \varepsilon^2 a^{(2)} + \ldots \right) (\partial_j \phi)^2$$
$$+ i\varepsilon^{-1} \left(a + \varepsilon a^{(1)} + \varepsilon^2 a^{(2)} + \ldots \right) \partial_j^2 \phi$$
$$+ 2i\varepsilon^{-1} \left(\partial_j a + \varepsilon \partial_j a^{(1)} + \varepsilon^2 \partial_j a^{(2)} + \ldots \right) \partial_j \phi$$
$$+ \partial_j^2 a + \varepsilon \partial_j^2 a^{(1)} + \varepsilon^2 \partial_j^2 a^{(2)} + \ldots \Big) e^{i\phi/\varepsilon}.$$

Ordering in powers of ε, we infer:

$$i\varepsilon \partial_t u^\varepsilon + \frac{\varepsilon^2}{2} \Delta u^\varepsilon \underset{\varepsilon \to 0}{\sim} \Bigg(-\left(\partial_t \phi + \frac{1}{2} |\nabla \phi|^2 \right) \left(a + \varepsilon a^{(1)} + \varepsilon^2 a^{(2)} + \ldots \right)$$
$$+ i\varepsilon \left(\partial_t a + \nabla \phi \cdot \nabla a + \frac{1}{2} a \Delta \phi \right)$$
$$+ i\varepsilon^2 \left(\partial_t a^{(1)} + \nabla \phi \cdot \nabla a^{(1)} + \frac{1}{2} a^{(1)} \Delta \phi - \frac{i}{2} \Delta a \right)$$
$$\vdots$$
$$+ i\varepsilon^{j+1} \left(\partial_t a^{(j)} + \nabla \phi \cdot \nabla a^{(j)} + \frac{1}{2} a^{(j)} \Delta \phi - \frac{i}{2} \Delta a^{(j-1)} \right)$$
$$+ \ldots \Bigg) e^{i\phi/\varepsilon}.$$

For the nonlinear term, we choose to compute only the first two terms:

$$|u^\varepsilon|^{2\sigma} u^\varepsilon \underset{\varepsilon \to 0}{\sim} \left(|a|^{2\sigma} a + \varepsilon \left(|a|^{2\sigma} a^{(1)} + 2\sigma \operatorname{Re}\left(\overline{a} a^{(1)} \right) |a|^{2\sigma-2} a \right) + \ldots \right) e^{i\phi/\varepsilon}.$$

To simplify the discussion, assume in the following lines that α is an integer, $\alpha \in \mathbb{N}$. Since we want to consider a leading order amplitude a which is not identically zero, it is natural to demand, for the term of order ε^0:

$$\partial_t \phi + \frac{1}{2} |\nabla \phi|^2 + V = \begin{cases} 0 & \text{if } \alpha > 0, \\ -\lambda |a|^{2\sigma} & \text{if } \alpha = 0. \end{cases} \tag{1.7}$$

For the term of order ε^1, we find:

$$\partial_t a + \nabla \phi \cdot \nabla a + \frac{1}{2} a \Delta \phi = \begin{cases} 0 & \text{if } \alpha > 1, \\ -i\lambda |a|^{2\sigma} a & \text{if } \alpha = 1, \\ -2i\lambda\sigma \operatorname{Re}\left(\overline{a} a^{(1)} \right) |a|^{2\sigma-2} a & \text{if } \alpha = 0. \end{cases} \tag{1.8}$$

Before giving a rigorous meaning to this approach, we comment on these cases. Intuitively, the larger the α, the smaller the influence of the nonlinearity: for large α, the nonlinearity is not expected to be relevant at leading order as $\varepsilon \to 0$. In terms of the problem (1.1)–(1.3), this means that small initial waves (large κ) evolve linearly at leading order: this corresponds to the general phenomenon that very small nonlinear waves behave linearly at leading order. Here, we see that if $\alpha > 1$, then ϕ and a solve equations which are independent of λ, hence of the nonlinearity. Since at leading order, we expect

$$u^\varepsilon(t, x) \underset{\varepsilon \to 0}{\sim} a(t, x) e^{i\phi(t,x)/\varepsilon},$$

this means that the leading order behavior of u^ε is linear. As a consequence, we also expect

$$u^\varepsilon(t, x) \underset{\varepsilon \to 0}{\sim} u_{\text{lin}}^\varepsilon(t, x),$$

where $u_{\text{lin}}^\varepsilon$ solves the linear problem

$$i\varepsilon \partial_t u_{\text{lin}}^\varepsilon + \frac{\varepsilon^2}{2} \Delta u_{\text{lin}}^\varepsilon = V u_{\text{lin}}^\varepsilon \quad ; \quad u_{\text{lin}}^\varepsilon(0, x) = u^\varepsilon(0, x) = a_0^\varepsilon(x) e^{i\phi_0(x)/\varepsilon}.$$

Decreasing the value of α, the critical threshold corresponds to $\alpha = 1$: the nonlinearity shows up in the equation for a, but not in the equation for ϕ. This régime is referred to as *weakly nonlinear geometric optics*. The term "weakly" means that the phase ϕ is determined independently of the nonlinearity: the equations for a and ϕ are decoupled. We will see that for $\alpha \geqslant 1$, the equation for a can be understood as a transport equation along the classical trajectories (rays of geometric optics) associated to ϕ, which in turn are determined by the initial phase ϕ_0 and the semi-classical Hamiltonian

$$\tau + \frac{|\xi|^2}{2} + V(t, x).$$

See Sec. 1.3.1 below.

The case $\alpha = 0$ is supercritical, and contains several difficulties. We point out two of those, which show that dealing with the supercritical case requires a different approach. First, the equation for the phase involves the amplitude a. But to solve the equation for a, it seems necessary to know $a^{(1)}$. One could continue the expansion in powers of ε at arbitrarily high order: no matter how many terms are included, the system is never closed. This aspect is a general feature of supercritical geometric optics (see also [Cheverry (2005, 2006); Cheverry and Guès (2007)]). The second difficulty concerns the *stability* analysis. We have claimed that the general approach consists in computing $\phi, a, a^{(1)}, \ldots$, so that

$$u_\ell^\varepsilon(t,x) := \left(a(t,x) + \varepsilon a^{(1)}(t,x) + \ldots + \varepsilon^\ell a^{(\ell)}(t,x) \right) e^{i\phi(t,x)/\varepsilon}$$

solves (1.6) up to a small error term. Typically (recall that $\alpha = 0$),

$$i\varepsilon \partial_t u_\ell^\varepsilon + \frac{\varepsilon^2}{2} \Delta u_\ell^\varepsilon = V u_\ell^\varepsilon + \lambda |u_\ell^\varepsilon|^{2\sigma} u_\ell^\varepsilon + \varepsilon^\ell r_\ell^\varepsilon,$$

where r_ℓ^ε is bounded in a space "naturally" associated to the study of (1.6). When working in spaces based on the conservation of the L^2 norm for nonlinear Schrödinger equations (see Sec. 1.4), we expect estimates in $L^\infty([0,T]; L^2(\mathbb{R}^n))$ for some $T > 0$ independent of $\varepsilon \in]0,1]$. Suppose that we have managed to construct such an approximate solution u_ℓ^ε. Assume for simplicity that u^ε and u_ℓ^ε coincide at time $t = 0$. Setting $w_\ell^\varepsilon = u^\varepsilon - u_\ell^\varepsilon$, we have:

$$i\varepsilon \partial_t w_\ell^\varepsilon + \frac{\varepsilon^2}{2} \Delta w_\ell^\varepsilon = V w_\ell^\varepsilon + \lambda \left(|u^\varepsilon|^{2\sigma} u^\varepsilon - |u_\ell^\varepsilon|^{2\sigma} u_\ell^\varepsilon \right) - \varepsilon^\ell r_\ell^\varepsilon.$$

Suppose that u^ε and u_ℓ^ε remain bounded in $L^\infty(\mathbb{R}^n)$ on the time interval $[0,T]$. Then we have:

$$\left| |u^\varepsilon(t,x)|^{2\sigma} u^\varepsilon(t,x) - |u_\ell^\varepsilon(t,x)|^{2\sigma} u_\ell^\varepsilon(t,x) \right| \leqslant C(T)|w_\ell^\varepsilon(t,x)|,$$

uniformly for $t \in [0,T]$ and $x \in \mathbb{R}^n$. Lemma 1.2 yields the formal estimate, for $t \in [0,T]$:

$$\varepsilon \|w_\ell^\varepsilon(t)\|_{L^2} \leqslant 2|\lambda| C(T) \int_0^t \|w_\ell^\varepsilon(\tau)\|_{L^2} d\tau + 2\varepsilon^\ell \int_0^t \|r_\ell^\varepsilon(\tau)\|_{L^2} d\tau.$$

Using Gronwall lemma, we infer:

$$\|w^\varepsilon(t)\|_{L^2} \leqslant C\varepsilon^{\ell-1} e^{Ct/\varepsilon}.$$

The exponential factor shows that this method may yield interesting results only up to time of the order $c\varepsilon |\log \varepsilon|^\theta$ for some $c, \theta > 0$. Note that in some functional analysis contexts, this may be satisfactory (see Sec. 5.1). However, in general, we wish to have a description of the solution of (1.6) on a time interval independent of ε.

In Chap. 9, we give a rather explicit example of a situation similar to the one considered above, where ℓ can be taken arbitrarily large, but w_ℓ^ε is not small in L^2, past the time where Gronwall lemma is satisfactory (see Sec. 9.1.3).

1.3   Linear Schrödinger equation

Before proceeding to the nonlinear analysis, we justify the above discussion in the linear case: we consider (1.6) with $\lambda = 0$, that is

$$i\varepsilon\partial_t u^\varepsilon + \frac{\varepsilon^2}{2}\Delta u^\varepsilon = V u^\varepsilon \quad ; \quad u^\varepsilon(0,x) = a_0^\varepsilon(x)e^{i\phi_0(x)/\varepsilon}. \tag{1.9}$$

The results presented here will also be useful in the study of the pointwise behavior of the solution u^ε to (1.1) in the nonlinear case (see Chap. 2 and Sec. 4.2.2). We invite the reader to consult [Robert (1987)] for results related to the semi-classical limit of Eq. (1.9) with a different point of view.

1.3.1   *The eikonal equation*

To cancel the ε^0 term, the first step consists in solving (1.7):

$$\partial_t\phi_{\text{eik}} + \frac{1}{2}|\nabla\phi_{\text{eik}}|^2 + V = 0 \quad ; \quad \phi_{\text{eik}}(0,x) = \phi_0(x). \tag{1.10}$$

This equation is called the *eikonal equation*. The term "eikonal" stems from the theory of geometric optics: the solution to this equation determines the set where light is propagated. In the case of the (linear) Schrödinger equation, we will see that a similar phenomenon occurs: the phase ϕ_{eik} determines the way the initial amplitude a_0 is transported (see Sec. 1.3.2). Equation (1.10) is also referred to as a *Hamilton–Jacobi equation*. It is usually solved locally in space and time in terms of the semi-classical Hamiltonian

$$H(t,x,\tau,\xi) = \tau + \frac{|\xi|^2}{2} + V(t,x), \quad (t,x,\tau,\xi) \in \mathbb{R}\times\mathbb{R}^n\times\mathbb{R}\times\mathbb{R}^n.$$

More general Hamilton–Jacobi equations are equations of the form

$$H(t,x,\partial_t\phi,\nabla\phi) = 0,$$

where H is a smooth real-valued function of its arguments. For the propagation of light in a medium of variable speed of propagation $c(x)$, we have

$$H(t,x,\tau,\xi) = \tau^2 - c(x)^2|\xi|^2.$$

The local resolution of such equations appears in many books (see e.g. [Dereziński and Gérard (1997); Grigis and Sjöstrand (1994); Evans (1998)]), so we shall only outline the usual approach. Since in our case $\partial_\tau H = 1$, the Hamiltonian flow is given by the system of ordinary differential equations

$$\begin{cases} \partial_t x(t,y) = \partial_\xi H = \xi(t,y) & ; \quad x(0,y) = y, \\ \partial_t \xi(t,y) = -\partial_x H = -\nabla_x V(t,x(t,y)) & ; \quad \xi(0,y) = \nabla\phi_0(y). \end{cases} \tag{1.11}$$

The projection of the solution (x, ξ) on the physical space, that is $x(t, y)$, is called *classical trajectory*, or *ray*. The Cauchy-Lipschitz Theorem yields:

Lemma 1.3. *Assume that V and ϕ_0 are smooth: $V \in C^\infty(\mathbb{R} \times \mathbb{R}^n; \mathbb{R})$ and $\phi_0 \in C^\infty(\mathbb{R}^n; \mathbb{R})$. Then for all $y \in \mathbb{R}^n$, there exists $T_y > 0$ and a unique solution to (1.11), $(x(t, y), \xi(t, y)) \in C^\infty([-T_y, T_y] \times \mathbb{R}^n; \mathbb{R}^n)^2$.*

The link with (1.10) appears in

Lemma 1.4. *Let ϕ_{eik} be a smooth solution to (1.10). Then necessarily,*

$$\nabla \phi_{\mathrm{eik}}(t, x(t, y)) = \xi(t, y),$$

as long as all the terms remain smooth.

Proof.   For ϕ_{eik} a smooth solution to (1.10), introduce the ordinary differential equation

$$\frac{d}{dt}\widetilde{x} = \nabla \phi_{\mathrm{eik}}(t, \widetilde{x}) \quad ; \quad \widetilde{x}\big|_{t=0} = y. \tag{1.12}$$

By the Cauchy-Lipschitz Theorem, (1.12) has a smooth solution $\widetilde{x} \in C^\infty([-\widetilde{T}_y, \widetilde{T}_y])$ for some $\widetilde{T}_y > 0$ possibly very small. Set

$$\widetilde{\xi}(t) := \nabla \phi_{\mathrm{eik}}(t, \widetilde{x}(t)).$$

We compute

$$\begin{aligned}
\frac{d}{dt}\widetilde{\xi} &= \nabla \partial_t \phi_{\mathrm{eik}}(t, \widetilde{x}(t)) + \nabla^2 \phi_{\mathrm{eik}}(t, \widetilde{x}(t)) \cdot \frac{d}{dt}\widetilde{x}(t) \\
&= \nabla \partial_t \phi_{\mathrm{eik}}(t, \widetilde{x}(t)) + \nabla^2 \phi_{\mathrm{eik}}(t, \widetilde{x}(t)) \cdot \nabla \phi_{\mathrm{eik}}(t, \widetilde{x}(t)) \\
&= \nabla \left(\partial_t \phi_{\mathrm{eik}} + \frac{1}{2}|\nabla \phi_{\mathrm{eik}}|^2 \right)(t, \widetilde{x}(t)) = -\nabla V(t, \widetilde{x}(t)).
\end{aligned}$$

We infer that $(\widetilde{x}, \widetilde{\xi})$ solves (1.11). The lemma then follows from uniqueness for (1.11).   □

Note that knowing $\nabla \phi_{\mathrm{eik}}$ suffices to get ϕ_{eik} itself, which is given by

$$\phi_{\mathrm{eik}}(t, x) = \phi_0(x) - \int_0^t \left(\frac{1}{2}|\nabla \phi_{\mathrm{eik}}(\tau, x)|^2 + V(\tau, x) \right) d\tau.$$

The above lemma and the Local Inversion Theorem yield

Lemma 1.5. *Let V and ϕ_0 smooth as in Lemma 1.3. Let $t \in [-T, T]$ and θ_0 an open set of $\mathbb{R}^n$. Denote*

$$\theta_t := \{x(t, y) \in \mathbb{R}^n, y \in \theta_0\} \quad ; \quad \theta := \{(t, x) \in [-T, T] \times \mathbb{R}^n, x \in \theta_t\}.$$

Suppose that for $t \in [-T, T]$, the mapping

$$\theta_0 \ni y \mapsto x(t, y) \in \theta_t$$

is bijective, and denote by $y(t, x)$ its inverse. Assume also that

$$\nabla_x y \in L^\infty_{\mathrm{loc}}(\theta).$$

Then there exists a unique function $\theta \ni (t, x) \mapsto \phi_{\mathrm{eik}}(t, x) \in \mathbb{R}$ that solves (1.10), and satisfies $\nabla^2_x \phi_{\mathrm{eik}} \in L^\infty_{\mathrm{loc}}(\theta)$. Moreover,

$$\nabla \phi_{\mathrm{eik}}(t, x) = \xi(t, y(t, x)). \tag{1.13}$$

Note that the existence time T may depend on the neighborhood θ_0. It actually does in general, as shown by the following example.

Example 1.6. Assume that $V \equiv 0$ and

$$\phi_0(x) = -\frac{1}{(2 + 2\delta)T_c} \left(|x|^2 + 1 \right)^{1+\delta}, \quad T_c > 0, \ \delta \geqslant 0.$$

For $\delta > 0$, integrating (1.11) yields:

$$x(t, y) = y + \int_0^t \xi(s, y) ds = y + \int_0^t \xi(0, y) ds = y - \frac{t}{T_c} \left(|y|^2 + 1 \right)^\delta y$$

$$= y \left(1 - \frac{t}{T_c} \left(|y|^2 + 1 \right)^\delta \right).$$

For $R > 0$, we see that the rays starting from the ball $\{ |y| = R \}$ meet at the origin at time

$$T_c(R) = \frac{T_c}{(R^2 + 1)^\delta}.$$

Since R is arbitrary, this shows that several rays can meet arbitrarily fast, thus showing that the above lemma cannot be applied uniformly in space.

Of course, the above issue would not appear if the space variable x belonged to a compact set instead of the whole space $\mathbb{R}^n$. To obtain a local time of existence with is independent of $y \in \mathbb{R}^n$, we have to make an extra assumption, in order to be able to apply a *global* inversion theorem.

Assumption 1.7 (Geometric assumption). *We assume that the potential and the initial phase are smooth, real-valued, and subquadratic:*

- *$V \in C^\infty(\mathbb{R} \times \mathbb{R}^n)$, and $\partial^\alpha_x V \in C(\mathbb{R}; L^\infty(\mathbb{R}^n))$ as soon as $|\alpha| \geqslant 2$.*
- *$\phi_0 \in C^\infty(\mathbb{R}^n)$, and $\partial^\alpha \phi_0 \in L^\infty(\mathbb{R}^n)$ as soon as $|\alpha| \geqslant 2$.*

As a consequence of this assumption on V, if $a_0^\varepsilon \in L^2$, then (1.9) has a unique solution $u^\varepsilon \in C(\mathbb{R}; L^2)$. See e.g. [Reed and Simon (1975)].

The following result can be found in [Schwartz (1969)], or in Appendix A of [Dereziński and Gérard (1997)].

Lemma 1.8. *Suppose that the function* $\mathbb{R}^n \ni y \mapsto x(y) \in \mathbb{R}^n$ *satisfies:*

$$|\det \nabla_y x| \geqslant C_0 > 0 \quad and \quad |\partial_y^\alpha x| \leqslant C, \ |\alpha| = 1, 2.$$

Then x *is bijective.*

We can then prove

Proposition 1.9. *Under Assumption 1.7, there exists* $T > 0$ *and a unique solution* $\phi_{\mathrm{eik}} \in C^\infty([-T, T] \times \mathbb{R}^n)$ *to (1.10). In addition, this solution is subquadratic:* $\partial_x^\alpha \phi_{\mathrm{eik}} \in L^\infty([-T, T] \times \mathbb{R}^n)$ *as soon as* $|\alpha| \geqslant 2$.

Proof. We know that we can solve (1.11) locally in time in the neighborhood of any $y \in \mathbb{R}^n$. In order to apply the above global inversion result, differentiate (1.11) with respect to y:

$$\begin{cases} \partial_t \partial_y x(t, y) = \partial_y \xi(t, y) & ; \ \partial_y x(0, y) = \mathrm{Id}, \\ \partial_t \partial_y \xi(t, y) = -\nabla_x^2 V(t, x(t, y)) \partial_y x(t, y) & ; \ \partial_y \xi(0, y) = \nabla^2 \phi_0(y). \end{cases} \quad (1.14)$$

Integrating (1.14) in time, we infer from Assumption 1.7 that for any $T > 0$, there exists C_T such that for $(t, y) \in [-T, T] \times \mathbb{R}^n$:

$$|\partial_y x(t, y)| + |\partial_y \xi(t, y)| \leqslant C_T + C_T \int_0^t \left(|\partial_y x(s, y)| + |\partial_y \xi(s, y)| \right) ds.$$

Gronwall lemma yields:

$$\|\partial_y x(t)\|_{L_y^\infty} + \|\partial_y \xi(t)\|_{L_y^\infty} \leqslant C'(T). \quad (1.15)$$

Similarly,

$$\|\partial_y^\alpha x(t)\|_{L_y^\infty} + \|\partial_y^\alpha \xi(t)\|_{L_y^\infty} \leqslant C(\alpha, T), \quad \forall \alpha \in \mathbb{N}^n, \ |\alpha| \geqslant 1. \quad (1.16)$$

Integrating the first line of (1.14) in time, we have:

$$\det \nabla_y x(t, y) = \det \left(\mathrm{Id} + \int_0^t \nabla_y \xi(s, y) \, ds \right).$$

We infer from (1.15) that for $t \in [-T, T]$, provided that $T > 0$ is sufficiently small, we can find $C_0 > 0$ such that:

$$|\det \nabla_y x(t, y)| \geqslant C_0, \quad \forall (t, y) \in [-T, T] \times \mathbb{R}^n. \quad (1.17)$$

Lemma 1.8 shows that we can invert $y \mapsto x(t, y)$ for $t \in [-T, T]$.

To apply Lemma 1.5 with $\theta_0 = \theta = \theta_t = \mathbb{R}^n$, we must check that $\nabla_x y \in L^\infty_{\text{loc}}(\mathbb{R}^n)$. Differentiate the relation

$$x(t, y(t, x)) = x$$

with respect to x:

$$\nabla_x y(t, x) \nabla_y x(t, y(t, x)) = \text{Id}.$$

Therefore, $\nabla_x y(t, x) = \nabla_y x(t, y(t, x))^{-1}$ as matrices, and

$$\nabla_x y(t, x) = \frac{1}{\det \nabla_y x(t, y)} \text{adj}(\nabla_y x(t, y(t, x))), \qquad (1.18)$$

where $\text{adj}(\nabla_y x)$ denotes the adjugate of $\nabla_y x$. We infer from (1.15) and (1.17) that $\nabla_x y \in L^\infty(\mathbb{R}^n)$ for $t \in [-T, T]$. Therefore, Lemma 1.5 yields a smooth solution ϕ_{eik} to (1.10); it is local in time and *global in space*: $\phi_{\text{eik}} \in C^\infty([-T, T] \times \mathbb{R}^n)$.

The fact that ϕ_{eik} is subquadratic as stated in Proposition 1.9 then stems from (1.13), (1.16), (1.17) and (1.18). $\qquad\square$

Note that Example 1.6 shows that the above result is essentially sharp: if Assumption 1.7 is not satisfied, then the above result fails to be true. Similarly, if we consider $V = V(x) = -x^4$ in space dimension $n = 1$, then, also due to an infinite speed of propagation, the Hamiltonian $-\partial_x^2 - x^4$ is not essentially self-adjoint (see Chap. 13, Sect. 6, Cor. 22 in [Dunford and Schwartz (1963)]). We now give some examples of cases where the phase ϕ_{eik} can be computed explicitly, which also show that in general, the above time T is necessarily finite.

Example 1.10 (Quadratic phase). *Resume Example 1.6, and consider the value $\delta = 0$. In that case, Assumption 1.7 is satisfied, and (1.10) is solved explicitly:*

$$\phi_{\text{eik}}(t, x) = \frac{|x|^2}{2(t - T_c)} - \frac{1}{2T_c}.$$

This shows that we can solve (1.10) globally in space, but only locally in time: as $t \to T_c$, ϕ_{eik} ceases to be smooth. A caustic reduced to a single point (the origin) is formed.

Remark 1.11. More generally, the space-time set where the map $y \mapsto x(t, y)$ ceases to be a diffeomorphism is called *caustic*. The behavior of the solution u^ε to (1.6) with $\lambda = 0$ is given for all time in terms of oscillatory integrals ([Duistermaat (1974); Maslov and Fedoriuk (1981)]). We present results concerning the asymptotic behavior of solutions to (1.6) with $\lambda \neq 0$ in the presence of point caustics in the second part of this book.

Example 1.12 (Harmonic potential). *When $\phi_0 \equiv 0$, and V is independent of time and quadratic, $V = V(x) = \frac{1}{2} \sum_{j=1}^{n} \omega_j^2 x_j^2$, we have:*

$$\phi_{\text{eik}}(t, x) = -\sum_{j=1}^{n} \frac{\omega_j}{2} x_j^2 \tan(\omega_j t).$$

This also shows that we can solve (1.10) globally in space, but locally in time only. Note that if we replace formally ω_j by $i\omega_j$, then V is turned into $-V$, and the trigonometric functions become hyperbolic functions: we can then solve (1.10) globally in space and time.

Example 1.13 (Plane wave). *If we assume $V \equiv 0$ and $\phi_0(x) = \xi_0 \cdot x$ for some $\xi_0 \in \mathbb{R}^n$, then we find:*

$$\phi_{\text{eik}}(t, x) = \xi_0 \cdot x - \frac{1}{2} |\xi_0|^2 t.$$

Also in this case, we can solve (1.10) globally in space and time.

1.3.2   The transport equations

To cancel the ε^1 term, the second step consists in solving (1.8):

$$\partial_t a + \nabla \phi_{\text{eik}} \cdot \nabla a + \frac{1}{2} a \Delta \phi_{\text{eik}} = 0 \quad ; \quad a(0, x) = a_0(x), \qquad (1.19)$$

where a_0 is given as the first term in the asymptotic expansion of the initial amplitude (1.4). The equation is a transport equation (see e.g. [Evans (1998)]), since the characteristics for the operator $\partial_t + \nabla \phi_{\text{eik}} \cdot \nabla$ do not meet for $t \in [-T, T]$, by construction. As a matter of fact, this equation can be solved rather explicitly, in terms of the geometric tools that we have used in the previous paragraph.

Introduce the *Jacobi's determinant*

$$J_t(y) = \det \nabla_y x(t, y),$$

where $x(t, y)$ is given by the Hamiltonian flow (1.11). Note that $J_0(y) = 1$ for all $y \in \mathbb{R}^n$. By construction, for $t \in [-T, T]$, the function $y \mapsto J_t(y)$ is uniformly bounded from above and from below:

$$\exists C > 0, \ \frac{1}{C} \leqslant J_t(y) \leqslant C, \quad \forall (t, y) \in [-T, T] \times \mathbb{R}^n.$$

Define the function A by

$$A(t, y) := a\left(t, x(t, y)\right) \sqrt{J_t(y)}.$$

Then since for $t \in [-T, T]$, $y \mapsto x(t, y)$ is a global diffeomorphism on $\mathbb{R}^n$, (1.19) is *equivalent* to the equation

$$\partial_t A(t, y) = 0 \quad ; \quad A(0, y) = a_0(y).$$

We obviously have $A(t, y) = a_0(y)$ for all $t \in [-T, T]$, and back to the function a, this yields

$$a(t, x) = \frac{1}{\sqrt{J_t(y(t, x))}} a_0(y(t, x)), \tag{1.20}$$

where $y(t, x)$ is the inverse map of $y \mapsto x(t, y)$.

Remark 1.14. The computations of Sec. 1.2 show that the amplitudes are given by

$$\partial_t a^{(j)} + \nabla \phi_{\text{eik}} \cdot \nabla a^{(j)} + \frac{1}{2} a^{(j)} \Delta \phi_{\text{eik}} = \frac{i}{2} \Delta a^{(j-1)} \quad ; \quad a^{(j)}_{|t=0} = a_j,$$

with the convention $a^{(-1)} = 0$ and $a^{(0)} = a$. For $j \geqslant 1$, this equation is the inhomogeneous analogue of (1.19). It can be solved by using the same change of variable as above. This shows that when ϕ_{eik} becomes singular (formation of a caustic), *all* the terms computed by this WKB analysis become singular in general. The WKB hierarchy ceases to be relevant at a caustic.

Proposition 1.15. *Let $s \geqslant 0$ and $a_0 \in H^s(\mathbb{R}^n)$. Then (1.19) has a unique solution $a \in C([-T; T]; H^s)$, where $T > 0$ is given by Proposition 1.9.*

Proof. Existence and uniqueness at the L^2 level stem from the above analysis, (1.20). To prove that an H^s regularity is propagated for $s > 0$, we could also use (1.20). We shall use another approach, which will be more natural in the nonlinear setting. To simplify the presentation, we assume $s \in \mathbb{N}$, and prove *a priori* estimates in H^s. Let $\alpha \in \mathbb{N}^n$, with $|\alpha| \leqslant s$. Applying ∂_x^α to (1.19), we find:

$$\partial_t \partial_x^\alpha a + \nabla \phi_{\text{eik}} \cdot \nabla \partial_x^\alpha a = [\nabla \phi_{\text{eik}} \cdot \nabla, \partial_x^\alpha] a - \frac{1}{2} \partial_x^\alpha (a \Delta \phi_{\text{eik}}) =: R_\alpha, \tag{1.21}$$

where $[P, Q] = PQ - QP$ denotes the commutator of the operators P and Q. Take the inner product of (1.21) with $\partial_x^\alpha a$, and consider the real part:

$$\frac{1}{2} \frac{d}{dt} \|\partial_x^\alpha a\|_{L^2}^2 + \text{Re} \int_{\mathbb{R}^n} \overline{\partial_x^\alpha a} \nabla \phi_{\text{eik}} \cdot \nabla \partial_x^\alpha a \leqslant \|R_\alpha\|_{L^2} \|a\|_{H^s}.$$

Notice that we have

$$\left| \text{Re} \int_{\mathbb{R}^n} \overline{\partial_x^\alpha a} \nabla \phi_{\text{eik}} \cdot \nabla \partial_x^\alpha a \right| = \frac{1}{2} \left| \int_{\mathbb{R}^n} \nabla \phi_{\text{eik}} \cdot \nabla |\partial_x^\alpha a|^2 \right|$$

$$= \frac{1}{2} \left| \int_{\mathbb{R}^n} |\partial_x^\alpha a|^2 \Delta \phi_{\text{eik}} \right| \leqslant C \|a\|_{H^s}^2,$$

since $\Delta\phi_{\mathrm{eik}} \in L^\infty([-T,T] \times \mathbb{R}^n)$ from Proposition 1.9. Summing over α such that $|\alpha| \leqslant s$, we infer:

$$\frac{d}{dt}\|a\|_{H^s}^2 \leqslant C\|a\|_{H^s}^2 + \|R_\alpha\|_{H^s}^2 .$$

To apply Gronwall's lemma, we need to estimate the last term. We use the fact that the derivatives of order at least two of ϕ_{eik} are bounded, from Proposition 1.9, to have:

$$\|R_\alpha\|_{L^2} \leqslant C\|a\|_{H^s} .$$

We can then conclude:

$$\|a\|_{L^\infty([-T,T];H^s)} \leqslant C\|a_0\|_{H^s} ,$$

which completes the proof of the proposition. $\qquad\square$

Let us examine what can be deduced at this stage, and see which rigorous meaning can be given to the relation $u^\varepsilon \sim ae^{i\phi_{\mathrm{eik}}/\varepsilon}$. Let

$$v_1^\varepsilon(t,x) := a(t,x)e^{i\phi_{\mathrm{eik}}(t,x)/\varepsilon}.$$

Proposition 1.16. *Let $s \geqslant 2$, $a_0 \in H^s(\mathbb{R}^n)$, and Assumption 1.7 be satisfied. Suppose that*

$$\|a_0^\varepsilon - a_0\|_{H^{s-2}} = \mathcal{O}\left(\varepsilon^\beta\right)$$

for some $\beta > 0$. Then there exists $C > 0$ independent of $\varepsilon \in]0,1]$ such that

$$\sup_{t\in[-T,T]} \|u^\varepsilon(t) - v_1^\varepsilon(t)\|_{L^2} \leqslant C\varepsilon^{\min(1,\beta)},$$

where T is given by Proposition 1.9. If in addition $s > n/2 + 2$, then

$$\sup_{t\in[-T,T]} \|u^\varepsilon(t) - v_1^\varepsilon(t)\|_{L^\infty} \leqslant C\varepsilon^{\min(1,\beta)}.$$

Proof. Let $w_1^\varepsilon := u^\varepsilon - v_1^\varepsilon$. By construction, is solves

$$i\varepsilon\partial_t w_1^\varepsilon + \frac{\varepsilon^2}{2}\Delta w_1^\varepsilon = Vw_1^\varepsilon - \frac{\varepsilon^2}{2}e^{i\phi_{\mathrm{eik}}/\varepsilon}\Delta a \quad ; \quad w_{1|t=0}^\varepsilon = a_0^\varepsilon - a_0. \qquad (1.22)$$

By Lemma 1.2, which can be made rigorous in the present setting (exercise), we have:

$$\sup_{t\in[-T,T]} \|w_1^\varepsilon(t)\|_{L^2} \leqslant \|a_0^\varepsilon - a_0\|_{L^2} + \frac{\varepsilon}{2}\int_{-T}^{T} \|\Delta a(\tau)\|_{L^2} d\tau \leqslant C\left(\varepsilon^\beta + \varepsilon\right),$$

where we have used the assumption on $a_0^\varepsilon - a_0$ and Proposition 1.15. This yields the first estimate of the proposition.

To prove the second estimate, we want to use the Sobolev embedding $H^s \subset L^\infty$ for $s > n/2$. A first idea could be to differentiate (1.22) with respect to space variables, and use Lemma 1.2. However, this direct approach fails, because the source term

$$\frac{\varepsilon^2}{2} e^{i\phi_{\text{eik}}/\varepsilon} \Delta a$$

is of order $\mathcal{O}(\varepsilon^2)$ in L^2, but of order $\mathcal{O}(\varepsilon^{2-s})$ in H^s, $s \geqslant 0$. This is due to the rapidly oscillatory factor $e^{i\phi_{\text{eik}}/\varepsilon}$. Moreover, under our assumptions, it is not guaranteed that $\nabla\phi_{\text{eik}}\Delta a \in C([-T,T]; L^2)$, since $\nabla\phi_{\text{eik}}$ may grow linearly with respect to the space variable, as shown by Examples 1.10 and 1.12. We therefore adopt a different point of view, relying on the remark:

$$\left| u^\varepsilon - v_1^\varepsilon \right| = \left| u^\varepsilon - a e^{i\phi_{\text{eik}}/\varepsilon} \right| = \left| u^\varepsilon e^{-i\phi_{\text{eik}}/\varepsilon} - a \right|.$$

Set $a^\varepsilon := u^\varepsilon e^{-i\phi_{\text{eik}}/\varepsilon}$. We check that it solves

$$\partial_t a^\varepsilon + \nabla\phi_{\text{eik}} \cdot \nabla a^\varepsilon + \frac{1}{2} a^\varepsilon \Delta\phi_{\text{eik}} = i\frac{\varepsilon}{2}\Delta a^\varepsilon \quad ; \quad a^\varepsilon_{|t=0} = a_0^\varepsilon.$$

Therefore, $r^\varepsilon = a^\varepsilon - a = w_1^\varepsilon e^{-i\phi_{\text{eik}}/\varepsilon}$ solves

$$\partial_t r^\varepsilon + \nabla\phi_{\text{eik}} \cdot \nabla r^\varepsilon + \frac{1}{2} r^\varepsilon \Delta\phi_{\text{eik}} = i\frac{\varepsilon}{2}\Delta r^\varepsilon + i\frac{\varepsilon}{2}\Delta a \quad ; \quad r^\varepsilon_{|t=0} = a_0^\varepsilon - a_0. \tag{1.23}$$

Note that this equation is very similar to the transport equation (1.19), with two differences. First, the presence of the operator $i\varepsilon\Delta$ acting on r^ε on the right hand side. Second, the source term $i\varepsilon\Delta a$, which makes the equation inhomogeneous.

We know by construction that $r^\varepsilon \in C([-T,T]; L^2)$, and we seek *a priori* estimates in $C([-T,T]; H^k)$. These are established along the same lines as in the proof of Proposition 1.15. We note that since the operator $i\Delta$ is skew-symmetric on H^s, the term $i\varepsilon\Delta r^\varepsilon$ vanishes from the energy estimates in H^s. Then, the source term is of order ε in $C([-T;T]; H^{s-2})$ from Proposition 1.15. We infer:

$$\sup_{t\in[-T,T]} \|r^\varepsilon(t)\|_{H^{s-2}} \lesssim \varepsilon^\beta + \varepsilon.$$

Note that this estimate, along with a standard continuation argument, shows that $a^\varepsilon \in C([-T;T]; H^{s-2})$ for $\varepsilon > 0$ sufficiently small. Since $s - 2 > n/2$, we deduce

$$\sup_{t\in[-T,T]} \|r^\varepsilon(t)\|_{L^\infty} \lesssim \varepsilon^\beta + \varepsilon,$$

which completes the proof of the proposition. $\qquad\square$

Before analyzing the accuracy of higher order approximate solutions, let us examine the candidate v_1^ε in the case of the examples given in Sec. 1.3.1.

Example 1.17 (Quadratic phase). *Resume Example 1.10. In this case, we compute, for $t < T_c$,*

$$a(t,x) = \left(\frac{T_c}{T_c - t}\right)^{n/2} a_0\left(\frac{T_c}{T_c - t}x\right).$$

As $t \to T_c$, not only ϕ_{eik} ceases to be smooth, but also a. This is a general feature of the formation of caustics: all the terms constructed by the usual WKB analysis become singular.

Example 1.18 (Harmonic potential). *Resume Example 1.12. If $|t|$ is sufficiently small so that ϕ_{eik} remains smooth on $[0,t]$, we find:*

$$a(t,x) = \prod_{j=1}^{n}\left(\frac{1}{\cos(\omega_j t)}\right)^{1/2} a_0\left(\frac{x_1}{\cos(\omega_1 t)}, \ldots, \frac{x_n}{\cos(\omega_n t)}\right).$$

Here again, ϕ_{eik} and a become singular simultaneously.

Example 1.19 (Plane wave). *If we assume $V \equiv 0$ and $\phi_0(x) = \xi_0 \cdot x$ for some $\xi_0 \in \mathbb{R}^n$, then we find:*

$$a(t,x) = a_0\left(x - \xi_0 t\right).$$

The initial amplitude is simply transported with constant velocity.

We can continue this analysis to arbitrary order:

Proposition 1.20. *Let $k \in \mathbb{N} \setminus \{0\}$ and $s \geqslant 2k+2$. Let $a_0, a_1, \ldots, a_k$ with $a_j \in H^{s-2j}(\mathbb{R}^n)$, and let Assumption 1.7 be satisfied. Suppose that*

$$\|a_0^\varepsilon - a_0 - \varepsilon a_1 - \ldots - \varepsilon^k a_k\|_{H^{s-2k-2}} = \mathcal{O}\left(\varepsilon^{k+\beta}\right)$$

for some $\beta > 0$. Then we can find $a^{(1)}, \ldots, a^{(k)}$, with

$$a^{(j)} \in C([-T,T]; H^{s-2j}),$$

such that if we set

$$v_{k+1}^\varepsilon = \left(a + \varepsilon a^{(1)} + \ldots + \varepsilon^k a^{(k)}\right)e^{i\phi_{\text{eik}}/\varepsilon},$$

there exists $C > 0$ independent of $\varepsilon \in]0,1]$ such that

$$\sup_{t\in[-T,T]}\left\|\left(u^\varepsilon(t) - v_{k+1}^\varepsilon(t)\right)e^{-i\phi_{\text{eik}}(t)/\varepsilon}\right\|_{H^{s-2k-2}} \leqslant C\varepsilon^{\min(k+1,k+\beta)},$$

where T is given by Proposition 1.9.

Proof. We simply sketch the proof, since it follows arguments which have been introduced above. First, the computations presented in Sec. 1.2 show that to cancel the term in ε^{j+1}, $1 \leqslant j \leqslant k$, we naturally impose:

$$\partial_t a^{(j)} + \nabla\phi_{\text{eik}} \cdot \nabla a^{(j)} + \frac{1}{2}a^{(j)}\Delta\phi_{\text{eik}} = \frac{i}{2}\Delta a^{(j-1)} \quad ; \quad a^{(j)}_{|t=0} = a_j.$$

This equation is the inhomogeneous analogue of (1.19). Using the same arguments as in the proof of Proposition 1.15, it is easy to see that it has a unique solution $a^{(j)} \in C([-T,T];L^2)$, whose spatial regularity is that of $a^{(j-1)}$, minus 2. Starting an induction with Proposition 1.15, we construct

$$a^{(j)} \in C([-T,T];H^{s-2j}).$$

To prove the error estimate, introduce $r_k^\varepsilon = a^\varepsilon - a - \varepsilon a^{(1)} - \ldots - \varepsilon^k a^{(k)}$, where we recall that $a^\varepsilon = u^\varepsilon e^{-i\phi_{\text{eik}}/\varepsilon}$. By construction, the remainder r_k^ε is in $C([-T;T];L^2)$ since $s \geqslant 2k+2$, and it solves:

$$\begin{cases} \partial_t r_k^\varepsilon + \nabla\phi_{\text{eik}} \cdot \nabla r_k^\varepsilon + \frac{1}{2}r_k^\varepsilon\Delta\phi_{\text{eik}} = i\frac{\varepsilon}{2}\Delta r_k^\varepsilon + i\frac{\varepsilon^{k+1}}{2}\Delta a^{(k)}, \\ r^\varepsilon_{|t=0} = a_0^\varepsilon - a_0 - \ldots - \varepsilon^k a_k. \end{cases}$$

We can then mimic the end of the proof of Proposition 1.16. □

To conclude, we see that we can construct an arbitrarily accurate (as $\varepsilon \to 0$) approximation of u^ε on $[-T,T]$, provided that the initial profiles a_j are sufficiently smooth. The goal now is to see how this approach can be adapted to a nonlinear framework.

1.4 Basic results in the nonlinear case

Before presenting a WKB analysis in the case $f \neq 0$ in (1.1), we recall a few important facts about the nonlinear Cauchy problem for (1.1). We shall simply gather classical results, which can be found for instance in [Cazenave and Haraux (1998); Cazenave (2003); Ginibre and Velo (1985a); Kato (1989); Tao (2006)]. Several notions of solutions are available. According to the cases, we will work with the notion of strong solutions (Chapters 2, 4 and 5), of weak solutions (Chapters 3 and 5) or of mild solutions (especially in the second part of this book).

In this section, one should think that the parameter $\varepsilon > 0$ is *fixed*. The dependence upon ε is discussed in the forthcoming sections.

1.4.1   Formal properties

Since V and f are real-valued, the L^2 norm of u^ε is formally independent of time:

$$\|u^\varepsilon(t)\|_{L^2} = \|u^\varepsilon(0)\|_{L^2}. \tag{1.24}$$

This can be seen from the proof of Lemma 1.2, with $F^\varepsilon = V + f\left(|u^\varepsilon|^2\right)$ and $R^\varepsilon = 0$. This relation yields an *a priori* bound for the L^2 norm of u^ε.

When the potential V is time-independent, $V = V(x)$, (1.1) has a Hamiltonian structure. Introduce

$$F(y) = \int_0^y f(\eta)d\eta.$$

The following energy is formally independent of time:

$$\begin{aligned}
E^\varepsilon(u^\varepsilon(t)) =& \frac{1}{2}\|\varepsilon\nabla u^\varepsilon(t)\|_{L^2}^2 + \int_{\mathbb{R}^n} F\left(|u^\varepsilon(t,x)|^2\right) dx \\
& + \int_{\mathbb{R}^n} V(x)|u^\varepsilon(t,x)|^2 dx.
\end{aligned} \tag{1.25}$$

We see that if E^ε is finite, and if $V \geqslant 0$ and $F \geqslant 0$, then this yields an *a priori* bound on $\|\varepsilon\nabla u^\varepsilon(t)\|_{L^2}$.

Example 1.21. If $V = V(x) \geqslant$ and $f(y) = \lambda y^\sigma$, then (1.25) becomes

$$E^\varepsilon = \frac{1}{2}\|\varepsilon\nabla u^\varepsilon(t)\|_{L^2}^2 + \frac{\lambda}{\sigma+1}\int_{\mathbb{R}^n} |u^\varepsilon(t,x)|^{2\sigma+2} dx + \int_{\mathbb{R}^n} V(x)|u^\varepsilon(t,x)|^2 dx.$$

If $\lambda \geqslant 0$ (*defocusing nonlinearity*), this yields an *a priori* bound on $\|\varepsilon\nabla u^\varepsilon(t)\|_{L^2}$. On the other hand, if $\lambda < 0$, then $\|\varepsilon\nabla u^\varepsilon(t)\|_{L^2}$ may become unbounded in finite time: this is the *finite time blow-up* phenomenon (see e.g. [Cazenave (2003); Sulem and Sulem (1999)]). Since the L^2 norm of u^ε is conserved, one can replace the assumption $V \geqslant 0$ with $V \geqslant -C$ for some $C > 0$, and leave the above discussion unchanged.

Example 1.22. If V is unbounded from below, the conservation of the energy does not seem to provide interesting informations. For instance, if $V(x) = -|x|^2$, then even in the linear case $f = 0$, the energy is not a positive energy functional (see [Carles (2003a)] though, for the nonlinear Cauchy problem).

1.4.2 *Strong solutions*

A remarkable fact is that if the external potential V is subquadratic in the sense of Assumption 1.7, then one can define a strongly continuous semi-group for the linear equation (1.9). As we have mentioned already, if no sign assumption is made on V, then Assumption 1.7 is essentially sharp: if $n = 1$ and $V(x) = -x^4$, then $-\partial_x^2 + V$ is not essentially self-adjoint on the set of test functions ([Dunford and Schwartz (1963)]). Under Assumption 1.7, one defines $U^\varepsilon(t, s)$ such that $u_{\text{lin}}^\varepsilon(t, x) = U^\varepsilon(t, s)\varphi^\varepsilon(x)$, where

$$i\varepsilon\partial_t u_{\text{lin}}^\varepsilon + \frac{\varepsilon^2}{2}\Delta u_{\text{lin}}^\varepsilon = V u_{\text{lin}}^\varepsilon \quad ; \quad u_{\text{lin}}^\varepsilon(s, x) = \varphi^\varepsilon(x).$$

Note that $U^\varepsilon(t, t) = \text{Id}$. The existence of $U^\varepsilon(t, s)$ is established in [Fujiwara (1979)], along with the following properties:

- The map $(t, s) \mapsto U^\varepsilon(t, s)$ is strongly continuous.
- $U^\varepsilon(t, s)^* = U^\varepsilon(t, s)^{-1}$.
- $U^\varepsilon(t, \tau)U^\varepsilon(\tau, s) = U^\varepsilon(t, s)$.
- $U^\varepsilon(t, s)$ is unitary on L^2: $\|U^\varepsilon(t, s)\varphi^\varepsilon\|_{L^2} = \|\varphi^\varepsilon\|_{L^2}$.

We construct strong solutions which are (at least) in $H^s(\mathbb{R}^n)$, for $s > n/2$. Recall that H^s is then an algebra, embedded into $L^\infty(\mathbb{R}^n)$. We shall also use the following version of Schauder's lemma:

Lemma 1.23 (Schauder's lemma). *Suppose that $G : \mathbb{C} \to \mathbb{C}$ is a smooth function, such that $G(0) = 0$. Then the map $u \mapsto G(u)$ sends $H^s(\mathbb{R}^n)$ to itself provided $s > n/2$. The map is uniformly Lipschitzean on bounded subsets of H^s.*

We refer to [Taylor (1997)] or [Rauch and Keel (1999)] for the proof of this result, as well as to the following refinement (*tame estimate*):

Lemma 1.24 (Moser's inequality). *Suppose that $G : \mathbb{C} \to \mathbb{C}$ is a smooth function, such that $G(0) = 0$. Then there exists $C : [0, \infty[\to [0, \infty[$ such that for all $u \in H^s(\mathbb{R}^n)$,*

$$\|G(u)\|_{H^s} \leqslant C\left(\|u\|_{L^\infty}\right)\|u\|_{H^s}.$$

For $k \in \mathbb{N}$, denote

$$\Sigma(k) = H^k \cap \mathcal{F}(H^k) = \{f \in H^k(\mathbb{R}^n) \ ; \ x \mapsto \langle x\rangle^k f(x) \in L^2(\mathbb{R}^n)\}.$$

Proposition 1.25. *Let V satisfy Assumption 1.7, and let $f \in C^\infty(\mathbb{R}_+; \mathbb{R})$. Let $k \in \mathbb{N}$, with $k > n/2$, and fix $\varepsilon \in\ ]0, 1]$.*

- *If $u_0^\varepsilon \in \Sigma(k)$, then there exist $T_-^\varepsilon, T_+^\varepsilon > 0$ and a unique maximal solution $u^\varepsilon \in C(]-T_-^\varepsilon, T_+^\varepsilon[; \Sigma(k))$ to (1.1), such that $u_{|t=0}^\varepsilon = u_0^\varepsilon$. It is maximal in the sense that if, say, $T_+^\varepsilon < \infty$, then*

$$\limsup_{t \to T_+^\varepsilon} \|u^\varepsilon(t)\|_{L^\infty(\mathbb{R}^n)} = +\infty. \tag{1.26}$$

- *Assume in addition that V is sub-linear: $\nabla V \in L_{\mathrm{loc}}^\infty(\mathbb{R}; L^\infty(\mathbb{R}^n))$. Let $s > n/2$ (not necessarily an integer). If $u_0^\varepsilon \in H^s(\mathbb{R}^n)$, then there exist $T_-^\varepsilon, T_+^\varepsilon > 0$ and a unique maximal solution $u^\varepsilon \in C(]-T_-^\varepsilon, T_+^\varepsilon[; H^s)$ to (1.1), such that $u_{|t=0}^\varepsilon = u_0^\varepsilon$. It is maximal in the sense that if, say, $T_+^\varepsilon < \infty$, then (1.26) holds. In particular, if $u_0^\varepsilon \in H^\infty$, then $u^\varepsilon \in C^\infty(]-T_-^\varepsilon, T_+^\varepsilon[; H^\infty)$.*

Proof. The proof follows arguments which are classical in the context of semilinear evolution equations. We indicate a few important facts, and refer to [Cazenave and Haraux (1998)] to fill the gaps.

The general idea consists in applying a fixed point argument on the Duhamel's formulation of (1.1) with associated initial datum u_0^ε:

$$u^\varepsilon(t) = U^\varepsilon(t,0)u_0^\varepsilon - i\varepsilon^{-1} \int_0^t U^\varepsilon(t,\tau) \left(f\left(|u^\varepsilon(\tau)|^2\right) u^\varepsilon(\tau) \right) d\tau. \tag{1.27}$$

We claim that for any $k \in \mathbb{N}$ and any $T > 0$,

$$\sup_{t \in [-T,T]} \|U^\varepsilon(t,0)u_0^\varepsilon\|_{\Sigma(k)} \leqslant C(k,T)\|u_0^\varepsilon\|_{\Sigma(k)}. \tag{1.28}$$

For $k = 0$, this is due to the fact that $U^\varepsilon(t,0)$ is unitary on $L^2(\mathbb{R}^n)$. For $k = 1$, notice the commutator identities

$$\left[\nabla, i\varepsilon\partial_t + \frac{\varepsilon^2}{2}\Delta - V\right] = -\nabla V \quad ; \quad \left[x, i\varepsilon\partial_t + \frac{\varepsilon^2}{2}\Delta - V\right] = -\varepsilon^2\nabla. \tag{1.29}$$

By Assumption 1.7, $|\nabla V(t,x)| \leqslant C(T) \langle x \rangle$ for $|t| \leqslant T$, and (1.28) follows for $k = 1$. For $k \geqslant 2$, the proof follows the same lines.

To estimate the nonlinear term, we can assume without loss of generality that $f(0) = 0$. Indeed, we can replace f with $f - f(0)$ and V with $V + f(0)$. Schauder's lemma shows that

$$u \mapsto f\left(|u|^2\right)u$$

sends $H^s(\mathbb{R}^n)$ (resp. $\Sigma(k)$) to itself, provided $s > n/2$ (resp. $k > n/2$), and the map is uniformly Lipschitzean on bounded subsets of $H^s(\mathbb{R}^n)$ (resp. $\Sigma(k)$). The existence and the uniqueness of a solution in the first part of the proposition follow easily. The notion of maximality is then a consequence of Lemma 1.24.

When V is sub-linear, notice that in view of the commutator identities (1.29), the estimate (1.28) can be replaced with

$$\sup_{t \in [-T,T]} \|U^\varepsilon(t,0)u_0^\varepsilon\|_{H^s} \leqslant C(s,T)\|u_0^\varepsilon\|_{H^s}.$$

This is straightforward if $s \in \mathbb{N}$, and follows by interpolation for general $s \geqslant 0$. The proof of the second part of the proposition then follows the same lines as the first part. Finally, if $u_0^\varepsilon \in H^\infty$, then u^ε is also smooth with respect to the time variable, $u^\varepsilon \in C^\infty(]-T_-^\varepsilon, T_+^\varepsilon[; H^\infty)$, by a bootstrap argument. $\qquad\square$

Note that the times T_-^ε and T_+^ε may very well go to zero as $\varepsilon \to 0$. The fact that we can bound these two quantities by $T > 0$ independent of $\varepsilon \in]0,1]$ is also a non-trivial information which will be provided by WKB analysis.

1.4.3 *Mild solutions*

Until the end of Sec. 1.4, to simplify the notations, we assume that the nonlinearity is homogeneous:

$$f(y) = \lambda y^\sigma, \quad \lambda \in \mathbb{R}, \sigma > 0.$$

In view of the conservations of mass (1.24) and energy (1.25), it is natural to look for solutions to (1.1) with initial data which are not necessarily as smooth as in Proposition 1.25. Typically, rather that (1.1), we rather consider its Duhamel's formulation, which now reads

$$u^\varepsilon(t) = U^\varepsilon(t,0)u_0^\varepsilon - i\lambda\varepsilon^{-1}\int_0^t U^\varepsilon(t,\tau)\left(|u^\varepsilon(\tau)|^{2\sigma}u^\varepsilon(\tau)\right)d\tau. \tag{1.30}$$

An extra property of U^ε was proved in [Fujiwara (1979)], which becomes interesting at this stage, that is, a dispersive estimate:

$$\|U^\varepsilon(t,0)U^\varepsilon(s,0)^*\varphi\|_{L^\infty(\mathbb{R}^n)} = \|U^\varepsilon(t,s)\varphi\|_{L^\infty(\mathbb{R}^n)} \leqslant \frac{C}{(\varepsilon|t-s|)^{n/2}}\|\varphi\|_{L^1(\mathbb{R}^n)},$$

provided that $|t-s| \leqslant \delta$, where C and $\delta > 0$ are independent of $\varepsilon \in]0,1]$. As a consequence, Strichartz estimates are available for U^ε (see e.g. [Keel and Tao (1998)]). Note that as $\varepsilon \to 0$, this dispersion estimate becomes worse and worse: the semi-classical limit $\varepsilon \to 0$ is sometimes referred to as *dispersionless limit*. Denoting

$$p = \frac{4\sigma + 4}{n\sigma},$$

(the pair $(p, 2\sigma + 2)$ is *admissible*, see Definition 7.4), we infer:

Proposition 1.26. *Let V satisfying Assumption 1.7.*
• *If $\sigma < 2/n$ and $u_0^\varepsilon \in L^2$, then (1.30) has a unique solution*

$$u^\varepsilon \in C(\mathbb{R}; L^2) \cap L_{\text{loc}}^p(\mathbb{R}; L^{2\sigma+2}),$$

and (1.24) holds for all $t \in \mathbb{R}$.
• *If $u_0^\varepsilon \in \Sigma(1)$ and $\sigma < 2/(n-2)$ when $n \geqslant 3$, then there exist $T_-^\varepsilon, T_+^\varepsilon > 0$ and a unique solution*

$$u^\varepsilon \in C(] - T_-^\varepsilon, T_+^\varepsilon[; \Sigma(1)) \cap L_{\text{loc}}^p(] - T_-^\varepsilon, T_+^\varepsilon[; W^{1,2\sigma+2})$$

to (1.30). Moreover, the mass (1.24) and the energy (1.25) do not depend on $t \in] - T_-^\varepsilon, T_+^\varepsilon[$.
• *If $V = V(x)$ is sub-linear, $u_0^\varepsilon \in H^1$ and $\sigma < 2/(n-2)$ when $n \geqslant 3$, then there exist $T_-^\varepsilon, T_+^\varepsilon > 0$ and a unique solution*

$$u^\varepsilon \in C(] - T_-^\varepsilon, T_+^\varepsilon[; H^1) \cap L_{\text{loc}}^p(] - T_-^\varepsilon, T_+^\varepsilon[; W^{1,2\sigma+2})$$

to (1.30). Moreover, the mass (1.24) does not depend on $t \in] - T_-^\varepsilon, T_+^\varepsilon[$. If the energy (1.25) is finite at time $t = 0$, then it is independent of $t \in] - T_-^\varepsilon, T_+^\varepsilon[$. If $\lambda \geqslant 0$, then we can take $T_-^\varepsilon = T_+^\varepsilon = \infty$, even if the energy is infinite.
• *If $V = 0$, $u_0^\varepsilon \in \Sigma(1)$ and $\sigma < 2/(n-2)$ when $n \geqslant 3$, then the following evolution law holds so long as $u^\varepsilon \in C_t\Sigma(1)$:*

$$\frac{d}{dt}\left(\frac{1}{2}\|(x + i\varepsilon t\nabla)\, u^\varepsilon\|_{L^2}^2 + \frac{\lambda t^2}{\sigma+1}\|u^\varepsilon\|_{L^{2\sigma+2}}^{2\sigma+2}\right) = $$
$$= \frac{\lambda t}{\sigma+1}(2 - n\sigma)\|u^\varepsilon\|_{L^{2\sigma+2}}^{2\sigma+2}. \tag{1.31}$$

In particular, if $\lambda \geqslant 0$, then $T_-^\varepsilon = T_+^\varepsilon = \infty$, and $u^\varepsilon \in C(\mathbb{R}; \Sigma(1))$.

Proof. The first point follows from the result of Y. Tsutsumi in the case $V = 0$ [Tsutsumi (1987)]. The proof relies on Strichartz estimates. The case $V \neq 0$ proceeds along the same lines, since local in time Strichartz estimates are available thanks to Assumption 1.7: the local in time result is made global thanks to the conservation of mass (1.24), since the local existence time depends only on the L^2 norm of the initial data.

The second point can be found in [Cazenave (2003)] in the case $V = 0$. To adapt it to the case $V \neq 0$, notice that (1.29) show that a closed family of estimates is available for u^ε, ∇u^ε and xu^ε. It is then possible to mimic the proof of the case $V = 0$. For the conservations of mass and energy, we refer to [Cazenave (2003)].

When V is sub-linear, it is possible to work in H^1 only, since

$$\left[\nabla, i\varepsilon\partial_t + \frac{\varepsilon^2}{2}\Delta - V\right] = -\nabla V$$

belongs to $L^\infty_{\text{loc}}(\mathbb{R}; L^\infty(\mathbb{R}^n))$. For the global existence result, rewrite formally the conservation of the energy as

$$\frac{d}{dt}\left(\frac{1}{2}\|\varepsilon\nabla u^\varepsilon(t)\|^2_{L^2} + \frac{\lambda}{\sigma+1}\|u^\varepsilon(t)\|^{2\sigma+2}_{L^{2\sigma+2}}\right) = -\frac{d}{dt}\int_{\mathbb{R}^n} V(x)|u^\varepsilon(t,x)|^2 dx$$

$$= -2\operatorname{Re}\int_{\mathbb{R}^n} V(x)\overline{u}^\varepsilon\partial_t u^\varepsilon dx = -2\operatorname{Im}\int_{\mathbb{R}^n} V(x)\overline{u}^\varepsilon\left(i\partial_t u^\varepsilon\right) dx$$

$$= \operatorname{Im}\int_{\mathbb{R}^n} V(x)\overline{u}^\varepsilon\varepsilon\Delta u^\varepsilon dx = -\operatorname{Im}\int_{\mathbb{R}^n} \overline{u}^\varepsilon\nabla V(x)\cdot\varepsilon\nabla u^\varepsilon dx.$$

We conclude thanks to Cauchy–Schwarz inequality, the conservation of mass and Gronwall lemma, that $\|\nabla u^\varepsilon(t)\|_{L^2}$ remains bounded on bounded time intervals. Therefore the solution is global in time. See [Carles (2008)] for details.

The identity of the last point follows from the *pseudo-conformal conservation law*, derived by J. Ginibre and G. Velo [Ginibre and Velo (1979)] for $\varepsilon = 1$. The case $\varepsilon \in\,]0, 1]$ is easily inferred *via* the scaling

$$(t, x) \mapsto \left(\frac{t}{\varepsilon}, \frac{x}{\varepsilon}\right).$$

Since from the previous point, $\varepsilon\nabla u^\varepsilon \in C(\mathbb{R}; L^2)$ and $u^\varepsilon \in C(\mathbb{R}; L^{2\sigma+2})$, this evolution law shows the *a priori* estimate $xu^\varepsilon \in L^\infty_{\text{loc}}\left(\mathbb{R}; L^2\right)$. □

1.4.4 *Weak solutions*

We will mention weak solutions only in the case $V = 0$, for a defocusing power-like nonlinearity. We therefore consider

$$i\varepsilon\partial_t u^\varepsilon + \frac{\varepsilon^2}{2}\Delta u^\varepsilon = |u^\varepsilon|^{2\sigma}u^\varepsilon \quad ; \quad u^\varepsilon_{|t=0} = u^\varepsilon_0. \tag{1.32}$$

Definition 1.27 (Weak solution). *Let* $u^\varepsilon_0 \in H^1 \cap L^{2\sigma+2}(\mathbb{R}^n)$. *A (global) weak solution to (1.32) is a function* $u^\varepsilon \in C(\mathbb{R}; \mathcal{D}') \cap L^\infty(\mathbb{R}; H^1 \cap L^{2\sigma+2})$ *solving (1.32) in* $\mathcal{D}'(\mathbb{R} \times \mathbb{R}^n) \cap C(\mathbb{R}; L^2)$, *and such that:*

- $\|u^\varepsilon(t)\|_{L^2} = \|u^\varepsilon_0\|_{L^2}$, $\forall t \in \mathbb{R}$.
- $E^\varepsilon(u^\varepsilon(t)) \leqslant E^\varepsilon(u^\varepsilon_0)$, $\forall t \in \mathbb{R}$.

Essentially, the energy conservation is replaced by an inequality, due to a limiting procedure and the use of Fatou's lemma in the construction of weak solutions.

Proposition 1.28 ([Ginibre and Velo (1985a)]). *Let $\sigma > 0$, $\varepsilon \in]0,1]$, and $u_0^\varepsilon \in H^1 \cap L^{2\sigma+2}(\mathbb{R}^n)$. Then (1.32) has a global weak solution. Moreover, if $\sigma < 2/(n-2)$, then this weak solution is unique, and coincides with the mild solution of the last point in Proposition 1.26.*

Chapter 2

Weakly Nonlinear Geometric Optics

In this chapter, we consider the initial value problem

$$i\varepsilon\partial_t u^\varepsilon + \frac{\varepsilon^2}{2}\Delta u^\varepsilon = Vu^\varepsilon + \varepsilon^\alpha f\left(|u^\varepsilon|^2\right)u^\varepsilon \quad ; \quad u^\varepsilon_{|t=0} = a^\varepsilon_0 e^{i\phi_0/\varepsilon}, \quad (2.1)$$

in the case $\alpha \geqslant 1$; note that α is not necessarily an integer. The formal analysis of Sec. 1.2 suggests that if $\alpha > 1$, then the nonlinearity f is not relevant at leading order in the limit $\varepsilon \to 0$. On the other hand, the value $\alpha = 1$ should be critical, and nonlinear effects are expected to influence the behavior of u^ε at leading order. We prove that this holds true. In the case $\alpha > 1$, this means that at leading order, u^ε is described as in Sec. 1.3. When $\alpha = 1$, we describe precisely the nonlinear effect at leading order: it consists of a *nonlinear phase shift*. In other words, the main nonlinear effect is a phase self-modulation.

In the last paragraph of this chapter (Sec. 2.5), we show a consequence of this analysis on the Cauchy problem without semi-classical parameter.

In addition to Assumption 1.7, the following assumption is made throughout all Chap. 2:

Assumption 2.1. We assume that the nonlinearity is smooth, and that the initial amplitude is bounded in the following sense:

- $f \in C^\infty(\mathbb{R};\mathbb{R})$, and $f(0) = 0$.
- There exists $s_0 > n/2$ such that $(a^\varepsilon_0)_\varepsilon$ is bounded in H^{s_0}.

As noticed in Sec. 1.4.2, the assumption $f(0) = 0$ comes for free, up to replacing V by $V + f(0)$ and f by $f - f(0)$. Note that we could also consider the equation

$$i\varepsilon\partial_t u^\varepsilon + \frac{\varepsilon^2}{2}\Delta u^\varepsilon = Vu^\varepsilon + f\left(\varepsilon^\alpha |u^\varepsilon|^2\right)u^\varepsilon \quad ; \quad u^\varepsilon_{|t=0} = a^\varepsilon_0 e^{i\phi_0/\varepsilon}.$$

31

Of course, when f is the identity (cubic nonlinearity), this is the same equation as (2.1). Otherwise, it is a different problem. Despite the appearance, this initial value problem is less general than (2.1) from the technical point of view. Indeed, recall that the WKB analysis considers times where the solution u^ε is of order $\mathcal{O}(1)$ (in L^∞) as $\varepsilon \to 0$. Therefore, since $\alpha \geqslant 1$, the Taylor expansion of f yields

$$f\left(\varepsilon^\alpha |u^\varepsilon|^2\right) u^\varepsilon \underset{\varepsilon \to 0}{\sim} f'(0)\varepsilon^\alpha |u^\varepsilon|^2 u^\varepsilon + \frac{f''(0)}{2}\varepsilon^{2\alpha}|u^\varepsilon|^4 u^\varepsilon + \ldots$$

Two cases can be distinguished. Either $\alpha > 1$, and the analysis for (2.1) will show that, in this case too, the nonlinearity is negligible at leading order before a caustic is formed (if any); or $\alpha = 1$, and only the cubic term $f'(0)\varepsilon |u^\varepsilon|^2 u^\varepsilon$ is expected to be relevant at leading order. In both cases, we leave it as an exercise to adapt the approach presented below, in order to justify these assertions.

2.1   Precised existence results

If $a_0^\varepsilon \in \Sigma(k) = H^k \cap \mathcal{F}(H^k)$ for some $k > n/2$, Proposition 1.25 shows that (2.1) has a unique solution $u^\varepsilon \in C(]-T_-^\varepsilon, T_+^\varepsilon[; \Sigma(k))$ for some $T_-^\varepsilon, T_+^\varepsilon > 0$. In this paragraph, we show that we may construct a strong solution u^ε by assuming only $a_0^\varepsilon \in H^s$ for some $s > n/2$. In addition, we show that u^ε remains bounded on $[-T, T]$, where $T > 0$ is given by Proposition 1.9, provided that a_0^ε in bounded in H^s as $\varepsilon \to 0$. As a corollary, we show that if a_0^ε is uniformly in $\Sigma(k)$ for some $k > n/2$, then $T_\pm^\varepsilon \geqslant T$.

As noticed during the proof of Proposition 1.16, it is more natural to work with

$$a^\varepsilon(t, x) = u^\varepsilon(t, x)e^{-i\phi_{\mathrm{eik}}(t,x)/\varepsilon}$$

than with u^ε directly. We check that (2.1) is equivalent to

$$\begin{cases} \partial_t a^\varepsilon + \nabla\phi_{\mathrm{eik}} \cdot \nabla a^\varepsilon + \dfrac{1}{2}a^\varepsilon \Delta\phi_{\mathrm{eik}} = i\dfrac{\varepsilon}{2}\Delta a^\varepsilon - i\varepsilon^{\alpha-1}f\left(|a^\varepsilon|^2\right)a^\varepsilon, \\ a^\varepsilon_{|t=0} = a_0^\varepsilon. \end{cases} \quad (2.2)$$

Proposition 2.2. *Let Assumptions 1.7 and 2.1 be satisfied. Then (2.2) has a unique solution $a^\varepsilon \in C([-T, T]; H^{s_0})$, where T is given by Proposition 1.9. Moreover, $(a^\varepsilon)_\varepsilon$ is bounded in $C([-T, T]; H^{s_0})$. If $(a_0^\varepsilon)_\varepsilon$ is bounded in H^s for some $s \geqslant s_0$, then $(a^\varepsilon)_\varepsilon$ is bounded in $C([-T, T]; H^s)$.*

Proof. There are at least two procedures to construct a solution to (2.2): an iterative scheme, or Galerkin methods. For the iterative scheme, we solve, for $j \geqslant 0$:

$$\begin{cases} \partial_t a_{j+1}^\varepsilon + \nabla \phi_{\text{eik}} \cdot \nabla a_{j+1}^\varepsilon + \dfrac{1}{2} a_{j+1}^\varepsilon \Delta \phi_{\text{eik}} = i \dfrac{\varepsilon}{2} \Delta a_{j+1}^\varepsilon - i \varepsilon^{\alpha-1} f\left(|a_j^\varepsilon|^2\right) a_j^\varepsilon, \\ a_{j+1|t=0}^\varepsilon = a_0^\varepsilon. \end{cases}$$

This is actually a linear Schrödinger equation: setting $u_j^\varepsilon := a_j^\varepsilon e^{i\phi_{\text{eik}}/\varepsilon}$, we see that the above equation is equivalent to:

$$i\varepsilon \partial_t u_{j+1}^\varepsilon + \frac{\varepsilon^2}{2} \Delta u_{j+1}^\varepsilon = V u_{j+1}^\varepsilon + \varepsilon^\alpha f\left(|u_j^\varepsilon|^2\right) u_j^\varepsilon \quad ; \quad u_{j+1|t=0}^\varepsilon = a_0^\varepsilon e^{i\phi_0/\varepsilon}.$$

Using Galerkin methods, we can mimic the mollification procedure presented for instance in [Alinhac and Gérard (1991); Majda (1984)]; roughly speaking, we solve an ordinary differential equation in H^s along characteristics by considering

$$\begin{cases} \partial_t a_h^\varepsilon + J_h \left(\nabla \phi_{\text{eik}} \cdot \nabla J_h a_h^\varepsilon\right) + \dfrac{1}{2} a_h^\varepsilon \Delta \phi_{\text{eik}} = i \dfrac{\varepsilon}{2} \Delta J_h^2 a_h^\varepsilon - i \varepsilon^{\alpha-1} f\left(|a_h^\varepsilon|^2\right) a_h^\varepsilon, \\ a_{h|t=0}^\varepsilon = a_0^\varepsilon, \end{cases}$$

where $J_h = \jmath(hD)$ is a Fourier multiplier, with $\jmath \in C_0^\infty(\mathbb{R}^n; \mathbb{R})$ equal to one in a neighborhood of the origin.

For both methods, the problem boils down to obtaining energy estimates for (2.2) in H^s, for all $s \geqslant s_0$. Let $s > n/2$. Applying the operator $\Lambda^s = (\text{Id} - \Delta)^{s/2}$ to (2.2), we find:

$$\partial_t \Lambda^s a^\varepsilon + \nabla \phi_{\text{eik}} \cdot \nabla \Lambda^s a^\varepsilon = i \frac{\varepsilon}{2} \Delta \Lambda^s a^\varepsilon - i \varepsilon^{\alpha-1} \Lambda^s \left(f\left(|a^\varepsilon|^2\right) a^\varepsilon\right) + R_s^\varepsilon, \quad (2.3)$$

where

$$R_s^\varepsilon = \left[\nabla \phi_{\text{eik}} \cdot \nabla, \Lambda^s\right] a^\varepsilon - \frac{1}{2} \Lambda^s \left(a^\varepsilon \Delta \phi_{\text{eik}}\right).$$

Take the inner product of (2.3) with $\Lambda^s a^\varepsilon$, and consider the real part: the first term of the right hand side of (2.3) vanishes, since $i\Delta$ is skew-symmetric, and we have:

$$\frac{1}{2} \frac{d}{dt} \|\Lambda^s a^\varepsilon\|_{L^2}^2 + \text{Re} \int_{\mathbb{R}^n} \overline{\Lambda^s a^\varepsilon} \nabla \phi_{\text{eik}} \cdot \nabla \Lambda^s a^\varepsilon \leqslant \varepsilon^{\alpha-1} \left\|f\left(|a^\varepsilon|^2\right) a^\varepsilon\right\|_{H^s} \|a^\varepsilon\|_{H^s}$$

$$+ \|R_s^\varepsilon\|_{L^2} \|a^\varepsilon\|_{H^s}.$$

Notice that we have

$$\left|\text{Re} \int_{\mathbb{R}^n} \overline{\Lambda^s a^\varepsilon} \nabla \phi_{\text{eik}} \cdot \nabla \Lambda^s a^\varepsilon\right| = \frac{1}{2} \left|\int_{\mathbb{R}^n} \nabla \phi_{\text{eik}} \cdot \nabla |\Lambda^s a^\varepsilon|^2\right|$$

$$= \frac{1}{2} \left|\int_{\mathbb{R}^n} |\Lambda^s a^\varepsilon|^2 \Delta \phi_{\text{eik}}\right| \leqslant C \|a^\varepsilon\|_{H^s}^2,$$

since $\Delta\phi_{\text{eik}} \in L^\infty([-T, T] \times \mathbb{R}^n)$ from Proposition 1.9. Moser's inequality (Lemma 1.24) yields, in view of Assumption 2.1:

$$\left\| f\left(|a^\varepsilon|^2\right) a^\varepsilon \right\|_{H^s} \leqslant C\left(\|a^\varepsilon\|_{L^\infty}\right) \|a^\varepsilon\|_{H^s}.$$

We infer:

$$\frac{d}{dt}\|a^\varepsilon\|_{H^s}^2 \leqslant C\left(\|a^\varepsilon\|_{L^\infty}\right) \|a^\varepsilon\|_{H^s}^2 + \|R_s^\varepsilon\|_{H^s} \|a^\varepsilon\|_{H^s}.$$

Note that the above locally bounded map $C(\cdot)$ can be taken independent of ε if and only if $\alpha \geqslant 1$. To apply Gronwall lemma, we need to estimate the last term: we use the fact that the derivatives of order at least two of ϕ_{eik} are bounded, from Proposition 1.9, to have:

$$\|R_s^\varepsilon\|_{L^2} \leqslant C\|a^\varepsilon\|_{H^s}.$$

We can then conclude by Gronwall lemma and a continuity argument:

$$\|a^\varepsilon\|_{L^\infty([-T,T];H^s)} \leqslant C\left(s, \|a_0^\varepsilon\|_{H^s}\right).$$

This yields boundedness in the "high" norm. Contraction in the "small" norm (that is, contraction in L^2) follows easily. Let a^ε and b^ε be solutions to (2.2), with initial data a_0^ε and b_0^ε respectively. Assume that a^ε and b^ε are bounded in $L^\infty([-T, T]; H^s)$ for some $s > n/2$. The difference $w^\varepsilon = a^\varepsilon - b^\varepsilon$ solves

$$\partial_t w^\varepsilon + \nabla\phi_{\text{eik}} \cdot \nabla w^\varepsilon + \frac{1}{2} w^\varepsilon \Delta\phi_{\text{eik}} = i\frac{\varepsilon}{2}\Delta w^\varepsilon - i\varepsilon^{\alpha-1}\left(f\left(|a^\varepsilon|^2\right) a^\varepsilon - f\left(|b^\varepsilon|^2\right) b^\varepsilon\right).$$

The above computations yield

$$\frac{d}{dt}\|w^\varepsilon(t)\|_{L^2}^2 \leqslant C\|w^\varepsilon(t)\|_{L^2}^2 + \left\| f\left(|a^\varepsilon|^2\right) a^\varepsilon - f\left(|b^\varepsilon|^2\right) b^\varepsilon \right\|_{L^2}^2,$$

where we have used Young's inequality

$$xy \leqslant \frac{1}{2}\left(x^2 + y^2\right), \quad \forall x, y \geqslant 0.$$

Denote $g(z) = f(|z|^2)z$. Using Taylor's formula, write

$$f\left(|a^\varepsilon|^2\right) a^\varepsilon - f\left(|b^\varepsilon|^2\right) b^\varepsilon = g\left(w^\varepsilon + b^\varepsilon\right) - g\left(b^\varepsilon\right)$$

$$= w^\varepsilon \int_0^1 \partial_z g\left(b^\varepsilon + \theta w^\varepsilon\right) d\theta + \overline{w}^\varepsilon \int_0^1 \partial_{\overline{z}} g\left(b^\varepsilon + \theta w^\varepsilon\right) d\theta.$$

We infer

$$\left\| f\left(|a^\varepsilon|^2\right) a^\varepsilon - f\left(|b^\varepsilon|^2\right) b^\varepsilon \right\|_{L^2}^2 \leqslant \|w^\varepsilon\|_{L^2}^2 \left\| \int_0^1 g'\left(b^\varepsilon + \theta w^\varepsilon\right) d\theta \right\|_{L^\infty}^2$$

$$\leqslant C\left(\|w^\varepsilon\|_{L^\infty}, \|b^\varepsilon\|_{L^\infty}\right) \|w^\varepsilon\|_{L^2}^2$$

$$\leqslant \widetilde{C}\left(\|a^\varepsilon\|_{L^\infty}, \|b^\varepsilon\|_{L^\infty}\right) \|w^\varepsilon\|_{L^2}^2.$$

This yields the contraction in the L^2 norm on small time intervals, hence the proposition. $\qquad\square$

Remark 2.3. It is not true in general that $u^\varepsilon \in C([-T,T]; H^s)$. Indeed, u^ε is the product of $a^\varepsilon \in C([-T,T]; H^s)$ and $e^{i\phi_{\text{eik}}/\varepsilon}$. If $xa^\varepsilon \notin C([-T,T]; L^2)$ and if ϕ_{eik} grows quadratically in space, then $u^\varepsilon \notin C([-T,T]; H^1)$. This phenomenon is geometric, not nonlinear: if $f = 0$ and if V is an harmonic potential, then u^ε may instantly cease to be in H^s for all $s > 0$ (but not $s = 0$). See Sec. 2.5.

Corollary 2.4. *Let Assumptions 1.7 and 2.1 be satisfied. If in addition $a_0^\varepsilon \in \Sigma(k)$ for an integer $k > n/2$, then $a^\varepsilon \in C([-T,T]; \Sigma(k))$. If $(a_0^\varepsilon)_\varepsilon$ is bounded in $\Sigma(k)$, then $(a^\varepsilon)_\varepsilon$ is bounded in $C([-T,T]; \Sigma(k))$.*

Proof. Proposition 2.2 shows that $(a^\varepsilon)_\varepsilon$ is bounded in $C([-T,T]; H^k)$. By multiplying (2.2) by x^β and using energy estimates as in the proof of Proposition 2.2, induction on $k' = |\beta|$ yields the corollary. $\square$

2.2 Leading order asymptotic analysis

To pass to the limit in (2.2), we assume:

There exist $s > \dfrac{n}{2} + 2$ and $a_0 \in H^s$, such that $a_0^\varepsilon \xrightarrow[\varepsilon \to 0]{} a_0$ in H^{s-2}. (2.4)

Since we deal with a general $\alpha \geqslant 1$, we keep the last term in (2.2), and consider

$$\begin{cases} \partial_t \widetilde{a}^\varepsilon + \nabla \phi_{\text{eik}} \cdot \nabla \widetilde{a}^\varepsilon + \dfrac{1}{2} \widetilde{a}^\varepsilon \Delta \phi_{\text{eik}} = -i\varepsilon^{\alpha-1} f\left(|\widetilde{a}^\varepsilon|^2\right) \widetilde{a}^\varepsilon, \\ \widetilde{a}^\varepsilon_{|t=0} = a_0. \end{cases} \quad (2.5)$$

The proof of Proposition 2.2 shows that (2.5) has a unique solution $\widetilde{a}^\varepsilon$, which is bounded in $C([-T,T]; H^s)$, where s appears in (2.4).

Proposition 2.5. *Let Assumptions 1.7 and 2.1 be satisfied, as well as (2.4). Then there exists $C > 0$ such that*

$$\|a^\varepsilon - \widetilde{a}^\varepsilon\|_{L^\infty([-T,T]; H^{s-2})} \leqslant C\left(\varepsilon + \|a_0^\varepsilon - a_0\|_{H^{s-2}}\right).$$

Proof. Set $w_a^\varepsilon = a^\varepsilon - \widetilde{a}^\varepsilon$. It solves

$$\begin{cases} \partial_t w_a^\varepsilon + \nabla \phi_{\text{eik}} \cdot \nabla w_a^\varepsilon + \dfrac{1}{2} w_a^\varepsilon \Delta \phi_{\text{eik}} = i\dfrac{\varepsilon}{2} \Delta w_a^\varepsilon + i\dfrac{\varepsilon}{2} \Delta \widetilde{a}^\varepsilon \\ \qquad\qquad\qquad\qquad\qquad\qquad\qquad - i\varepsilon^{\alpha-1}\left(g\left(a^\varepsilon\right) - g\left(\widetilde{a}^\varepsilon\right)\right), \\ w_a^\varepsilon{}_{|t=0} = a_0^\varepsilon - a_0, \end{cases}$$

where $g(z) = f(|z|^2)z$. Note that the term $i\Delta w_a^\varepsilon$ vanishes from the energy estimates. The term $i\varepsilon \Delta \widetilde{a}^\varepsilon$ is viewed as a source term of order $\mathcal{O}(\varepsilon)$

in $C([-T,T]; H^{s-2})$. Since $s - 2 > n/2$, g is uniformly Lipschitzean on bounded sets of H^{s-2} (Lemma 1.23), and the same computations as in the proof of Proposition 2.2 show that we can apply Gronwall lemma, which yields the proposition. $\qquad\square$

Proposition 2.5 shows that the leading order asymptotic behavior of u^ε is given by

$$\left\| u^\varepsilon - \widetilde{a}^\varepsilon e^{i\phi_{\mathrm{eik}}/\varepsilon} \right\|_{L^\infty([-T,T]; L^2 \cap L^\infty)} \xrightarrow[\varepsilon \to 0]{} 0.$$

Note also that in the case $\alpha = 1$, $\widetilde{a}^\varepsilon$ *does not* depend on ε. Therefore, the formal computations of Sec. 1.2 are justified at leading order when $\alpha = 1$. It turns out that (2.5) can be solved rather explicitly, in terms of the geometric objects introduced in Sec. 1.3, as shown in the next paragraph, where the case $\alpha > 1$ is also analyzed more precisely.

2.3 Interpretation

Recall from Sec. 1.3 that classical trajectories are given by the Hamiltonian system (1.11):

$$\begin{cases} \partial_t x(t,y) = \xi(t,y) & ; \quad x(0,y) = y, \\ \partial_t \xi(t,y) = -\nabla_x V(t, x(t,y)) & ; \quad \xi(0,y) = \nabla\phi_0(y). \end{cases}$$

For $t \in [-T,T]$, $y \mapsto x(t,y)$ is a diffeomorphism on $\mathbb{R}^n$, and the Jacobi's determinant

$$J_t(y) = \det\nabla_y x(t,y)$$

is uniformly bounded from above and from below:

$$\exists C > 0, \quad \frac{1}{C} \leqslant J_t(y) \leqslant C, \quad \forall (t,y) \in [-T,T] \times \mathbb{R}^n. \tag{2.6}$$

Denote

$$\widetilde{A}^\varepsilon(t,y) := \widetilde{a}^\varepsilon(t, x(t,y)) \sqrt{J_t(y)}.$$

For $t \in [-T,T]$, (2.5) is equivalent to:

$$\partial_t \widetilde{A}^\varepsilon = -i\varepsilon^{\alpha-1} f\left(J_t(y)^{-1} \left| \widetilde{A}^\varepsilon \right|^2 \right) \widetilde{A}^\varepsilon \quad ; \quad \widetilde{A}^\varepsilon(0,y) = a_0(y).$$

Despite the appearances, this ordinary differential equation along the rays of geometrical optics is a *linear* equation. Indeed, since f is real-valued, it is of the form

$$\partial_t A = i\mathcal{V}A, \quad \mathcal{V} \in \mathbb{R}.$$

This implies $\partial_t |A|^2 = 0$. In our case, $\partial_t |\widetilde{A}^\varepsilon|^2 = 0$, hence

$$\widetilde{A}^\varepsilon(t,y) = a_0(y) \exp\left(-i\varepsilon^{\alpha-1} \int_0^t f\left(J_s(y)^{-1} |a_0(y)|^2 \right) ds \right).$$

Back to the function $\widetilde{a}^\varepsilon$, we have:

$$\widetilde{a}^\varepsilon(t,x) = a(t,x) e^{i\varepsilon^{\alpha-1} G(t,x)},$$

where a is given in the linear case by (1.20), that is

$$a(t,x) = \frac{1}{\sqrt{J_t(y(t,x))}} a_0\left(y(t,x) \right),$$

and the phase shift G is given by:

$$G(t,x) = -\int_0^t f\left(J_s(y(t,x))^{-1} |a_0(y(t,x))|^2 \right) ds. \tag{2.7}$$

If $\alpha > 1$, then

$$\widetilde{a}^\varepsilon(t,x) - a(t,x) = \mathcal{O}\left(\varepsilon^{\alpha-1} \right),$$

and no nonlinear effect is present at leading order. On the other hand, in the case $\alpha = 1$, we see that the leading order nonlinear effect is described by the function G: a phase shift generated by a nonlinear mechanism. In the context of laser physics, this phenomenon is known as *phase self-modulation* (see e.g. [Zakharov and Shabat (1971); Boyd (1992)]). Note that this function G does not affect the convergence of the main two quadratic quantities:

Position density: $\quad \rho^\varepsilon = |u^\varepsilon|^2 = |a^\varepsilon|^2,$

Current density: $\quad J^\varepsilon = \varepsilon \operatorname{Im}\left(\overline{u}^\varepsilon \nabla u^\varepsilon \right) = |a^\varepsilon|^2 \nabla \phi_{\text{eik}} + \varepsilon \operatorname{Im}\left(\overline{a}^\varepsilon \nabla a^\varepsilon \right).$

We refer to Chap. 3 for further discussions on these quantities.

One may wonder if this approach could be extended to some values $\alpha < 1$. To have a simple ansatz as above, we would like to remove the Laplacian in the limit $\varepsilon \to 0$ in (2.2). We find:

$$\widetilde{a}^\varepsilon(t,x) = a(t,x) e^{i\varepsilon^{\alpha-1} G(t,x)},$$

where a and G are given by the same expressions as above. Now recall that in Proposition 2.2, we prove that a^ε is bounded in H^s, uniformly for $\varepsilon \in]0,1]$; this property is used to approximate a^ε by $\widetilde{a}^\varepsilon$. But when $\alpha < 1$, $\widetilde{a}^\varepsilon$ is no longer uniformly bounded in H^s, because what was a phase modulation for $\alpha \geqslant 1$ is now a rapid oscillation. This is another hint that when $\alpha < 1$, the approach must be modified. The analogous study in the case $\alpha = 0$ is presented in Chap. 4. See also Sec. 4.2.3 for the case $0 < \alpha < 1$.

2.4　Higher order asymptotic analysis

To compute the next term in the asymptotic expansion of a^ε, we assume that $\alpha \in \mathbb{N} \setminus \{0\}$, and

$$\exists s > \frac{n}{2} + 4, a_0 \in H^s, a_1 \in H^{s-2} / \; \|a_0^\varepsilon - a_0 - \varepsilon a_1\|_{H^{s-4}} = o(\varepsilon). \qquad (2.8)$$

Define $a^{(0)}$ by

$$a^{(0)}(t,x) = \begin{cases} a(t,x) & \text{if } \alpha \geqslant 2, \\ a(t,x)e^{iG(t,x)} & \text{if } \alpha = 1. \end{cases}$$

Define the first corrector $a^{(1)}$ in the WKB analysis as the solution to

$$\partial_t a^{(1)} + \nabla\phi_{\text{eik}} \cdot \nabla a^{(1)} + \frac{1}{2}a^{(1)}\Delta\phi_{\text{eik}} = \frac{i}{2}\Delta a^{(0)} + S_\alpha, \qquad (2.9)$$

with initial datum $a^{(1)}(0,x) = a_1(x)$, where

$$S_\alpha = \begin{cases} 0 & \text{if } \alpha \geqslant 3, \\ -if\left(\left|a^{(0)}\right|^2\right)a^{(0)} & \text{if } \alpha = 2, \\ -if\left(\left|a^{(0)}\right|^2\right)a^{(1)} - 2if'\left(\left|a^{(0)}\right|^2\right)a^{(0)}\,\text{Re}\left(\overline{a}^{(0)}a^{(1)}\right) & \text{if } \alpha = 1. \end{cases}$$

Note that in all the cases, (2.9) is a linear equation. Since ϕ_{eik} is smooth and $a^{(0)} \in C([-T,T]; H^s)$, the regularity of $a^{(1)}$ is given by the regularity of its initial datum and of its source term $i\Delta a^{(0)} \in C([-T,T]; H^{s-2})$. Since $s > n/2$, the term $\|S_\alpha\|_{H^s}$ is controlled by $1 + \|a^{(1)}\|_{H^s}$ in all the cases, so $a^{(1)} \in C([-T,T]; H^{s-2})$. Using the same approach as above, the following result is left as an exercise:

Proposition 2.6. *Let Assumptions 1.7 and 2.1 be satisfied, as well as (2.9). Then there exists $C > 0$ such that*

$$\left\|a^\varepsilon - a^{(0)} - \varepsilon a^{(1)}\right\|_{L^\infty([-T,T];H^{s-4})} \leqslant C\left(\varepsilon + \|a_0^\varepsilon - a_0 - \varepsilon a_1\|_{H^{s-4}}\right).$$

We see that this analysis can be continued to arbitrary order, with essentially the loss of two derivatives at each step, like in the linear case. To justify the asymptotics at order $j \geqslant 0$, we assume $s > n/2 + 2j + 2$. We leave out the computations here, for the higher order analysis bears no new difficulty or interest.

2.5 An application: Cauchy problem in Sobolev spaces for nonlinear Schrödinger equations with potential

In this paragraph, we set $\varepsilon = 1$, and consider the Cauchy problem

$$i\partial_t u + \frac{1}{2}\Delta u = Vu + f\left(|u|^2\right)u \quad ; \quad u_{|t=0} = a_0. \tag{2.10}$$

We assume $a_0 \in H^s$ for $s > n/2$, $f \in C^\infty(\mathbb{R};\mathbb{R})$, and $V = V(x)$ to simplify, satisfying Assumption 1.7. We consider Eq. (2.10) as a weakly nonlinear equation ($\alpha = 1$), even though $\varepsilon = 1$. We have seen in Sec. 1.4.2 that under Assumption 1.7, it is fairly natural to work in the space

$$\Sigma(k) = H^k \cap \mathcal{F}(H^k) = \{f \in H^k(\mathbb{R}^n) \; ; \; x \mapsto \langle x \rangle^k f(x) \in L^2(\mathbb{R}^n)\},$$

for some $k > n/2$. When $\nabla V \in L^\infty(\mathbb{R}^n)$, $\Sigma(k)$ can be replaced by H^k. We now address the same question when ∇V is unbounded. The typical example of such an occurrence under Assumption 1.7 is when V is an harmonic potential, say $V(x) = |x|^2$. In the linear case $f = 0$, it is well-known that V acts as a rotation in the phase space. Therefore, a_0 must have similar properties on the x-side and on the ξ-side in order for the Sobolev regularity to be propagated. In particular, if $a_0 \in H^\infty$ with $x \mapsto \langle x \rangle a_0(x) \notin L^2$, then $u(t, \cdot)$ ceases to be in H^s for $s \geqslant 1$ as soon as $t > 0$. It is natural to expect a similar phenomenon in the nonlinear case. We show that this is so, with a proof which is valid both for the linear case and the nonlinear case ([Carles (2008)]).

First, introduce ϕ_{eik} solution to the eikonal equation

$$\partial_t \phi_{\text{eik}} + \frac{1}{2}\left|\nabla \phi_{\text{eik}}\right|^2 + V = 0 \quad ; \quad \phi_{\text{eik}}(0, x) = 0.$$

From Proposition 1.9, there exist $T > 0$ and a unique $\phi_{\text{eik}} \in C^\infty([-T, T] \times \mathbb{R}^n)$ solution to the above equation. Proposition 2.2 shows that there exists $a \in C([-T, T]; H^\infty)$ such that $u = ae^{i\phi_{\text{eik}}}$ solves (2.10). The function a solves:

$$\partial_t a + \nabla \phi_{\text{eik}} \cdot \nabla a + \frac{1}{2}a\Delta\phi_{\text{eik}} = \frac{i}{2}\Delta a - if\left(|a|^2\right)a \quad ; \quad a_{|t=0} = a_0. \tag{2.11}$$

Since in particular, $u \in C([-T, T]; L^2 \cap L^\infty)$, u is the unique such solution: if $v \in C([-T, T]; L^2 \cap L^\infty)$ solves Eq. (2.10), then $w = u - v$ solves

$$i\partial_t w + \frac{1}{2}\Delta w = Vw + f\left(|v + w|^2\right)(v + w) - f\left(|v|^2\right)v \; ; \; w_{|t=0} = 0.$$

Lemma 1.2 shows that for $t > 0$,

$$\|w(t)\|_{L^2} \leqslant \int_0^t \left\| f\left(|v+w|^2\right)(v+w) - f\left(|v|^2\right)v \right\|_{L^2} d\tau$$

$$\leqslant C \int_0^t \|w(\tau)\|_{L^2} \, d\tau.$$

Gronwall lemma implies $w \equiv 0$.

Proposition 2.7. *Let $n \geqslant 1$, and f be smooth, $f \in C^\infty(\mathbb{R}_+; \mathbb{R})$. Assume that V is super-linear, and that there exist $0 < k(\leqslant 1)$ and $C > 0$ such that*

$$|\nabla V(x)| \leqslant C \langle x \rangle^k, \quad \forall x \in \mathbb{R}^n,$$

and $\omega, \omega' \in \mathbb{S}^{n-1}$ such that

$$|\omega \cdot \nabla V(x)| \geqslant c \left| \omega' \cdot x \right|^k \text{ as } |x| \to \infty, \text{ for some } c > 0. \tag{2.12}$$

Then there exists $a_0 \in H^\infty$ such that for arbitrarily small $t > 0$ and all $s > 0$, the solution $u(t, \cdot)$ to (2.10) fails to be in $H^s(\mathbb{R}^n)$.

Example 2.8. For V, we may consider any non-trivial quadratic form, or $V(x) = \pm \langle x' \rangle^a$, with $1 < a \leqslant 2$, for some decomposition $x = (x', x'') \in \mathbb{R}^n$.

Remark 2.9. No assumption is made on the growth of the nonlinearity at infinity: the above result reveals a *geometric phenomenon*, and not an ill-posedness result like in Chap. 5.

To prove the above result, recall that $u = ae^{i\phi_{\text{eik}}}$, where a is given by (2.11). Consider only positive times, and define b as the solution on $[0, T]$ to:

$$\partial_t b + \nabla \phi_{\text{eik}} \cdot \nabla b + \frac{1}{2} b \Delta \phi_{\text{eik}} = -if\left(|b|^2\right) b \quad ; \quad b_{|t=0} = a_0. \tag{2.13}$$

As noticed in Sec. 2.3,

$$b(t, x) = \frac{1}{\sqrt{J_t(y(t,x))}} a_0\left(y(t,x)\right) e^{-i\beta(t,x)},$$

$$\text{where } \beta(t,x) = \int_0^t f\left(J_s\left(y(t,x)\right)^{-1} |a_0\left(y(t,x)\right)|^2\right) ds,$$

$x(t, y)$ is given by Eq. (1.11) with $\phi_0 = 0$, and the Jacobi's determinant is defined by

$$J_t(y) = \det \nabla_y x(t, y).$$

Observe that $b \in C([0,T]; H^\infty)$. Let $r = a - b$: $r \in C([0,T]; H^\infty)$. For $1 \leqslant j \leqslant n$, $x_j r$ solves:

$$\begin{cases} \partial_t(x_j r) + \nabla\phi_{\text{eik}} \cdot \nabla(x_j r) + \dfrac{1}{2} x_j r \Delta\phi_{\text{eik}} = \dfrac{i}{2}\Delta(x_j r) + r\partial_j\phi_{\text{eik}} - i\partial_j r \\[2mm] \qquad\qquad + \dfrac{i}{2} x_j \Delta b - i x_j \left(f\left(|b+r|^2\right)(b+r) - f\left(|b|^2\right) b \right), \\[2mm] \qquad x_j r_{|t=0} = 0. \end{cases}$$

The fundamental theorem of calculus yields:

$$\begin{aligned} x_j \left(f\left(|a|^2\right) a - f\left(|b|^2\right) b \right) &= x_j \left(f\left(|b+r|^2\right)(b+r) - f\left(|b|^2\right) b \right) \\ &= x_j r \int_0^1 \partial_z g\left(b + sr\right) ds + x_j \bar{r} \int_0^1 \partial_{\bar{z}} g\left(b + sr\right) ds, \end{aligned}$$

where $g(z) = f(|z|^2)z$. In particular, we know that

$$\int_0^1 \partial_z g\left(b + sr\right) ds, \int_0^1 \partial_{\bar{z}} g\left(b + sr\right) ds \in C \cap L^\infty([0,T] \times \mathbb{R}^d).$$

Energy estimates as in §2.1 show that:

$$\|xr\|_{L^\infty([0,t]; L^2)} \leqslant C \left(1 + \|x\Delta b\|_{L^1([0,t]; L^2)}\right).$$

We must make sure that the last term is, or can be chosen, finite. We shall demand $x\Delta b \in L^\infty([0,T]; L^2)$. In view of (2.6), this requirement is met as soon as $a_0 \in H^\infty(\mathbb{R}^n)$ is such that $x\Delta a_0, xa_0|\nabla a_0|^2 \in L^2(\mathbb{R}^n)$:

If $a_0 \in H^\infty$ is such that $x\Delta a_0, xa_0|\nabla a_0|^2 \in L^2(\mathbb{R}^n)$, then: $a = b + r$, with $b, r \in C([0,T]; H^\infty)$, and $xr \in C([0,T]; L^2)$. $\qquad$ (2.14)

We now prove that for small times, $\nabla\phi_{\text{eik}}(t,x)$ can be approximated by $-t\nabla V(x)$.

Lemma 2.10. *Assume that there exist $0 \leqslant k \leqslant 1$ and $C > 0$ such that*

$$|\nabla V(x)| \leqslant C \langle x \rangle^k, \quad \forall x \in \mathbb{R}^n.$$

Then there exist $T_0, C_0 > 0$ such that

$$|\nabla\phi_{\text{eik}}(t,x) + t\nabla V(x)| \leqslant C_0 t^2 \langle x \rangle^k, \quad \forall t \in [0, T_0].$$

Proof. We infer from Proposition 1.9 that

$$\begin{aligned} |\partial_t \nabla\phi_{\text{eik}}(t,x) + \nabla V(x)| &\leqslant \|\nabla^2\phi_{\text{eik}}(t)\|_{L^\infty} |\nabla\phi_{\text{eik}}(t,x)| \\ &\lesssim |\nabla\phi_{\text{eik}}(t,x)|. \end{aligned} \qquad (2.15)$$

From Eq. (1.11), we also have

$$|\nabla\phi_{\text{eik}}(t,x)| = |\xi(t,y(t,x))| = \left|\int_0^t \nabla V\left(x(s,y(t,x))\right) ds\right|$$

$$\lesssim \int_0^t |\nabla V(y(t,x))|\, ds + \int_0^t |x(s,y(t,x)) - y(t,x)|\, ds.$$

We claim that

$$|x(t,y) - y| \lesssim t^2 \langle y \rangle^k. \qquad (2.16)$$

Indeed, we have from (1.11),

$$|x(t,y) - y| = \left|\int_0^t \partial_t x(s,y) ds\right| = \left|\int_0^t \int_0^s \nabla V\left(x(s',y)\right) ds' ds\right|$$

$$= \left|\int_0^t (t-s')\nabla V\left(x(s',y)\right) ds'\right|$$

$$= \left|\int_0^t (t-s)\nabla V(y)\, ds + \int_0^t (t-s)\left(\nabla V\left(x(s,y)\right) - \nabla V(y)\right) ds\right|$$

$$\lesssim t^2 \langle y \rangle^k + \int_0^t (t-s)|x(s,y) - y|\, ds,$$

and (2.16) follows from Gronwall lemma. We infer that for $t > 0$ sufficiently small,

$$|y(t,x) - x| \lesssim t^2 \langle x \rangle^k,$$

and therefore,

$$|\nabla\phi_{\text{eik}}(t,x)| \lesssim \int_0^t |\nabla V(y(t,x))|\, ds + \int_0^t |x(s,y(t,x)) - y(t,x)|\, ds$$

$$\lesssim \int_0^t |\nabla V(x)|\, ds + \int_0^t |x - y(t,x)|\, ds$$

$$+ \int_0^t |x(s,y(t,x)) - y(t,x)|\, ds$$

$$\lesssim t \langle x \rangle^k + t^3 \langle x \rangle^k + \int_0^t s^2 \langle y(t,x) \rangle^k\, ds$$

$$\lesssim t \langle x \rangle^k + t^3 \langle x \rangle^k + t^3 \left(\langle x \rangle^k + t^{2k} \langle x \rangle^{2k}\right).$$

Then (2.15) yields

$$|\partial_t \nabla\phi_{\text{eik}}(t,x) + \nabla V(x)| \lesssim t \langle x \rangle^k,$$

Lemma 2.10 follows by integration in time.   □

We infer that for $t > 0$ small enough,

$$|\omega \cdot \nabla \phi_{\text{eik}}(t, x)| \gtrsim t|\omega \cdot \nabla V(x)|. \qquad (2.17)$$

To complete the proof of Proposition 2.7, consider

$$a_0(x) = \frac{1}{\langle x \rangle^{n/2} \log(2 + |x|^2)}. \qquad (2.18)$$

As is easily checked, a_0 meets the requirements of the first line of (2.14). Denote

$$v = be^{i\phi_{\text{eik}}} \quad ; \quad w = re^{i\phi_{\text{eik}}}.$$

Obviously, $u = v + w$. From (2.14) and (2.17), $v(t, \cdot) \in L^2(\mathbb{R}^n) \setminus H^1(\mathbb{R}^n)$ for $t > 0$ sufficiently small. On the other hand, $w(t, \cdot) \in H^1(\mathbb{R}^n)$ for all $t \in [0, T]$, hence $u(t, \cdot) \in L^2(\mathbb{R}^n) \setminus H^1(\mathbb{R}^n)$ for $0 < t \ll 1$.

We now just have to see that the same holds if we replace $H^1(\mathbb{R}^n)$ with $H^s(\mathbb{R}^n)$ for $0 < s < 1$. We use the following characterization of $H^s(\mathbb{R}^n)$ (see e.g. [Chemin (1998)]): for $\varphi \in L^2(\mathbb{R}^n)$ and $0 < s < 1$,

$$\varphi \in H^s(\mathbb{R}^n) \iff \iint_{\mathbb{R}^n \times \mathbb{R}^n} \frac{|\varphi(x + y) - \varphi(x)|^2}{|y|^{n+2s}} dx dy < \infty.$$

Since $w(t, \cdot) \in H^1$ for all $t \in [0, T]$, we shall prove that $v(t, \cdot) \in L^2 \setminus H^s$ for t sufficiently small. We prove that for $0 < t \ll 1$ independent of $s \in]0, 1[$,

$$I := \int_{|y| \leqslant 1} \int_{x \in \mathbb{R}^n} \frac{|v(t, x + y) - v(t, x)|^2}{|y|^{n+2s}} dx dy = \infty.$$

To apply a fractional Leibnitz rule, write

$$v(t, x + y) - v(t, x) = (b(t, x + y) - b(t, x)) e^{i\phi_{\text{eik}}(t, x+y)}$$
$$+ \left(e^{i\phi_{\text{eik}}(t, x+y)} - e^{i\phi_{\text{eik}}(t, x)}\right) b(t, x).$$

In view of the inequality $|\alpha - \beta|^2 \geqslant \alpha^2/2 - \beta^2$, we have:

$$|v(t, x + y) - v(t, x)|^2 \geqslant \frac{1}{2} \left|\left(e^{i\phi_{\text{eik}}(t, x+y)} - e^{i\phi_{\text{eik}}(t, x)}\right) b(t, x)\right|^2$$
$$- |b(t, x + y) - b(t, x)|^2.$$

We can leave out the last term, since $b(t, \cdot) \in H^\infty$ for $t \in [0, T]$:

$$\iint_{\mathbb{R}^n \times \mathbb{R}^n} \frac{|b(t, x + y) - b(t, x)|^2}{|y|^{n+2s}} dx dy < \infty, \quad \forall t \in [0, T].$$

We now want to prove

$$\int_{|y| \leqslant 1} \int_{x \in \mathbb{R}^n} |b(t, x)|^2 \frac{\left|\sin\left(\frac{\phi_{\text{eik}}(t, x+y) - \phi_{\text{eik}}(t, x)}{2}\right)\right|^2}{|y|^{n+2s}} dx dy = \infty.$$

Proposition 1.9 yields:

$$(\partial_t + \nabla\phi_{\text{eik}} \cdot \nabla) \nabla^2\phi_{\text{eik}} \in L^\infty ([0,T] \times \mathbb{R}^n)^{n^2} \quad ; \quad \nabla^2\phi_{\text{eik}|t=0} = 0.$$

Therefore,

$$\left\| \nabla^2\phi_{\text{eik}}(t,\cdot) \right\|_{L^\infty(\mathbb{R}^n)^{n^2}} = \mathcal{O}(t) \quad \text{as } t \to 0.$$

We infer:

$$\phi_{\text{eik}}(t, x+y) - \phi_{\text{eik}}(t,x) = y \cdot \nabla\phi_{\text{eik}}(t,x) + \mathcal{O}(t|y|^2), \quad \text{uniformly for } x \in \mathbb{R}^n,$$

and

$$\sin\left(\frac{\phi_{\text{eik}}(t, x+y) - \phi_{\text{eik}}(t,x)}{2} \right) = \sin\left(\frac{y \cdot \nabla\phi_{\text{eik}}(t,x)}{2} \right) \cos\left(\mathcal{O}(t|y|^2) \right)$$

$$+ \cos\left(\frac{y \cdot \nabla\phi_{\text{eik}}(t,x)}{2} \right) \sin\left(\mathcal{O}(t|y|^2) \right).$$

The second term is $\mathcal{O}(t|y|^2)$. Using the estimate $|\alpha - \beta|^2 \geqslant \alpha^2/2 - \beta^2$ again, we see that the integral corresponding to the second term is finite, and can be left out. To prove that

$$I' = \int_{|y| \leqslant 1} \int_{x \in \mathbb{R}^n} |b(t,x)|^2 \frac{\left| \sin\left(\frac{y \cdot \nabla\phi_{\text{eik}}(t,x)}{2} \right) \right|^2}{|y|^{n+2s}} dx dy = \infty \quad \text{for } 0 < t \ll 1,$$

we can localize y in a small conic neighborhood of $\omega\mathbb{R} \cap \{|y| \leqslant 1\}$:

$$\mathcal{V}_\epsilon = \{|y| \leqslant 1 \ ; \ |y - (y \cdot \omega)\omega| \leqslant \epsilon|y|\}, \quad 0 < \epsilon \ll 1.$$

For $0 < \epsilon, t \ll 1$, (2.17) yields:

$$\left| \sin\left(\frac{y \cdot \nabla\phi_{\text{eik}}(t,x)}{2} \right) \right| \gtrsim t|y \cdot \omega| \times |\omega \cdot \nabla V(x)|, \quad y \in \mathcal{V}_\epsilon.$$

Introduce a conic localization for x close to ω', excluding the origin:

$$\mathcal{U}_\epsilon = \{|x| \geqslant 1 \ ; \ |x - (x \cdot \omega')\omega'| \leqslant \epsilon|x|\}.$$

Change the variable in the y-integral: for t and ϵ sufficiently small, and $x \in \mathcal{U}_\epsilon$, set

$$y' = \omega \cdot \nabla\phi_{\text{eik}}(t,x)y.$$

This change of variable is admissible, from (2.12) and (2.17). We infer, for $0 < \epsilon, t \ll 1$:

$$I' \geqslant \int_{y \in \mathcal{V}_\epsilon} \int_{x \in \mathbb{R}^n} |b(t,x)|^2 \frac{\left| \sin\left(\frac{y \cdot \nabla\phi_{\text{eik}}(t,x)}{2} \right) \right|^2}{|y|^{n+2s}} dx dy$$

$$\gtrsim \int_{x \in \mathcal{U}_\epsilon} |b(t,x)|^2 |\omega \cdot \nabla\phi_{\text{eik}}(t,x)|^{2s} \left(\int_{y \in |\omega \cdot \nabla\phi_{\text{eik}}(t,x)|\mathcal{V}_\epsilon} \frac{dy}{|y|^{n+2s-2}} \right) dx$$

$$\gtrsim \int_{x \in \mathcal{U}_\epsilon} |b(t,x)|^2 |\omega \cdot \nabla\phi_{\text{eik}}(t,x)|^{2s} \left(\int_{y \in ct\mathcal{V}_\epsilon} \frac{dy}{|y|^{n+2s-2}} \right) dx.$$

The assumption (2.12), the expression of b and the choice (2.18) for a_0 then show that for $0 < t \ll 1$, $I' = \infty$. This completes the proof of Proposition 2.7.

Chapter 3

Convergence of Quadratic Observables *via* Modulated Energy Functionals

3.1 Presentation

In this chapter, we turn to what has appeared as a supercritical case in Sec. 1.2: the case $\alpha = 0$. We consider the case of a defocusing power-like nonlinearity, with no external potential

$$i\varepsilon \partial_t u^\varepsilon + \frac{\varepsilon^2}{2}\Delta u^\varepsilon = |u^\varepsilon|^{2\sigma} u^\varepsilon \quad ; \quad u^\varepsilon_{|t=0} = a_0^\varepsilon e^{i\phi_0/\varepsilon}. \tag{3.1}$$

We discuss in Chap. 4 how to take the presence of an external potential into account. As noticed in Sec. 1.2, an aspect of supercriticality is that the cascade of equations (1.7), (1.8), etc., is not closed. However, we remark that the transport equation (1.8) reads, in this case:

$$\partial_t a + \nabla \phi \cdot \nabla a + \frac{1}{2} a \Delta \phi = -2i\sigma \operatorname{Re}\left(\bar{a} a^{(1)}\right) |a|^{2\sigma-2} a \text{ if } \alpha = 0.$$

Suppose that we know ϕ (which is not straightforward at all, since (1.7) now contains $|a|^{2\sigma}$). Then the above equation shares an interesting property with (2.5): after changing the space variable to follow the characteristics associated to $\nabla \phi$, this equation is of the form

$$\partial_t A = i\mathcal{V}A, \quad \mathcal{V} \in \mathbb{R}.$$

In particular, $\partial_t |A|^2 = 0$, which yields the identity which can be checked directly:

$$\partial_t |a|^2 + \nabla \phi \cdot \nabla |a|^2 + |a|^2 \Delta \phi = 0. \tag{3.2}$$

Setting $(\rho, v) := (|a|^2, \nabla \phi)$, we see that (1.7)–(1.8) *implies*:

$$\begin{cases} \partial_t v + v \cdot \nabla v + \nabla(\rho^\sigma) = 0 & ; \quad v_{|t=0} = \nabla \phi_0, \\ \partial_t \rho + \operatorname{div}(\rho v) = 0 & ; \quad \rho_{|t=0} = |a_0|^2, \end{cases} \tag{3.3}$$

45

where we have naturally assumed that $a_0^\varepsilon \to a_0$ as $\varepsilon \to 0$. The above system is an *isentropic, compressible Euler equation*, where the pressure law is such that $\rho\nabla(\rho^\sigma) = \nabla p$, that is

$$p(\rho) = \frac{\sigma}{\sigma+1}\rho^{\sigma+1}.$$

It is remarkable that this system is *quasi-linear*, while for fixed $\varepsilon > 0$, (3.1) is a *semi-linear* equation. We may say that this increase in the non-linear aspect of the equations we consider is due to the fact that (3.1) is supercritical as far as WKB analysis is concerned.

In this chapter, we do not prove any asymptotics for the wave function u^ε. This will be done in Chap. 4. We present here an approach that makes it possible to establish the convergence of two quadratic observables which are of particular interest in Physics. We have introduced in Sec. 2.3:

Position density: $\rho^\varepsilon = |u^\varepsilon|^2,$

Current density: $J^\varepsilon = \varepsilon\,\mathrm{Im}\,(\overline{u}^\varepsilon \nabla u^\varepsilon)\,.$

The above discussion suggests that we should have

$$\rho^\varepsilon \underset{\varepsilon\to 0}{\longrightarrow} \rho \quad ; \quad J^\varepsilon \underset{\varepsilon\to 0}{\longrightarrow} \rho v.$$

A rigorous proof of such convergences is a consequence of the analysis presented in this chapter.

Compare the above convergence to the result of the previous chapter. In (2.1), assume that $V \equiv 0$ and $\alpha \geqslant 1$. Then Proposition 2.5 shows that

$$\rho^\varepsilon \underset{\varepsilon\to 0}{\longrightarrow} \underline{\rho} \text{ in } L^\infty\left([-T,T];L^1(\mathbb{R}^n)\right),$$

$$J^\varepsilon \underset{\varepsilon\to 0}{\longrightarrow} \underline{\rho}\underline{v} \text{ in } L^\infty\left([-T,T];L^1_{\mathrm{loc}}(\mathbb{R}^n)\right),$$

where $(\underline{\rho}, \underline{v}) = (|a|^2, \nabla\phi_{\mathrm{eik}})$ solves

$$\begin{cases} \partial_t\underline{v} + \underline{v}\cdot\nabla\underline{v} = 0 \quad ; \quad \underline{v}_{|t=0} = \nabla\phi_0, \\ \partial_t\underline{\rho} + \mathrm{div}\,(\underline{\rho}\underline{v}) = 0 \quad ; \quad \underline{\rho}_{|t=0} = |a_0|^2. \end{cases}$$

This is a pressure-less Euler system. A non-trivial pressure law appears, when the semi-classical limit is considered, in a supercritical case only.

Note that for linear problems, and the nonlinear problem of the Schrödinger–Poisson system, the convergence of the above two quadratic observables can be determined thanks to the study of the Wigner measure associated to u^ε. See e.g. [Gérard *et al.* (1997); Lions and Paul (1993)], and Sec. 3.4 below. For the nonlinear case (3.1), another tool was introduced by

Y. Brenier [Brenier (2000)], inspired by the notion of dissipative solution in fluid mechanics [Lions (1996)]. In a context very similar to (3.1), this technique has been used initially by F. Lin and P. Zhang [Lin and Zhang (2005)].

The first idea to understand the modulated energy functional is that since the rapid oscillations (at scale ε) in u^ε are expected to be described by ϕ, $u^\varepsilon e^{-i\phi/\varepsilon}$ should not be ε-oscillatory; it should even converge strongly as ε goes to zero, while u^ε does not. Recall that the energy associated to (3.1) is

$$E^\varepsilon = \frac{1}{2}\|\varepsilon\nabla u^\varepsilon\|_{L^2}^2 + \frac{1}{\sigma+1}\|u^\varepsilon\|_{L^{2\sigma+2}}^{2\sigma+2}.$$

It is therefore natural to replace the first part of the energy (*kinetic energy*) with

$$\frac{1}{2}\left\|\varepsilon\nabla\left(u^\varepsilon e^{-i\phi/\varepsilon}\right)\right\|_{L^2}^2 = \frac{1}{2}\left\|(\varepsilon\nabla - iv)\,u^\varepsilon\right\|_{L^2}^2.$$

To understand how to treat the second part of the energy (*potential energy*), introduce a more general notation, as in Sec. 1.4: for the equation

$$i\varepsilon\partial_t u^\varepsilon + \frac{\varepsilon^2}{2}\Delta u^\varepsilon = f\left(|u^\varepsilon|^2\right)u^\varepsilon, \tag{3.4}$$

setting

$$F(y) = \int_0^y f(\eta)d\eta,$$

the potential energy is

$$\int_{\mathbb{R}^n} F\left(|u^\varepsilon(t,x)|^2\right)dx.$$

The above discussion shows that $|u^\varepsilon|^2$ is expected to be well approximated by ρ, where (ρ, v) solves an Euler equation whose pressure law is related to f. The Taylor expansion of the potential energy yields

$$F\left(|u^\varepsilon|^2\right) = F(\rho) + \left(|u^\varepsilon|^2 - \rho\right)f(\rho) + \mathcal{O}\left(\left(|u^\varepsilon|^2 - \rho\right)^2\right).$$

In the modulated energy functional, we subtract the first two terms of this Taylor expansion. This leads to the definition:

$$\begin{aligned}
H^\varepsilon(t) = &\frac{1}{2}\left\|(\varepsilon\nabla - iv)\,u^\varepsilon(t)\right\|_{L^2}^2 \\
&+ \int_{\mathbb{R}^n}\left(F\left(|u^\varepsilon|^2\right) - F(\rho) - \left(|u^\varepsilon|^2 - \rho\right)f(\rho)\right)(t,x)dx.
\end{aligned} \tag{3.5}$$

We recover the form suggested in Remark 1, (2) in [Lin and Zhang (2005)].

Note that the energy functional E^ε is non-negative because we consider a defocusing nonlinearity. We will see that thanks to convexity arguments, the modulated energy functional H^ε is non-negative as well. If instead of (3.1), we considered its focusing counterpart

$$i\varepsilon\partial_t u^\varepsilon + \frac{\varepsilon^2}{2}\Delta u^\varepsilon = -|u^\varepsilon|^{2\sigma}u^\varepsilon \quad ; \quad u^\varepsilon_{|t=0} = a^\varepsilon_0 e^{i\phi_0/\varepsilon},$$

then the energy functionals would not be signed any more, and the energy estimates presented below would fail. We refer to Sec. 4.5 for the case of focusing nonlinearities.

3.2   Formal computation

Let (ρ, v) solve (3.3) on some time interval $[-T, T]$. Introduce the hydrodynamic variables:

$$\rho^\varepsilon = |u^\varepsilon|^2 \quad ; \quad J^\varepsilon = \text{Im}\left(\varepsilon \overline{u}^\varepsilon \nabla u^\varepsilon\right).$$

For $y \geqslant 0$, denote

$$f(y) = y^\sigma \quad ; \quad F(y) = \int_0^y f(z)dz = \frac{1}{\sigma+1}y^{\sigma+1} \quad ;$$

$$G(y) = \int_0^y zf'(z)dz = yf(y) - F(y) = \frac{\sigma}{\sigma+1}y^{\sigma+1}.$$

We check that $(\rho^\varepsilon, J^\varepsilon)$ satisfies, for $\sigma \geqslant 1$:

$$\begin{cases} \partial_t \rho^\varepsilon + \text{div}\, J^\varepsilon = 0. \\ \partial_t J^\varepsilon_j + \dfrac{\varepsilon^2}{4}\sum_k \partial_k \left(4\,\text{Re}\,\partial_j \overline{u}^\varepsilon \partial_k u^\varepsilon - \partial^2_{jk}\rho^\varepsilon\right) + \partial_j G(\rho^\varepsilon) = 0. \end{cases} \quad (3.6)$$

Split the modulated energy functional (3.5) into two parts, and denote

$$K^\varepsilon(t) = \frac{1}{2}\int_{\mathbb{R}^n} |(\varepsilon\nabla - iv(t,x))\, u^\varepsilon(t,x)|^2\, dx$$

$$= \frac{1}{2}\int |\varepsilon\nabla u^\varepsilon|^2 + \frac{1}{2}\int |v|^2\,|u^\varepsilon|^2 - \int J^\varepsilon \cdot v.$$

Using the conservation of energy, we find

$$\frac{d}{dt}K^\varepsilon = -\frac{d}{dt}\int F(\rho^\varepsilon) + \frac{1}{2}\int |v|^2\partial_t\rho^\varepsilon + \int \rho^\varepsilon v \cdot \partial_t v$$

$$- \int J^\varepsilon \cdot \partial_t v - \int v \cdot \partial_t J^\varepsilon.$$

The first term on the right hand side is compensated by the first term of the potential energy in the modulated energy functional. In view of (3.6), the second term of the right had side is controlled by an integration by parts. The third and fourth are controlled directly, but the last term is not. The advantage of the modulated energy functional is to exactly cancel out this term: integrations by parts, which are studied in more detail below, yield

$$\frac{d}{dt} H^\varepsilon(t) = \mathcal{O}\left(K^\varepsilon + \varepsilon^2\right) - \int_{\mathbb{R}^n} \left(G(\rho^\varepsilon) - G(\rho) - (\rho^\varepsilon - \rho)G'(\rho)\right) \operatorname{div} v \, dx.$$

At first glance, we cannot apply Gronwall lemma directly yet: the right hand side does not involve H^ε. However, we get by thanks to convexity arguments. We check that there exists $c > 0$ such that

$$H^\varepsilon(t) \geqslant K^\varepsilon(t) + c \int_{\mathbb{R}^n} (\rho^\varepsilon - \rho)^2 \left((\rho^\varepsilon)^{\sigma-1} + \rho^{\sigma-1}\right) dx.$$

On the other hand, we have

$$|G(\rho^\varepsilon) - G(\rho) - (\rho^\varepsilon - \rho)G'(\rho)| \leqslant C(\rho^\varepsilon - \rho)^2 \left((\rho^\varepsilon)^{\sigma-1} + \rho^{\sigma-1}\right).$$

Setting

$$\widetilde{H}^\varepsilon(t) = K^\varepsilon(t) + c \int_{\mathbb{R}^n} (\rho^\varepsilon - \rho)^2 \left((\rho^\varepsilon)^{\sigma-1} + \rho^{\sigma-1}\right) dx,$$

we have therefore:

$$\widetilde{H}^\varepsilon(t) \leqslant \widetilde{H}^\varepsilon(0) + C \int_0^t \left(\widetilde{H}^\varepsilon(s) + \varepsilon^2\right) ds.$$

We check that if $a_0^\varepsilon = a_0 + \mathcal{O}(\varepsilon)$,

$$\widetilde{H}^\varepsilon(0) = \frac{1}{2} \|\varepsilon \nabla a_0^\varepsilon\|_{L^2}^2 + c \int_{\mathbb{R}^n} (|a_0^\varepsilon|^2 - |a_0|^2)^2 \left(|a_0^\varepsilon|^{2\sigma-2} + |a_0|^{2\sigma-2}\right) dx$$

$$= \mathcal{O}(\varepsilon^2).$$

We infer by Gronwall lemma that $\widetilde{H}^\varepsilon(t) = \mathcal{O}(\varepsilon^2)$ so long as it is defined, which can be rewritten:

$$\|(\varepsilon\nabla - iv)u^\varepsilon\|_{L^\infty([-T,T];L^2)}^2 + \left\|(\rho^\varepsilon - \rho)^2 \left((\rho^\varepsilon)^{\sigma-1} + \rho^{\sigma-1}\right)\right\|_{L^\infty([-T,T];L^1)} =$$

$$= \mathcal{O}(\varepsilon^2).$$

This suffices to establish the strong convergence of position and current densities.

3.3   Justification

Two aspects have to be pointed out in view of a rigorous justification of the above computations. First, we have to make sure that the limiting Euler equation possesses a unique, sufficiently smooth, solution. We will see that this is rather straightforward in the cubic case $\sigma = 1$, but demands a non-trivial argument when $\sigma \geqslant 2$. Next, the above integrations by part require sufficient regularity on the solution u^ε. Even if we know from Proposition 1.25 that u^ε remains smooth on some time interval $]-T_-^\varepsilon, T_+^\varepsilon[$, it may happen that T_-^ε or T_+^ε goes to zero as $\varepsilon \to 0$. In general, we only know that u^ε exists as a global weak solution, from Proposition 1.28.

3.3.1   *The Cauchy problem for* (3.3)

For a more general nonlinearity like in (3.4), the same arguments as above lead to the following Euler equation:

$$\begin{cases} \partial_t v + v \cdot \nabla v + \nabla \left(f\left(\rho\right)\right) = 0 & ; \quad v_{|t=0} = \nabla \phi_0, \\ \partial_t \rho + \operatorname{div}\left(\rho v\right) = 0 & ; \quad \rho_{|t=0} = |a_0|^2. \end{cases} \tag{3.7}$$

When $f' > 0$, this system enters the framework of symmetric, hyperbolic, quasi-linear equations. It is classical that if the initial data are in $H^s(\mathbb{R}^n)$ for some $s > n/2 + 1$, then there exists $T > 0$ such that (3.7) has a unique solution $(\rho, v) \in C([-T, T]; H^s)^2$. We refer for instance to [Majda (1984)] or [Taylor (1997)]. Moreover, tame estimates show that the time of existence $T > 0$ can be chosen independent of $s > n/2 + 1$.

In the homogeneous case $f(y) = y^\sigma$, $\sigma \in \mathbb{N} \setminus \{0\}$, which we have in mind, $f' > 0$ only in the cubic case $\sigma = 1$. For $\sigma \geqslant 2$, f' possesses zeroes, and this causes a lack of (strict) hyperbolicity in (3.7). This corresponds to the presence of vacuum in fluid dynamics. From the analytical point of view, such lack of hyperbolicity may cause a loss of regularity in energy estimates; see e.g. [Cicognani and Colombini (2006a,b)] and references therein. We overcome this issue, and in view of Chap. 4, we prove:

Lemma 3.1. *Let* $\sigma \in \mathbb{N} \setminus \{0\}$, $s > n/2 + 1$, $\phi_0 \in H^{s+1}$ *and* $a_0 \in H^s$. *Consider*

$$\begin{cases} \partial_t v + v \cdot \nabla v + \nabla \left(|a|^{2\sigma} \right) = 0 & ; \quad v_{|t=0} = \nabla \phi_0. \\ \partial_t a + v \cdot \nabla a + \dfrac{1}{2} a \operatorname{div} v = 0 & ; \quad a_{|t=0} = a_0. \end{cases} \tag{3.8}$$

There exists $T_-, T_+ > 0$ *such that Eq. (3.8) has a unique maximal solution* $(v, a) \in C(]-T_-, T_+[; H^s \times H^{s-1})$. *It is maximal in the sense that if, say,*

$T_+ < \infty$, *then*

$$\|(v,a)(t)\|_{W^{1,\infty}(\mathbb{R}^n)} \xrightarrow[t \to T_+]{} \infty.$$

In addition, if $\phi_0, a_0 \in H^\infty$, *then* $v, a \in C^\infty(]-T_-, T_+[; H^\infty)$. *Setting* $\rho = |a|^2$, *we infer that* (3.3) *has a unique solution* $(v, \rho) \in C(]-T_-, T_+[; H^s \times H^{s-1})$.

Remark 3.2. This shows that the possible loss of regularity due to the lack of hyperbolicity for Eq. (3.3) remains limited.

Remark 3.3. The backward and forward lifespans T_- and T_+ are finite for all compactly supported initial data, as shown in [Makino *et al.* (1986)] (see also [Chemin (1990)]): singularities appear in finite time.

Proof. Adapting the idea of [Makino *et al.* (1986)], consider the unknown $(v, u) = (v, a^\sigma)$. Even though the map $a \mapsto a^\sigma$ is not bijective, this will suffice to prove the lemma. The pair (v, u) solves:

$$\begin{cases} \partial_t v + v \cdot \nabla v + \nabla \left(|u|^2\right) = 0 & ; \quad v_{|t=0} = \nabla \phi_0 \in H^s(\mathbb{R}^n), \\ \partial_t u + v \cdot \nabla u + \dfrac{\sigma}{2} u \operatorname{div} v = 0 & ; \quad u_{|t=0} = a_0^\sigma \in H^s(\mathbb{R}^n). \end{cases} \tag{3.9}$$

This system is hyperbolic symmetric, with a constant symmetrizer. Therefore, there exist $T_-, T_+ > 0$ and a unique maximal solution $(v, u) \in C(]-T_-, T_+[; H^s)^2$. The notion of maximality follows from Moser's inequality (Lemma 1.24). Now that v is known, we define a as the solution of the linear transport equation

$$\partial_t a + v \cdot \nabla a + \frac{1}{2} a \operatorname{div} v = 0 \quad ; \quad a_{|t=0} = a_0.$$

The function a has the regularity announced in Lemma 3.1. We check that a^σ solves the second equation in (3.9). Since v is a smooth coefficient, by uniqueness for this linear equation, we have $u = a^\sigma$. Therefore, (v, a) solves Eq. (3.8). Note that the local existence times T_-, T_+ may be chosen independent of $s > n/2 + 1$, thanks to tame estimates. $\qquad \square$

3.3.2 *Rigorous estimates for the modulated energy*

Fix $T > 0$ such that $T < \min(T_-, T_+)$, where T_- and T_+ are given by Lemma 3.1. We stop counting the derivatives, and assume that the initial data are in H^∞:

Theorem 3.4. *Let $n \geqslant 1$, $\sigma \geqslant 1$ be an integer, and $\phi_0, a_0 \in H^\infty$. Let $(v, \rho) \in C([-T, T]; H^\infty)^2$ given by Lemma 3.1. Assume that there exists $s > n/2$ such that*

$$\|a_0^\varepsilon - a_0\|_{H^s} = \mathcal{O}(\varepsilon).$$

Denote $\rho^\varepsilon = |u^\varepsilon|^2$. Then we have the following estimate:

$$\|(\varepsilon\nabla - iv)u^\varepsilon\|_{L^\infty([-T,T];L^2)}^2 + \left\|(\rho^\varepsilon - \rho)^2 \left((\rho^\varepsilon)^{\sigma-1} + \rho^{\sigma-1}\right)\right\|_{L^\infty([-T,T];L^1)}$$
$$= \mathcal{O}(\varepsilon^2).$$

Remark 3.5. The above quantities are well-defined for weak solutions, so the above quantities are well-defined, since u^ε is (at least) a global weak solution.

Proof.   Recall that in general, the integrations by parts mentioned in Sec. 3.2 do not make sense for all $t \in [-T, T]$, since we consider weak solutions only. To make the above approach rigorous, we work on a sequence of global strong solutions, converging to a weak solution. For $(\delta_m)_m$ a sequence of positive numbers going to zero, introduce the saturated non-linearity, defined for $y \geqslant 0$:

$$f_m(y) = \frac{y^\sigma}{1 + (\delta_m y)^\sigma}.$$

Note that f_m is a symbol of degree 0. For fixed m and $\varepsilon > 0$, we have a global mild solution $u_m^\varepsilon \in C(\mathbb{R}; H^1)$ to:

$$i\varepsilon\partial_t u_m^\varepsilon + \frac{\varepsilon^2}{2}\Delta u_m^\varepsilon = f_m\left(|u_m^\varepsilon|^2\right) u_m^\varepsilon \quad ; \quad u_m^\varepsilon(0, x) = a_0^\varepsilon(x)e^{i\phi_0(x)/\varepsilon}. \quad (3.10)$$

As $m \to \infty$, the sequence $(u_m^\varepsilon)_m$ converges to a weak solution of (3.1) (see [Ginibre and Velo (1985a); Lebeau (2005)]). For $y \geqslant 0$, introduce also

$$F_m(y) = \int_0^y f_m(z)dz \quad ; \quad G_m(y) = \int_0^y zf_m'(z)dz = yf_m(y) - F_m(y).$$

The mass and energy associated to u_m^ε are conserved:

$$M_m^\varepsilon(t) = \int |u_m^\varepsilon(t, x)|^2 dx \equiv \|a_0\|_{L^2}^2.$$

$$E_m^\varepsilon(t) = \frac{1}{2}\|\varepsilon\nabla u_m^\varepsilon(t)\|_{L^2}^2 + \int_{\mathbb{R}^n} F_m\left(|u_m^\varepsilon(t, x)|^2\right) dx \equiv E_m^\varepsilon(0).$$

Moreover, the solution is in $H^2(\mathbb{R}^n)$ for all time: $u_m^\varepsilon \in C(\mathbb{R}; H^2)$. To see this, we use an idea due to T. Kato [Kato (1987, 1989)], and consider $\partial_t u_m^\varepsilon$. It solves

$$\left(i\varepsilon\partial_t + \frac{\varepsilon^2}{2}\Delta\right)\partial_t u_m^\varepsilon = 2f_m'\left(|u_m^\varepsilon|^2\right) \text{Re}\left(\overline{u}_m^\varepsilon\partial_t u_m^\varepsilon\right)u_m^\varepsilon + f_m\left(|u_m^\varepsilon|^2\right)\partial_t u_m^\varepsilon.$$

To compute the initial data for $\partial_t u_m^\varepsilon$, we use (3.10):

$$\partial_t u_m^\varepsilon{}_{|t=0} = \frac{i}{\varepsilon}\left(\frac{\varepsilon^2}{2}\Delta u_m^\varepsilon - f_m\left(|u_m^\varepsilon|^2\right)u_m^\varepsilon\right)\Big|_{t=0} \in H^\infty.$$

Note that since f_m is a symbol of degree 0, there exists $C_m > 0$ independent of u_m^ε such that

$$|u_m^\varepsilon|^2 f_m'\left(|u_m^\varepsilon|^2\right) + f_m\left(|u_m^\varepsilon|^2\right) \leqslant C_m.$$

Energy estimates (see Lemma 1.2) then show that $\partial_t u_m^\varepsilon \in C(\mathbb{R}; L^2)$. Using (3.10) and the boundedness of f_m, we infer $\Delta u_m^\varepsilon \in C(\mathbb{R}; L^2)$.

We consider the hydrodynamic variables:

$$\rho_m^\varepsilon = |u_m^\varepsilon|^2 \quad ; \quad J_m^\varepsilon = \mathrm{Im}\left(\varepsilon \overline{u}_m^\varepsilon \nabla u_m^\varepsilon\right).$$

From the above discussion, we have:

$$\rho_m^\varepsilon(t) \in W^{2,1}(\mathbb{R}^n) \text{ and } J_m^\varepsilon(t) \in W^{1,1}(\mathbb{R}^n), \quad \forall t \in \mathbb{R}. \tag{3.11}$$

The analogue of (3.6) is:

$$\begin{cases} \partial_t \rho_m^\varepsilon + \mathrm{div}\, J_m^\varepsilon = 0. \\ \partial_t (J_m^\varepsilon)_j + \dfrac{\varepsilon^2}{4}\sum_k \partial_k (4\,\mathrm{Re}\,\partial_j \overline{u}_m^\varepsilon \partial_k u_m^\varepsilon - \partial_{jk}^2 \rho_m^\varepsilon) + \partial_j G_m(\rho_m^\varepsilon) = 0. \end{cases} \tag{3.12}$$

Introduce the modulated energy functional "adapted to (3.10)":

$$H_m^\varepsilon(t) = \frac{1}{2}\int_{\mathbb{R}^n} |(\varepsilon\nabla - iv)u_m^\varepsilon|^2\, dx$$
$$+ \int_{\mathbb{R}^n} (F_m(\rho_m^\varepsilon) - F_m(\rho) - (\rho_m^\varepsilon - \rho)f_m(\rho))\, dx.$$

Notice that this functional is not exactly adapted to (3.10), since the limiting quantities (as $\varepsilon \to 0$) ρ and v are constructed with the nonlinearity f and not the nonlinearity f_m. We also distinguish the kinetic part:

$$K_m^\varepsilon(t) = \frac{1}{2}\int_{\mathbb{R}^n} |(\varepsilon\nabla - iv)u_m^\varepsilon|^2\, dx.$$

Thanks to the conservation of energy for u_m^ε, we have:

$$\frac{d}{dt}K_m^\varepsilon = -\frac{d}{dt}\int F_m(\rho_m^\varepsilon) + \frac{1}{2}\int |v|^2 \partial_t \rho_m^\varepsilon + \int \rho_m^\varepsilon v \cdot \partial_t v$$
$$- \int J_m^\varepsilon \cdot \partial_t v - \int v \cdot \partial_t J_m^\varepsilon.$$

Using Lemma 3.1, Eqs. (3.11) and (3.12), (licit) integrations by parts yield:

$$\frac{d}{dt}K_m^\varepsilon = -\frac{d}{dt}\int F_m(\rho_m^\varepsilon) - \frac{1}{2}\int |v|^2 \operatorname{div} J_m^\varepsilon - \sum_{j,k}\rho_m^\varepsilon v_j v_k \partial_j v_k$$
$$-\int \rho_m^\varepsilon \nabla f(\rho)\cdot v + \int (v\cdot\nabla v)\cdot J_m^\varepsilon + \int \nabla f(\rho)\cdot J_m^\varepsilon$$
$$-\sum_{j,k}\int \partial_k v_j \operatorname{Re}(\varepsilon\partial_j \overline{u}_m^\varepsilon \varepsilon \partial_k u_m^\varepsilon) - \frac{\varepsilon^2}{4}\int \nabla(\operatorname{div} v)\cdot\nabla\rho_m^\varepsilon + \int \rho_m^\varepsilon v\cdot\nabla f_m(\rho^\varepsilon).$$

Proceeding as in [Lin and Zhang (2005)], we have:

$$\varepsilon^2 \int \operatorname{div}(\nabla v)\cdot\nabla\rho_m^\varepsilon = \varepsilon\int \operatorname{div}(\nabla v)\cdot(\overline{u}_m^\varepsilon \varepsilon\nabla u_m^\varepsilon + u_m^\varepsilon \varepsilon\nabla\overline{u}_m^\varepsilon)$$
$$= \varepsilon\int \operatorname{div}(\nabla v)\cdot\left(\overline{u}_m^\varepsilon(\varepsilon\nabla - iv)u_m^\varepsilon + u_m^\varepsilon\overline{(\varepsilon\nabla - iv)u^\varepsilon}_m\right)$$
$$= \mathcal{O}\left(K_m^\varepsilon + \varepsilon^2\right),$$

where we have used the conservation of mass and Young's inequality

$$ab \leqslant \frac{1}{2}\left(a^2 + b^2\right), \quad \forall a, b \geqslant 0.$$

From now on, we use the convention that the constant associated to the notation $\mathcal{O}$ is independent of m and ε. Treating the term involving $\partial_k v_j \operatorname{Re}(\varepsilon\partial_j \overline{u}_m^\varepsilon \varepsilon \partial_k u_m^\varepsilon)$ in a similar fashion, simplifications yield:

$$\frac{d}{dt}K_m^\varepsilon = \mathcal{O}\left(K_m^\varepsilon + \varepsilon^2\right) - \frac{d}{dt}\int F_m(\rho_m^\varepsilon) + \int \nabla f_m(\rho_m^\varepsilon)\rho_m^\varepsilon v$$
$$-\int \nabla f(\rho)\cdot(\rho_m^\varepsilon v - J_m^\varepsilon).$$

Similar computations for $H_m^\varepsilon - K_m^\varepsilon$ yield:

$$\frac{d}{dt}H_m^\varepsilon = \mathcal{O}\left(K_m^\varepsilon + \varepsilon^2\right) - \int (G_m(\rho_m^\varepsilon) - G_m(\rho) - (\rho_m^\varepsilon - \rho)G_m'(\rho))\operatorname{div} v$$
$$+\int \nabla(f(\rho) - f_m(\rho))\cdot(J_m^\varepsilon - \rho_m^\varepsilon v).$$

Note that $f(\rho) - f_m(\rho) \to 0$ in $L^\infty([0,T]; W^{1,\infty})$ as $m \to \infty$. We can thus write:

$$\frac{d}{dt}H_m^\varepsilon = \mathcal{O}\left(K_m^\varepsilon + \varepsilon^2\right) + o_{m\to\infty}(1)$$
$$-\int (G_m(\rho_m^\varepsilon) - G_m(\rho) - (\rho_m^\varepsilon - \rho)G_m'(\rho))\operatorname{div} v. \tag{3.13}$$

We check that there exists C independent of m such that

$$|G_m(\rho_m^\varepsilon) - G_m(\rho) - (\rho_m^\varepsilon - \rho)G_m'(\rho)| \leqslant C(\rho_m^\varepsilon - \rho)^2\left(\theta_m(\rho_m^\varepsilon) + \theta_m(\rho)\right),$$

where we have set, for $y \geqslant 0$,

$$\theta(y) = \frac{y^{\sigma-1}}{1+y^{\sigma}} \quad ; \quad \theta_m(y) = \frac{y^{\sigma-1}}{1+(\delta_m y)^{\sigma}}.$$

Easy computations show that there exists $K > 0$ such that

$$\frac{1}{K}\left(\theta(a) + \theta(b)\right) \leqslant \theta(a+b) \leqslant K\left(\theta(a) + \theta(b)\right), \quad \forall a, b \geqslant 0.$$

Since the numerator of θ_m is homogeneous, we infer that the above estimate remains true when θ is replaced by θ_m, with the same constant K (independent of m). Therefore, there exists $c > 0$ independent of m, such that:

$$H_m^\varepsilon(t) \geqslant \widetilde{H}_m^\varepsilon(t) := K_m^\varepsilon(t) + c \int \left(\rho_m^\varepsilon - \rho\right)^2 \left(\theta_m(\rho_m^\varepsilon) + \theta_m(\rho)\right).$$

The above computations yield

$$\frac{d}{dt}\widetilde{H}_m^\varepsilon(t) \leqslant C\left(\widetilde{H}_m^\varepsilon(t) + \varepsilon^2\right) + o_{m\to\infty}(1),$$

where C is independent of ε and m. By assumption,

$$\widetilde{H}_m^\varepsilon(0) = \frac{1}{2}\|\varepsilon\nabla a_0^\varepsilon\|_{L^2}^2 + c\int\left(|a_0^\varepsilon|^2 - |a_0|^2\right)^2 \left(\theta_m(|a_0^\varepsilon|^2) + \theta_m(|a_0|^2)\right)dx$$
$$= \mathcal{O}\left(\varepsilon^2\right).$$

Using Gronwall lemma, we infer

$$\sup_{t\in[-T,T]} \widetilde{H}_m^\varepsilon(t) \leqslant C\varepsilon^2 + o_{m\to\infty}(1),$$

for some constant C independent of m. Letting $m \to \infty$, Fatou's lemma yields

$$\|(\varepsilon\nabla - iv)u^\varepsilon(t)\|_{L^2}^2 + \int_{\mathbb{R}^n}(\rho^\varepsilon - \rho)^2\left((\rho^\varepsilon)^{\sigma-1} + \rho^{\sigma-1}\right)dx = \mathcal{O}\left(\varepsilon^2\right),$$

uniformly for $t \in [-T, T]$. This completes the proof of Theorem 3.4. $\quad\square$

3.4 Convergence of quadratic observables

We infer the convergence of quadratic observables from Theorem 3.4. Recall a few basic facts about the Wigner measures. For $(x, \xi) \in \mathbb{R}^n \times \mathbb{R}^n$, the Wigner transform of $u^\varepsilon \in L_t^\infty L_x^2$ is defined by

$$w^\varepsilon(t, x, \xi) = (2\pi)^{-n}\int_{\mathbb{R}^n} u^\varepsilon\left(t, x - \varepsilon\frac{\eta}{2}\right)\overline{u}^\varepsilon\left(t, x + \varepsilon\frac{\eta}{2}\right)e^{i\eta\cdot\xi}d\eta,$$

The position and current densities can be recovered from w^ε, by

$$\rho^\varepsilon(t,x) = |u^\varepsilon(t,x)|^2 = \int_{\mathbb{R}^n} w^\varepsilon(t,x,\xi)d\xi,$$

$$J^\varepsilon(t,x) = \operatorname{Im}\left(\varepsilon \overline{u}^\varepsilon \nabla u^\varepsilon\right)(t,x) = \int_{\mathbb{R}^n} \xi w^\varepsilon(t,x,\xi)d\xi.$$

A measure μ is a Wigner measure associated to u^ε (there is no uniqueness in general) if, up to extracting a subsequence, w^ε converges to μ as $\varepsilon \to 0$. Note that μ is a non-negative measure on the phase space. We refer to [Burq (1997)] and references therein for various results on Wigner measures, as well as applications to several problems.

Recall that if X and Y are two Banach spaces, $X + Y$ is equipped with the norm

$$\|u\|_{X+Y} = \inf\left\{\|u_1\|_X + \|u_2\|_Y \quad ; \quad u = u_1 + u_2,\ u_1 \in X,\ u_2 \in Y\right\}.$$

Corollary 3.6. *Under the assumptions of Theorem 3.4, the position and current densities converge strongly on $[-T, T]$ as $\varepsilon \to 0$:*

$$|u^\varepsilon|^2 \xrightarrow[\varepsilon \to 0]{} |a|^2 \qquad\qquad \text{in } C\left([-T,T]; L^{\sigma+1}(\mathbb{R}^n)\right).$$

$$\operatorname{Im}\left(\varepsilon \overline{u}^\varepsilon \nabla u^\varepsilon\right) \xrightarrow[\varepsilon \to 0]{} |a|^2 v \qquad \text{in } C\left([-T,T]; L^{\sigma+1}(\mathbb{R}^n) + L^1(\mathbb{R}^n)\right).$$

In particular, there is only one Wigner measure associated to $(u^\varepsilon)_\varepsilon$, and it is given by

$$\mu(t, dx, d\xi) = |a(t,x)|^2 dx \otimes \delta\left(\xi - v(t,x)\right).$$

Remark 3.7. The above convergences imply the following local L^1 convergences:

$$|u^\varepsilon|^2 \xrightarrow[\varepsilon \to 0]{} |a|^2 \qquad\qquad \text{in } C\left([-T,T]; L^1(|x| \leqslant R)\right),$$

$$\operatorname{Im}\left(\varepsilon \overline{u}^\varepsilon \nabla u^\varepsilon\right) \xrightarrow[\varepsilon \to 0]{} |a|^2 v \qquad \text{in } C\left([-T,T]; L^1(|x| \leqslant R)\right), \quad \forall R \geqslant 0.$$

Proof.   The second part of the estimate in Theorem 3.4 yields

$$\sup_{t \in [-T,T]} \int_{\mathbb{R}^n} \left(|u^\varepsilon(t,x)|^2 - |a(t,x)|^2\right)^2 \left(|u^\varepsilon(t,x)|^{2\sigma-2} + |a(t,x)|^{2\sigma-2}\right)^2 dx =$$

$$= \mathcal{O}\left(\varepsilon^2\right).$$

Therefore, since there exists C_σ such that

$$|\alpha - \beta|^{\sigma+1} \leqslant C_\sigma(\alpha - \beta)^2 \left(\alpha^{\sigma-1} + \beta^{\sigma-1}\right), \quad \forall \alpha, \beta \geqslant 0,$$

we infer:

$$\sup_{t \in [T,T]} \int_{\mathbb{R}^n} \left| |u^\varepsilon(t,x)|^2 - |a(t,x)|^2 \right|^{\sigma+1} dx = \mathcal{O}\left(\varepsilon^2\right).$$

This yields the first part of the corollary, along with a bound on the rate of convergence as $\varepsilon \to 0$. For the current density, write

$$\operatorname{Im}\left(\varepsilon \overline{u}^\varepsilon \nabla u^\varepsilon\right) = \operatorname{Im}\left(\overline{u}^\varepsilon \left(\varepsilon \nabla - iv\right) u^\varepsilon\right) + |u^\varepsilon|^2 v.$$

Since $v \in L^\infty([-T,T] \times \mathbb{R}^n)$, we have

$$|u^\varepsilon|^2 v \xrightarrow[\varepsilon \to 0]{} |a|^2 v \quad \text{in } C([-T,T]; L^{\sigma+1}).$$

On the other hand, Cauchy–Schwarz inequality and Theorem 3.4 yield

$$\operatorname{Im}\left(\overline{u}^\varepsilon \left(\varepsilon \nabla - iv\right) u^\varepsilon\right) = \mathcal{O}(\varepsilon) \quad \text{in } C([-T,T]; L^1).$$

This completes the proof of the corollary. □

To conclude this paragraph, we point out a phenomenon which is typical of the supercritical WKB régime, and which is the key point at the origin of the instability mechanisms presented in Chap. 5. Suppose that no rapid oscillation is present in the initial datum for u^ε:

$$\phi_0 = 0.$$

Then $v_{|t=0} = 0$ in (3.8). The equation for v at time $t = 0$ then yields:

$$\partial_t v_{|t=0} = -\nabla \left(|a_0|^{2\sigma}\right).$$

Thus, at least for $t > 0$ independent of ε but sufficiently small,

$$v(t,\cdot) = -t\nabla \left(|a_0|^{2\sigma}\right) + \mathcal{O}\left(t^2\right).$$

Note also that for a non-trivial $a_0 \in L^2(\mathbb{R}^n)$, $\nabla \left(|a_0|^{2\sigma}\right) \not\equiv 0$. Therefore, even if no rapid oscillation is present initially, u^ε becomes *instantaneously* ε-oscillatory.

Chapter 4

Pointwise Description of the Wave Function

In the previous chapter, we have established the convergence of quadratic observables thanks to a modulated energy functional, in the case of a defocusing nonlinearity, with no external potential. In this chapter, we study the asymptotic behavior of the wave function itself, as $\varepsilon \to 0$, in the same supercritical WKB régime: we consider

$$i\varepsilon\partial_t u^\varepsilon + \frac{\varepsilon^2}{2}\Delta u^\varepsilon = Vu^\varepsilon + f\left(|u^\varepsilon|^2\right)u^\varepsilon \quad ; \quad u^\varepsilon_{|t=0} = a^\varepsilon_0 e^{i\phi_0/\varepsilon}. \quad (4.1)$$

Recall that the standard WKB approach meets the problem of closing the cascade of equations. This problem was eluded in the previous section, since we have noticed that the system relating the phase ϕ (or equivalently, its gradient v) and the *modulus* of the leading order profile a is closed (Euler equation). To study the wave function u^ε itself, this does not suffice: we will see in particular that $\mathcal{O}(\varepsilon)$ perturbations of the initial profile a^ε_0 affect the wave function u^ε at *leading order*, through a term of modulus one (phase modulation).

We first discuss several possibilities to adapt the WKB method to this case. In particular, we explain why the case of a focusing nonlinearity is extremely different (Sec. 4.5). For instance, the framework of Sobolev spaces is not well adapted to this case. For defocusing nonlinearities, we provide a pointwise description of u^ε as $\varepsilon \to 0$ in two cases:

- Defocusing nonlinearity, which is *cubic at the origin*: $f' > 0$.
- Smooth, *homogeneous*, defocusing nonlinearity: $f(y) = y^\sigma$, $\sigma \in \mathbb{N}$.

4.1   Several possible approaches

To overcome the absence of closure in the regular WKB analysis, a possibility consists in trying to write the exact solution u^ε as the product of an amplitude and of a rapidly oscillatory factor:

$$u^\varepsilon = a^\varepsilon e^{i\Phi^\varepsilon/\varepsilon}, \tag{4.2}$$

where a^ε *and* Φ^ε depend on ε. If one can construct a^ε and Φ^ε such that u^ε can be represented as above, then we recover a WKB-like expansion as soon as a^ε and Φ^ε have asymptotic expansions as $\varepsilon \to 0$.

A standard approach is to assume that a_0^ε is real-valued, and to seek a real-valued amplitude a^ε; see e.g. [Landau and Lifschitz (1967)]. The phase Φ^ε is real-valued too. Plugging (4.2) into Eq. (4.1), and separating real and imaginary parts, we find (adding tildes to avoid confusion):

$$\begin{cases} \widetilde{a}^\varepsilon \left(\partial_t \widetilde{\Phi}^\varepsilon + \frac{1}{2}|\nabla \widetilde{\Phi}^\varepsilon|^2 + V + f\left(|\widetilde{a}^\varepsilon|^2\right) \right) = \frac{\varepsilon^2}{2}\Delta \widetilde{a}^\varepsilon \quad ; \quad \widetilde{\Phi}^\varepsilon_{|t=0} = \phi_0, \\ \partial_t \widetilde{a}^\varepsilon + \nabla \widetilde{\Phi}^\varepsilon \cdot \widetilde{\nabla} a^\varepsilon + \frac{1}{2}\widetilde{a}^\varepsilon \Delta \widetilde{\Phi}^\varepsilon = 0 \qquad\qquad ; \quad \widetilde{a}^\varepsilon_{|t=0} = a_0^\varepsilon. \end{cases} \tag{4.3}$$

The problem in seeking a solution $(\widetilde{\Phi}^\varepsilon, \widetilde{a}^\varepsilon)$ to the above system, with $\widetilde{a}^\varepsilon(t, \cdot) \in L^2(\mathbb{R}^n)$, is the meaning of the first equation at the zeroes of $\widetilde{a}^\varepsilon$ (that is, the zeroes of u^ε).

Another possibility consists in allowing a^ε to be complex-valued. In that case, we have an extra degree of freedom in imposing the system solved by $(\Phi^\varepsilon, a^\varepsilon)$. The choice proposed by E. Grenier [Grenier (1998)] (in the case $V \equiv 0$) consists in considering

$$\begin{cases} \partial_t \Phi^\varepsilon + \frac{1}{2}|\nabla \Phi^\varepsilon|^2 + V + f\left(|a^\varepsilon|^2\right) = 0 \quad ; \quad \Phi^\varepsilon_{|t=0} = \phi_0, \\ \partial_t a^\varepsilon + \nabla \Phi^\varepsilon \cdot \nabla a^\varepsilon + \frac{1}{2}a^\varepsilon \Delta \Phi^\varepsilon = i\frac{\varepsilon}{2}\Delta a^\varepsilon \quad ; \quad a^\varepsilon_{|t=0} = a_0^\varepsilon. \end{cases} \tag{4.4}$$

We will see in the next section that this approach is very efficient when $f' > 0$, that is, for a defocusing nonlinearity which is cubic at the origin.

In Eq. (4.3) as well as in Eq. (4.4), assuming that Φ^ε and a^ε are bounded in, say, $C([-T, T]; H^s)$ for some sufficiently large s, it is natural to expect

$$\Phi^\varepsilon \xrightarrow[\varepsilon \to 0]{} \Phi, \quad a^\varepsilon \xrightarrow[\varepsilon \to 0]{} a,$$

where (Φ, a) solves

$$\begin{cases} \partial_t \Phi + \frac{1}{2}|\nabla \Phi|^2 + V + f\left(|a|^2\right) = 0 \quad ; \quad \Phi_{|t=0} = \phi_0, \\ \partial_t a + \nabla \Phi \cdot \nabla a + \frac{1}{2}a\Delta \Phi = 0 \qquad\quad ; \quad a_{|t=0} = a_0. \end{cases} \tag{4.5}$$

Note that when $V = 0$, we see that $(\nabla\Phi, a)$ solves the Euler type system (3.8). This leads to the second approach which we present here. This approach makes it possible to treat the case of defocusing nonlinearities which are not cubic at the origin, but homogeneous of degree $2\sigma + 1$, with $\sigma \in \mathbb{N} \setminus \{0\}$. The idea, already present in Chap. 3, is to say that if Eq. (4.5) has some rigorous meaning, then at least the rapid oscillations of u^ε should be described by Φ. We point out at this stage that this is the most reasonable thing to expect. In general, a does not suffice to describe the (complex-valued) amplitude of the wave function u^ε, unless, for instance, $a_0 \in \mathbb{R}$ and $a_1 \in i\mathbb{R}$ (e.g. $a_1 = 0$), where

$$a_0^\varepsilon = a_0 + \varepsilon a_1 + o(\varepsilon) \quad \text{in } H^s(\mathbb{R}^n).$$

See Sec. 4.2.1 below. Once Φ is determined, change the unknown function u^ε — and the notation (4.2) — to

$$a^\varepsilon := u^\varepsilon e^{i\Phi/\varepsilon}.$$

The idea is that this process should filter out all the rapid oscillations, so that a^ε is bounded in Sobolev spaces, and converges strongly (while u^ε does not, as soon as Φ is not trivial). With this definition for a^ε, Eq. (4.1) is *equivalent* to

$$\partial_t a^\varepsilon + \nabla\Phi \cdot \nabla a^\varepsilon + \frac{1}{2} a^\varepsilon \Delta\Phi = i\frac{\varepsilon}{2}\Delta a^\varepsilon - \frac{i}{\varepsilon}\left(f\left(|a^\varepsilon|^2\right) - f\left(|a|^2\right)\right) a^\varepsilon.$$

The major difficulty to prove that a^ε is bounded and converges strongly in Sobolev spaces, is the singular factor $1/\varepsilon$ in front of the last term. Note already that it is reasonable to hope that this singularity is "artificial", since we expect $|a^\varepsilon|^2 = |a|^2 + \mathcal{O}(\varepsilon)$ (this is already suggested by the results of Chap. 3). We prove that this is so in the homogeneous case $f(y) = y^\sigma$, $\sigma \in \mathbb{N}$, in Sec. 4.3. Note that the results in the case $f' > 0$ as well as in the case $f(y) = y^\sigma$, $\sigma \in \mathbb{N}$, show that the presence of vacuum (zeroes of u^ε) is not a real problem, but barely a technical difficulty (a non-trivial one, though).

Finally, we discuss the case of a focusing nonlinearity ($f' < 0$) in Sec. 4.5.

4.2 E. Grenier's idea

In this section, we explain the approach based on Eq. (4.4). We first consider the case $V = 0$ for some initial phase $\phi_0 \in H^s$ for large s, in Sec. 4.2.1. We then show how to adapt the approach to the case where V and ϕ_0 are smooth and subquadratic, that is, under Assumption 1.7, in Sec. 4.2.2.

4.2.1　*Without external potential*

In this paragraph, we assume $V = 0$, and we recall the approach of Grenier (1998). To study Eq. (4.4), we introduce an intermediary system, in terms of the amplitude a^ε and the "velocity" $v^\varepsilon := \nabla \Phi^\varepsilon$. The second equation in Eq. (4.4) can directly be expressed in terms of a^ε and v^ε. Differentiating the first equation in (4.4) with respect to x, we find:

$$
\begin{cases}
\partial_t v^\varepsilon + v^\varepsilon \cdot \nabla v^\varepsilon + 2f'\left(|a^\varepsilon|^2\right)\operatorname{Re}\left(\overline{a}^\varepsilon \nabla a^\varepsilon\right) = 0 \quad ; \quad v^\varepsilon_{|t=0} = \nabla \phi_0, \\
\partial_t a^\varepsilon + v^\varepsilon \cdot \nabla a^\varepsilon + \dfrac{1}{2}a^\varepsilon \operatorname{div} v^\varepsilon = i\dfrac{\varepsilon}{2}\Delta a^\varepsilon \qquad ; \quad a^\varepsilon_{|t=0} = a^\varepsilon_0.
\end{cases}
\tag{4.6}
$$

The important remark made by E. Grenier is to notice that if $f' > 0$, the above system is hyperbolic symmetric, perturbed by a skew-symmetric term. To make this fact more explicit, separate the real and imaginary parts of a^ε, to consider the unknown

$$
\mathbf{u}^\varepsilon = \begin{pmatrix} \operatorname{Re} a^\varepsilon \\ \operatorname{Im} a^\varepsilon \\ v^\varepsilon_1 \\ \vdots \\ v^\varepsilon_n \end{pmatrix} = \begin{pmatrix} a^\varepsilon_1 \\ a^\varepsilon_2 \\ v^\varepsilon_1 \\ \vdots \\ v^\varepsilon_n \end{pmatrix} \in \mathbb{R}^{n+2}.
$$

In terms of this unknown function, Eq. (4.6) reads

$$
\partial_t \mathbf{u}^\varepsilon + \sum_{j=1}^n A_j(\mathbf{u}^\varepsilon)\partial_j \mathbf{u}^\varepsilon = \frac{\varepsilon}{2}L\mathbf{u}^\varepsilon,
\tag{4.7}
$$

where the matrices $A_j \in \mathcal{M}_{n+2}(\mathbb{R})$ are given by:

$$
A(\mathbf{u}, \xi) = \sum_{j=1}^n A_j(\mathbf{u})\xi_j = \begin{pmatrix} v \cdot \xi & 0 & \frac{1}{2}a_1 {}^t\xi \\ 0 & v \cdot \xi & \frac{1}{2}a_2 {}^t\xi \\ 2f'a_1 \xi & 2f'a_2 \xi & v \cdot \xi I_n \end{pmatrix},
$$

where f' stands for $f'(|a_1|^2 + |a_2|^2)$. The linear operator L is given by

$$
L = \begin{pmatrix} 0 & -\Delta & 0 & \cdots & 0 \\ \Delta & 0 & 0 & \cdots & 0 \\ 0 & 0 & & 0_{n \times n} & \end{pmatrix}.
$$

The important remark is that even though L is a differential operator of order two, it causes no loss of regularity in the energy estimates, since it is skew-symmetric. The other important fact is that the left hand side of Eq. (4.7) is hyperbolic symmetric (or symmetrizable), provided $f' > 0$. Let

$$
S = \begin{pmatrix} I_2 & 0 \\ 0 & \frac{1}{4f'}I_n \end{pmatrix}.
\tag{4.8}
$$

This matrix is symmetric and positive if (and only if) $f' > 0$, and SA is symmetric,

$$SA(\mathbf{u}, \xi) \in S_{n+2}(\mathbb{R}), \quad \forall (\mathbf{u}, \xi) \in \mathbb{R}^{n+2} \times \mathbb{R}^n.$$

Theorem 4.1 ([Grenier (1998)]). *Let $f \in C^\infty(\mathbb{R}_+; \mathbb{R})$ with $f(0) = 0$ and $f' > 0$. Let $s > 2 + n/2$. Assume that $\phi_0 \in H^{s+1}$, and that a_0^ε is uniformly bounded in H^s for $\varepsilon \in\,]0, 1]$. There exist $T > 0$ independent of $\varepsilon \in\,]0, 1]$ and $s > n/2 + 2$, and $u^\varepsilon = a^\varepsilon e^{i\Phi^\varepsilon/\varepsilon}$ solution to (4.1) on $[-T, T]$. Moreover, a^ε and Φ^ε are bounded in $C([-T, T]; H^s)$ and $C([-T, T]; H^{s+1})$ respectively, uniformly in $\varepsilon \in\,]0, 1]$.*

Remark 4.2. The assumption $f(0) = 0$ is not really one. Indeed, considering $u^\varepsilon e^{itf(0)/\varepsilon}$ instead of u^ε turns f into $f - f(0)$ in Eq. (4.1).

Proof. We first prove that (4.6) has a unique solution $(v^\varepsilon, a^\varepsilon)$ in $C([-T, T]; H^s)^2$, uniformly in $\varepsilon \in\,]0, 1]$. The main step to prove this fact consists in obtaining *a priori* estimates, so we shall detail this part only. For $s > n/2 + 2$, we bound

$$\langle S\Lambda^s \mathbf{u}^\varepsilon, \Lambda^s \mathbf{u}^\varepsilon \rangle,$$

(scalar product in $L^2(\mathbb{R}^{n+2})$) by computing its time derivative:

$$\frac{d}{dt} \langle S\Lambda^s \mathbf{u}^\varepsilon, \Lambda^s \mathbf{u}^\varepsilon \rangle = \langle \partial_t S\Lambda^s \mathbf{u}^\varepsilon, \Lambda^s \mathbf{u}^\varepsilon \rangle + 2 \langle S\partial_t \Lambda^s \mathbf{u}^\varepsilon, \Lambda^s \mathbf{u}^\varepsilon \rangle,$$

since S is symmetric. For the first term, we consider the lower $n \times n$ block:

$$\langle \partial_t S\Lambda^s \mathbf{u}^\varepsilon, \Lambda^s \mathbf{u}^\varepsilon \rangle \leqslant \left\| \frac{1}{f'} \partial_t \left(f' \left(|a_1^\varepsilon|^2 + |a_2^\varepsilon|^2 \right) \right) \right\|_{L^\infty} \langle S\Lambda^s \mathbf{u}^\varepsilon, \Lambda^s \mathbf{u}^\varepsilon \rangle.$$

Since a_0^ε is bounded in $H^s(\mathbb{R}^n) \subset L^\infty(\mathbb{R}^n)$, there exists C_0 independent of $\varepsilon \in\,]0, 1]$ such that

$$\|a_0^\varepsilon\|_{L^\infty} \leqslant C_0.$$

So long as $\|\mathbf{u}^\varepsilon\|_{L^\infty} \leqslant 2C_0$, we have:

$$f' \left(|a_1^\varepsilon|^2 + |a_2^\varepsilon|^2 \right) \geqslant \inf \left\{ f'(y) \,;\, 0 \leqslant y \leqslant 4C_0^2 \right\} = \delta_n > 0,$$

where δ_n is now fixed, since f' is continuous with $f' > 0$. Note that this property implies that there exists $C > 0$ such that

$$\frac{1}{C} I_{n+2} \leqslant S \leqslant C I_{n+2}, \tag{4.9}$$

in the sense of symmetric matrices. We infer,

$$\left\| \frac{1}{f'} \partial_t \left(f' \left(|a_1^\varepsilon|^2 + |a_2^\varepsilon|^2 \right) \right) \right\|_{L^\infty} \leqslant C \left\| \partial_t \left(|a_1^\varepsilon|^2 + |a_2^\varepsilon|^2 \right) \right\|_{L^\infty}$$

$$\leqslant C \| \partial_t a^\varepsilon \|_{L^\infty} \leqslant C \| \mathbf{u}^\varepsilon \|_{H^s},$$

where we have used (4.7), the assumption $s > n/2 + 2$, and Sobolev embeddings. Note that we need to assume $s > n/2 + 2$ instead of the more standard assumption $s > n/2 + 1$ for quasi-linear systems, because the operator L is of second order. If the symmetrizer S is constant (in the case of an exactly cubic, defocusing nonlinearity), we can assume simply $s > n/2 + 1$.

For the second term we use

$$\langle S\partial_t \Lambda^s \mathbf{u}^\varepsilon, \Lambda^s \mathbf{u}^\varepsilon \rangle = \frac{\varepsilon}{2} \langle SL(\Lambda^s \mathbf{u}^\varepsilon), \Lambda^s \mathbf{u}^\varepsilon \rangle - \Big\langle S\Lambda^s \Big(\sum_{j=1}^n A_j(\mathbf{u}^\varepsilon)\partial_j \mathbf{u}^\varepsilon \Big), \Lambda^s \mathbf{u}^\varepsilon \Big\rangle.$$

We notice that SL is a skew-symmetric second order operator, so the first term is zero. For the second term, write

$$\Big\langle S\Lambda^s \Big(\sum_{j=1}^n A_j(\mathbf{u}^\varepsilon)\partial_j \mathbf{u}^\varepsilon \Big), \Lambda^s \mathbf{u}^\varepsilon \Big\rangle = \sum_{j=1}^n \Big\langle SA_j(\mathbf{u}^\varepsilon)\partial_j \Lambda^s \mathbf{u}^\varepsilon, \Lambda^s \mathbf{u}^\varepsilon \Big\rangle$$

$$+ \Big\langle S\Big(\sum_{j=1}^n [\Lambda^s, A_j(\mathbf{u}^\varepsilon)\partial_j]\mathbf{u}^\varepsilon \Big), \Lambda^s \mathbf{u}^\varepsilon \Big\rangle. \quad (4.10)$$

Since the matrices $SA_j(\mathbf{u}^\varepsilon)$ are symmetric, we have

$$\Big\langle SA_j(\mathbf{u}^\varepsilon)\partial_j \Lambda^s \mathbf{u}^\varepsilon, \Lambda^s \mathbf{u}^\varepsilon \Big\rangle = \Big\langle \partial_j \Lambda^s \mathbf{u}^\varepsilon, SA_j(\mathbf{u}^\varepsilon)\Lambda^s \mathbf{u}^\varepsilon \Big\rangle$$

$$= -\Big\langle \Lambda^s \mathbf{u}^\varepsilon, \partial_j \left(SA_j(\mathbf{u}^\varepsilon) \right) \Lambda^s \mathbf{u}^\varepsilon \Big\rangle$$

$$- \Big\langle \Lambda^s \mathbf{u}^\varepsilon, SA_j(\mathbf{u}^\varepsilon)\Lambda^s \partial_j \mathbf{u}^\varepsilon \Big\rangle$$

$$= -\frac{1}{2}\Big\langle \Lambda^s \mathbf{u}^\varepsilon, \partial_j \left(SA_j(\mathbf{u}^\varepsilon) \right) \Lambda^s \mathbf{u}^\varepsilon \Big\rangle.$$

This yields

$$\Big| \Big\langle SA_j(\mathbf{u}^\varepsilon)\partial_j \Lambda^s \mathbf{u}^\varepsilon, \Lambda^s \mathbf{u}^\varepsilon \Big\rangle \Big| \leqslant \| \partial_j \left(SA_j(\mathbf{u}^\varepsilon) \right) \|_{L^\infty} \| \mathbf{u}^\varepsilon \|_{H^s}^2$$

$$\leqslant C \left(\| \mathbf{u}^\varepsilon \|_{H^s} \right) \| \mathbf{u}^\varepsilon \|_{H^s}^2,$$

where we have used Schauder's lemma (Lemma 1.23) and the assumption $s > n/2 + 1$. Usual estimates on commutators (see e.g. [Majda (1984); Taylor (1997)]), and Eq. (4.9), yield finally:

$$\frac{d}{dt} \langle S\Lambda^s \mathbf{u}^\varepsilon, \Lambda^s \mathbf{u}^\varepsilon \rangle \leqslant C \left(\| \mathbf{u}^\varepsilon \|_{H^s} \right) \langle S\Lambda^s \mathbf{u}^\varepsilon, \Lambda^s \mathbf{u}^\varepsilon \rangle,$$

for $s > n/2 + 2$. Gronwall lemma along with a continuity argument (to make sure that $\| \mathbf{u}^\varepsilon \|_{L^\infty} \leqslant 2C_0$) show that we can find $T > 0$ independent of ε, such that Eq. (4.4) has a unique solution $(v^\varepsilon, a^\varepsilon)$ in $C([-T, T]; H^s)^2$, uniformly in $\varepsilon \in]0, 1]$.

The fact that T can be chosen independent of $s > n/2 + 2$ follows from tame estimates (see Lemma 1.24).

Finally, once v^ε is known, we can proceed in two ways to conclude. Either remark that v^ε is irrotational ($\nabla \times v^\varepsilon \equiv 0$), so there exists $\widetilde{\Phi}^\varepsilon$ such that $v^\varepsilon = \nabla \widetilde{\Phi}^\varepsilon$; up to adding a function $F = F(t)$ of time only, $\Phi^\varepsilon = \widetilde{\Phi}^\varepsilon + F$ solves the first equation in (4.4). The other possibility is to define directly Φ^ε as

$$\Phi^\varepsilon(t,x) = \phi_0(x) - \int_0^t \left(\frac{1}{2}|v^\varepsilon(\tau,x)|^2 + f\left(|a^\varepsilon(\tau,x)|^2\right) \right) d\tau.$$

We check

$$\partial_t \left(\nabla \Phi^\varepsilon - v^\varepsilon\right) = \nabla \partial_t \Phi^\varepsilon - \partial_t v^\varepsilon = 0.$$

Since $\nabla \Phi^\varepsilon$ and v^ε have the same initial data, we infer that $\nabla \Phi^\varepsilon = v^\varepsilon$, and $(\Phi^\varepsilon, a^\varepsilon)$ solves Eq. (4.4). Since $v^\varepsilon, a^\varepsilon \in C([-T,T]; H^s)$ and $s > n/2$, we have directly $\Phi^\varepsilon \in C([-T,T]; L^2)$ (this is where we need $f(0) = 0$), and we conclude $\Phi^\varepsilon \in C([-T,T]; H^{s+1})$. $\qquad\square$

Once we have constructed the solution to Eq. (4.4), the next step is to study the asymptotic behavior of $(\Phi^\varepsilon, a^\varepsilon)$ as $\varepsilon \to 0$. In view of Eq. (4.4), it is natural to consider Eq. (4.5) in the case $V = 0$:

$$\begin{cases} \partial_t \Phi + \dfrac{1}{2}|\nabla \Phi|^2 + f\left(|a|^2\right) = 0 \quad ; \quad \Phi_{|t=0} = \phi_0, \\[2mm] \partial_t a + \nabla \Phi \cdot \nabla a + \dfrac{1}{2}a\Delta \Phi = 0 \quad ; \quad a_{|t=0} = a_0. \end{cases} \tag{4.11}$$

The proof of Theorem 4.1 shows that if $a_0 \in H^s$ (and $\phi_0 \in H^{s+1}$) for some $s > n/2 + 1$ (there is no second order operator in the analogue of Eq. (4.6) with $\varepsilon = 0$), then Eq. (4.11) has a unique solution

$$(\Phi, a) \in C([-T,T]; H^{s+1} \times H^s).$$

The error estimate between $(\Phi^\varepsilon, a^\varepsilon)$ and (Φ, a) is given by:

Proposition 4.3. *Let* $f \in C^\infty(\mathbb{R}_+; \mathbb{R})$ *with* $f(0) = 0$ *and* $f' > 0$. *Let* $s > 2 + n/2$. *Assume that* $\phi_0 \in H^{s+3}$, $a_0 \in H^{s+2}$ *and*

$$\|a_0^\varepsilon - a_0\|_{H^s} \xrightarrow[\varepsilon \to 0]{} 0.$$

Then for $T > 0$ *given by Theorem 4.1, there exists* $C > 0$ *independent of* ε *such that*

$$\|\Phi^\varepsilon - \Phi\|_{L^\infty([-T,T]; H^{s+1})} + \|a^\varepsilon - a\|_{L^\infty([-T,T]; H^s)} \leqslant C\left(\varepsilon + \|a_0^\varepsilon - a_0\|_{H^s}\right).$$

Proof.   Resume the above notation $\mathbf{u}^\varepsilon$, and introduce $\mathbf{u}$, its counterpart associated to (Φ, a). We know that $\mathbf{u} \in C([-T, T]; H^{s+2})$. Consider the error $\mathbf{w}^\varepsilon = \mathbf{u}^\varepsilon - \mathbf{u}$. It solves

$$\partial_t \mathbf{w}^\varepsilon + \sum_{j=1}^n \left(A_j(\mathbf{u}^\varepsilon)\partial_j \mathbf{u}^\varepsilon - A_j(\mathbf{u})\partial_j \mathbf{u} \right) = \frac{\varepsilon}{2} L \mathbf{u}^\varepsilon.$$

Rewrite this equation as:

$$\partial_t \mathbf{w}^\varepsilon + \sum_{j=1}^n A_j(\mathbf{u}^\varepsilon)\partial_j \mathbf{w}^\varepsilon = -\sum_{j=1}^n \left(A_j(\mathbf{u}^\varepsilon) - A_j(\mathbf{u}) \right)\partial_j \mathbf{u} + \frac{\varepsilon}{2} L \mathbf{w}^\varepsilon + \frac{\varepsilon}{2} L \mathbf{u}.$$

The operator on the left hand side is the same operator as in Eq. (4.7). It is symmetrized by S, defined in Eq. (4.8). This means that we keep the symmetrizer associated to $\mathbf{u}^\varepsilon$. We do not consider the symmetrizer associated to $\mathbf{u}$. The term $L\mathbf{w}^\varepsilon$ is not present in the energy estimates, since it is skew-symmetric. The term $\varepsilon L\mathbf{u}$ is considered as a source term: it is of order ε, uniformly in $C([-T, T]; H^s)$. Finally, the first term on the right hand side is a semi-linear perturbation:

$$\begin{aligned}
\left\| \left(A_j(\mathbf{u}^\varepsilon) - A_j(\mathbf{u}) \right)\partial_j \mathbf{u} \right\|_{H^s} &\leqslant \left\| \left(A_j(\mathbf{u}^\varepsilon) - A_j(\mathbf{u}) \right) \right\|_{H^s} \left\| \mathbf{u} \right\|_{H^{s+1}} \\
&\leqslant C \left\| \left(A_j(\mathbf{w}^\varepsilon + \mathbf{u}) - A_j(\mathbf{u}) \right) \right\|_{H^s} \\
&\leqslant C \left(\left\| \mathbf{w}^\varepsilon \right\|_{L^\infty}, \left\| \mathbf{u} \right\|_{L^\infty} \right) \left\| \mathbf{w}^\varepsilon \right\|_{H^s},
\end{aligned}$$

where we have used Moser's inequality. Finally, we know that $\mathbf{w}^\varepsilon$ is bounded in $L^\infty([-T, T] \times \mathbb{R}^n)$, as the difference of two bounded terms. With the same approach as in the proof of Theorem 4.1, we end up with:

$$\frac{d}{dt} \langle S\Lambda^s \mathbf{w}^\varepsilon, \Lambda^s \mathbf{w}^\varepsilon \rangle \leqslant C \left(\varepsilon + \left\| \mathbf{w}^\varepsilon \right\|_{H^s}^2 \right) \leqslant C \left(\varepsilon + \langle S\Lambda^s \mathbf{w}^\varepsilon, \Lambda^s \mathbf{w}^\varepsilon \rangle \right).$$

Gronwall lemma yields:

$$\left\| \nabla \Phi^\varepsilon - \nabla \Phi \right\|_{L^\infty([-T,T];H^s)} + \left\| a^\varepsilon - a \right\|_{L^\infty([-T,T];H^s)} \leqslant C \left(\varepsilon + \left\| a_0^\varepsilon - a_0 \right\|_{H^s} \right).$$

We infer

$$\left\| \partial_t (\Phi^\varepsilon - \Phi) \right\|_{H^s} \leqslant C \left(\varepsilon + \left\| a_0^\varepsilon - a_0 \right\|_{H^s} \right),$$

and since Φ^ε and Φ coincide at time $t = 0$,

$$\left\| \Phi^\varepsilon(t) - \Phi(t) \right\|_{H^{s+1}} \leqslant Ct \left(\varepsilon + \left\| a_0^\varepsilon - a_0 \right\|_{H^s} \right), \quad t \in [-T, T]. \qquad (4.12)$$

This completes the proof of the proposition. $\square$

At this stage of the study, it is tempting to consider $ae^{i\Phi/\varepsilon}$ as a decent approximation for u^ε. In general, this approximation is interesting for very small time only:

$$\left|u^\varepsilon - ae^{i\Phi/\varepsilon}\right| = \left|a^\varepsilon e^{i\Phi^\varepsilon/\varepsilon} - ae^{i\Phi/\varepsilon}\right|$$

$$\leqslant |a^\varepsilon - a| + |a|\left|e^{i\Phi^\varepsilon/\varepsilon} - e^{i\Phi/\varepsilon}\right|$$

$$\leqslant C\left(\varepsilon + \|a_0^\varepsilon - a_0\|_{H^s}\right) + C\left|\sin\left(\frac{\Phi^\varepsilon - \Phi}{2\varepsilon}\right)\right|.$$

In view of (4.12), we infer:

$$\left\|u^\varepsilon(t) - a(t)e^{i\Phi(t)/\varepsilon}\right\|_{L^\infty} \leqslant o_{\varepsilon\to 0}(1) + \mathcal{O}(t) + \mathcal{O}\left(\frac{t\|a_0^\varepsilon - a_0\|_{H^s}}{\varepsilon}\right).$$

The best we can expect in general is, provided $\|a_0^\varepsilon - a_0\|_{H^s} = \mathcal{O}(\varepsilon)$,

$$\left\|u^\varepsilon(t) - a(t)e^{i\Phi(t)/\varepsilon}\right\|_{L^\infty} = o_{\varepsilon\to 0}(1) + \mathcal{O}(t). \tag{4.13}$$

This shows that because the phase is divided by ε, we will obtain a good approximation for u^ε only if we know the asymptotic behavior of Φ^ε as $\varepsilon \to 0$ up to a remainder which is at least $o(\varepsilon)$. The above computation shows that it is reasonable to require the same thing about a_0^ε, because of the coupling because phase and amplitude.

We therefore seek an asymptotic expansion for $(\Phi^\varepsilon, a^\varepsilon)$. The formal approach is the same as the one presented in Sec. 1.2: we plug an asymptotic expansion of the form

$$(\Phi^\varepsilon, a^\varepsilon) = (\Phi, a) + \varepsilon\left(\Phi^{(1)}, a^{(1)}\right) + \varepsilon^2\left(\Phi^{(2)}, a^{(2)}\right) + \cdots$$

into Eq. (4.4), and we identify the powers of ε. Of course, we also assume

$$a_0^\varepsilon = a_0 + \varepsilon a_1 + \varepsilon^2 a_2 + \ldots$$

The term in ε^0 yields Eq. (4.5). For the term in ε^1, we obtain:

$$\begin{cases} \partial_t\Phi^{(1)} + \nabla\Phi\cdot\nabla\Phi^{(1)} + 2f'\left(|a|^2\right)\text{Re}\left(\overline{a}a^{(1)}\right) = 0, \\ \partial_t a^{(1)} + \nabla\Phi\cdot\nabla a^{(1)} + \nabla\Phi^{(1)}\cdot\nabla a + \frac{1}{2}a^{(1)}\Delta\Phi + \frac{1}{2}a\Delta\Phi^{(1)} = \frac{i}{2}\Delta a, \\ \Phi^{(1)}_{|t=0} = 0 \quad ; \quad a^{(1)}_{|t=0} = a_1. \end{cases}$$

To solve this system, introduce $v^{(1)} = \nabla\Phi^{(1)}$ (and $v = \nabla\Phi$):

$$\begin{cases} \partial_t v^{(1)} + v\cdot\nabla v^{(1)} + 2\nabla\left(f'\left(|a|^2\right)\text{Re}\left(\overline{a}a^{(1)}\right)\right) = -v^{(1)}\cdot\nabla v, \\ \partial_t a^{(1)} + v\cdot\nabla a^{(1)} + \frac{1}{2}a\,\text{div}\,v^{(1)} = -v^{(1)}\cdot\nabla a - \frac{1}{2}a^{(1)}\Delta\Phi + \frac{i}{2}\Delta a, \quad (4.14) \\ v^{(1)}_{|t=0} = 0 \quad ; \quad a^{(1)}_{|t=0} = a_1. \end{cases}$$

The left hand side is a *linear* hyperbolic symmetric operator, applied to $(v^{(1)}, a^{(1)})$. The right hand side consists of terms which are linear in $(v^{(1)}, a^{(1)})$, plus the source $i\Delta a$. We infer the following existence lemma, whose easy proof is left out.

Lemma 4.4. *Let $s > n/2 + 2$. Assume that $\phi_0 \in H^{s+3}$, $a_0 \in H^{s+2}$ and $a_1 \in H^s$. Then Eq. (4.14) has a unique solution*

$$(v^{(1)}, a^{(1)}) \in C([-T; T]; H^s)^2.$$

With this lemma, we can find $(\Phi^{(1)}, a^{(1)})$ as the second term of the asymptotic expansion for $(\Phi^\varepsilon, a^\varepsilon)$. Set

$$(\Phi_1^\varepsilon, a_1^\varepsilon) = (\Phi, a) + \varepsilon \left(\Phi^{(1)}, a^{(1)} \right).$$

By construction, it solves Eq. (4.4), up to a source term of order $\mathcal{O}(\varepsilon^2)$:

$$\begin{cases} \partial_t \Phi_1^\varepsilon + \dfrac{1}{2}|\nabla\Phi_1^\varepsilon|^2 + f\left(|a_1^\varepsilon|^2\right) = -\dfrac{\varepsilon^2}{2}|\nabla\Phi^{(1)}|^2 \\ \qquad\qquad\qquad - \varepsilon^2 \displaystyle\int_0^1 h''\left(a + \theta\varepsilon a^{(1)}\right) \cdot a^{(1)} \cdot a^{(1)} d\theta, \\ \partial_t a_1^\varepsilon + \nabla\Phi_1^\varepsilon \cdot \nabla a_1^\varepsilon + \dfrac{1}{2} a_1^\varepsilon \Delta\Phi_1^\varepsilon = i\dfrac{\varepsilon}{2}\Delta a_1^\varepsilon - \varepsilon^2 \nabla\Phi^{(1)} \cdot \nabla a^{(1)} \\ \qquad\qquad\qquad - \dfrac{\varepsilon^2}{2} a^{(1)}\Delta\Phi^{(1)} - i\dfrac{\varepsilon^2}{2}\Delta a^{(1)}, \\ \Phi_1^\varepsilon{}_{|t=0} = \phi_0 \quad ; \quad a_1^\varepsilon{}_{|t=0} = a_0 + \varepsilon a_1. \end{cases}$$

We have used the notation $h(z) = f(|z|^2)$, and the last term in the equation for Φ_1^ε is an obvious formal notation. Mimicking the proof of Proposition 4.3, the following result is left as an exercise:

Proposition 4.5. *Let $f \in C^\infty(\mathbb{R}_+; \mathbb{R})$ with $f(0) = 0$ and $f' > 0$. Let $s > 2 + n/2$. Assume that $\phi_0 \in H^{s+5}$, $a_0 \in H^{s+4}$, $a_1 \in H^{s+2}$, and*

$$\|a_0^\varepsilon - a_0 - \varepsilon a_1\|_{H^s} = o(\varepsilon) \quad \text{as } \varepsilon \to 0.$$

Then for $T > 0$ given by Theorem 4.1, there exists $C > 0$ independent of ε such that

$$\left\| \Phi^\varepsilon - \Phi - \varepsilon\Phi^{(1)} \right\|_{L^\infty([-T,T];H^{s+1})} + \left\| a^\varepsilon - a - \varepsilon a^{(1)} \right\|_{L^\infty([-T,T];H^s)} \leqslant$$
$$\leqslant C \left(\varepsilon^2 + \|a_0^\varepsilon - a_0 - \varepsilon a^{(1)}\|_{H^s} \right).$$

Despite the notations, it seems unadapted to consider $\Phi^{(1)}$ as being part of the phase. Indeed, we infer from the above proposition

$$\left\| u^\varepsilon - ae^{i\Phi^{(1)}} e^{i\Phi/\varepsilon} \right\|_{L^\infty([-T,T];L^2 \cap L^\infty)} \leqslant \left\| a^\varepsilon - a \right\|_{L^\infty([-T,T];L^2 \cap L^\infty)}$$

$$+ \|a\|_{L^\infty([-T,T];L^2 \cap L^\infty)} \left\| e^{i\Phi^\varepsilon/\varepsilon} - e^{i\Phi^{(1)}} e^{i\Phi/\varepsilon} \right\|_{L^\infty([-T,T] \times \mathbb{R}^n)}$$

$$\leqslant \mathcal{O}(\varepsilon) + \mathcal{O} \left(\frac{\|a_0^\varepsilon - a_0 - \varepsilon a^{(1)}\|_{H^s}}{\varepsilon} \right)$$

$$\leqslant o(1).$$

If in addition $\|a_0^\varepsilon - a_0 - \varepsilon a^{(1)}\|_{H^s} = \mathcal{O}(\varepsilon^2)$ (as is usual in WKB analysis), we find

$$\left\| u^\varepsilon - ae^{i\Phi^{(1)}} e^{i\Phi/\varepsilon} \right\|_{L^\infty([-T,T];L^2 \cap L^\infty)} = \mathcal{O}(\varepsilon).$$

Since $\Phi^{(1)}$ depends on a_1 while a does not, we retrieve the fact that in supercritical régimes, the leading order amplitude in WKB methods depends on the initial first corrector a_1. This phenomenon was called *ghost effect* in the context of gas dynamics [Sone *et al.* (1996)]: the corrector a_1 vanishes in the limit $\varepsilon \to 0$ at time $t = 0$, but plays a non-negligible role for $t > 0$.

Remark 4.6. The term $e^{i\Phi^{(1)}}$ does not appear in the Wigner measure of $ae^{i\Phi^{(1)}} e^{i\Phi/\varepsilon}$. Thus, from the point of view of Wigner measures, the asymptotic behavior of the exact solution is described by the Euler-type system (3.7). We also recover the result of Chap. 3

The above procedure can be pursued to arbitrary order, and we leave it at this stage. To conclude this paragraph, we examine more closely the relevance of the term $\Phi^{(1)}$. From the equation, we find

$$\Phi^{(1)}_{|t=0} = 0 \quad ; \quad \partial_t \Phi^{(1)}_{|t=0} = -2f' \left(|a_0|^2\right) \operatorname{Re}\left(\bar{a}_0 a_1\right).$$

So if $\operatorname{Re}\left(\bar{a}_0 a_1\right) \not\equiv 0$, $\Phi^{(1)}$ is non-trivial for $t > 0$. Note that even if $a_1 = 0$, then in general, $\Phi^{(1)}$ is non-trivial for $t > 0$. Indeed, if $a_1 = 0$, we have

$$a^{(1)}_{|t=0} = 0 \quad ; \quad \partial_t a^{(1)}_{|t=0} = \frac{i}{2} \Delta a_0,$$

and therefore

$$\Phi^{(1)}_{|t=0} = \partial_t \Phi^{(1)}_{|t=0} = 0 \quad ; \quad \partial_t^2 \Phi^{(1)}_{|t=0} = -f' \left(|a_0|^2\right) \operatorname{Im}\left(\bar{a}_0 \Delta a_0\right).$$

If a_0 has a constant argument, then $\Phi^{(1)} \equiv 0$, as shown below.

With the case of Eq. (4.3) in mind, assume that $a_0 e^{i\theta}$ is real-valued some some *constant* $\theta \in \mathbb{R}$; then so is $ae^{i\theta}$, from Eq. (4.5). In that case, we check that $\left(\Phi^{(1)}, \operatorname{Re}\left(\overline{a}a^{(1)}\right)\right)$ solves an *homogeneous* linear system, since then

$$\partial_t \operatorname{Re}\left(\overline{a}a^{(1)}\right) + \nabla\Phi \cdot \operatorname{Re}\left(\overline{a}a^{(1)}\right) = -\frac{1}{2}\operatorname{div}\left(|a|^2\nabla\Phi^{(1)}\right) - \operatorname{Re}\left(\overline{a}a^{(1)}\right)\Delta\Phi.$$

By uniqueness, if $a_0 e^{i\theta}$ is real-valued and $a_1 e^{i\theta}$ is purely imaginary (e.g. $a_1 = 0$), then $\Phi^{(1)} \equiv 0$. Note however that if $a_1 e^{i\theta} \notin i\mathbb{R}$ is non-trivial (e.g. $a_1 e^{i\theta} \in \mathbb{R}$), then $\Phi^{(1)}$ is non-trivial.

4.2.2   *With an external potential*

When $V \neq 0$, the first idea consists in trying the same arguments as above. Obviously, if $V \in H^\infty$, then the previous approach can be repeated. Even if we assume only $\nabla V \in H^\infty$, we can construct $(v^\varepsilon, a^\varepsilon)$ in Sobolev spaces, and then Φ^ε is not necessarily in L^2, which is not a big issue.

This approach is essentially perturbative. Its main drawback is that it is incompatible with the case when V is an harmonic potential for instance, a case motivated by Physics. We seek a solution to Eq. (4.4), with

$$\Phi^\varepsilon = \phi_{\text{eik}} + \phi^\varepsilon,$$

where ϕ_{eik} was constructed in Sec. 1.3.1, and ϕ^ε belongs to some Sobolev space. This idea is very naïve, since the equations at stake are nonlinear (note that even when $f = 0$, the eikonal equation is a nonlinear equation). However, it turns out to be fruitful, essentially because ϕ_{eik} is subquadratic with respect to the space variable. For the sake of readability, we rewrite Assumption 1.7 and the main result of Sec. 1.3.1:

Assumption 4.7 (Geometric assumption). *We assume that the potential and the initial phase are smooth, real-valued, and subquadratic:*

- $V \in C^\infty(\mathbb{R} \times \mathbb{R}^n)$, *and* $\partial_x^\alpha V \in C(\mathbb{R}; L^\infty(\mathbb{R}^n))$ *as soon as* $|\alpha| \geqslant 2$.
- $\phi_0 \in C^\infty(\mathbb{R}^n)$, *and* $\partial^\alpha \phi_0 \in L^\infty(\mathbb{R}^n)$ *as soon as* $|\alpha| \geqslant 2$.

Proposition 4.8. *Under Assumption 4.7, there exists $T > 0$ and a unique solution $\phi_{\text{eik}} \in C^\infty\left([-T, T] \times \mathbb{R}^n\right)$ to*

$$\partial_t \phi_{\text{eik}} + \frac{1}{2}\left|\nabla\phi_{\text{eik}}\right|^2 + V = 0 \quad ; \quad \phi_{\text{eik}}(0, x) = \phi_0(x).$$

In addition, this solution is subquadratic: $\partial_x^\alpha \phi_{\text{eik}} \in L^\infty([-T, T] \times \mathbb{R}^n)$ as soon as $|\alpha| \geqslant 2$.

In terms of the unkown function $(\phi^\varepsilon, a^\varepsilon)$, Eq. (4.4) reads:

$$
\begin{cases}
\partial_t \phi^\varepsilon + \nabla \phi_{\text{eik}} \cdot \nabla \phi^\varepsilon + \dfrac{1}{2} |\nabla \phi^\varepsilon|^2 + f\left(|a^\varepsilon|^2\right) = 0, \\[2mm]
\partial_t a^\varepsilon + \nabla \phi_{\text{eik}} \cdot \nabla a^\varepsilon + \nabla \phi^\varepsilon \cdot \nabla a^\varepsilon + \dfrac{1}{2} a^\varepsilon \Delta \phi_{\text{eik}} + \dfrac{1}{2} a^\varepsilon \Delta \phi^\varepsilon = i\dfrac{\varepsilon}{2} \Delta a^\varepsilon, \\[2mm]
\phi^\varepsilon_{|t=0} = 0 \quad ; \quad a^\varepsilon_{|t=0} = a^\varepsilon_0.
\end{cases}
$$

By construction, the potential V has disappeared from the equation. To prove the analogue of Theorem 4.1, we stop counting the derivatives, and we distinguish two cases, whether the nonlinearity is exactly cubic or not:

Assumption 4.9 (Analytical assumption). *We assume that a^ε_0 is bounded in H^s for all $s \geqslant 0$, and $f \in C^\infty(\mathbb{R}_+; \mathbb{R})$ with $f(0) = 0$ and $f' > 0$. Moreover,*

- *The nonlinearity is exactly cubic, $f(y) = \lambda y$ for some $\lambda > 0$, or*
- *The first momentum of a^ε_0 is bounded in Sobolev spaces: xa^ε_0 is bounded in H^s for all $s \geqslant 0$.*

The above distinction makes it possible to refine a result in [Carles (2007c)]:

Theorem 4.10. *Let Assumptions 4.7 and 4.9 be satisfied. There exist $T_* > 0$ independent of $\varepsilon \in]0, 1]$, and $u^\varepsilon = a^\varepsilon e^{i(\phi_{\text{eik}} + \phi^\varepsilon)/\varepsilon}$ solution to (4.1) on $[-T_*, T_*]$. Moreover, a^ε and ϕ^ε are bounded in $C([-T_*, T_*]; H^\infty)$. In addition, in the second case of Assumption 4.9, xa^ε and $x\nabla \phi^\varepsilon$ are bounded in $L^\infty([-T_*, T_*]; H^s)$ for all $s \geqslant 0$.*

Proof. The proof proceeds along the same lines as the proof of Theorem 4.1, so we point out the main differences. We only work with times such that $|t| \leqslant T$, so that ϕ_{eik} remains smooth. Like in the previous paragraph, we introduce $v^\varepsilon = \nabla \phi^\varepsilon$, and the notations

$$
\mathbf{u}^\varepsilon = \begin{pmatrix} \operatorname{Re} a^\varepsilon \\ \operatorname{Im} a^\varepsilon \\ v^\varepsilon_1 \\ \vdots \\ v^\varepsilon_n \end{pmatrix} = \begin{pmatrix} a^\varepsilon_1 \\ a^\varepsilon_2 \\ v^\varepsilon_1 \\ \vdots \\ v^\varepsilon_n \end{pmatrix}, \quad L = \begin{pmatrix} 0 & -\Delta & 0 & \dots & 0 \\ \Delta & 0 & 0 & \dots & 0 \\ 0 & 0 & & 0_{n \times n} & \end{pmatrix},
$$

and
$$
A(\mathbf{u}, \xi) = \sum_{j=1}^{n} A_j(\mathbf{u})\xi_j = \begin{pmatrix} v \cdot \xi & 0 & \frac{a_1}{2} {}^t\xi \\ 0 & v \cdot \xi & \frac{a_2}{2} {}^t\xi \\ 2f'a_1 \xi & 2f'a_2 \xi & v \cdot \xi I_n \end{pmatrix},
$$

where f' stands for $f'(|a_1|^2 + |a_2|^2)$. Instead of (4.7), we now have a system of the form

$$\partial_t \mathbf{u}^\varepsilon + \sum_{j=1}^n A_j(\mathbf{u}^\varepsilon)\partial_j \mathbf{u}^\varepsilon + \nabla\phi_{\text{eik}} \cdot \nabla \mathbf{u}^\varepsilon + M\left(\nabla^2\phi_{\text{eik}}\right)\mathbf{u}^\varepsilon = \frac{\varepsilon}{2}L\mathbf{u}^\varepsilon, \quad (4.15)$$

where the matrix M is smooth and locally bounded. The quasi-linear part of the above equation is the same as in Eq. (4.7), and involves the matrices A_j. In particular, we keep the same symmetrizer S given by (4.8). The term $\nabla\phi_{\text{eik}} \cdot \nabla\mathbf{u}^\varepsilon$ has a semi-linear contribution, as we see below. The term corresponding to the matrix M can obviously be considered as a source term, since ϕ_{eik} is subquadratic.

For $s > n/2 + 2$, we still have

$$\frac{d}{dt}\left\langle S\Lambda^s\mathbf{u}^\varepsilon, \Lambda^s\mathbf{u}^\varepsilon\right\rangle = \left\langle \partial_t S\Lambda^s\mathbf{u}^\varepsilon, \Lambda^s\mathbf{u}^\varepsilon\right\rangle + 2\left\langle S\partial_t\Lambda^s\mathbf{u}^\varepsilon, \Lambda^s\mathbf{u}^\varepsilon\right\rangle.$$

Two cases must then be distinguished, which explain the two cases in Assumption 4.9: if f' is constant, then so is the symmetrizer S. Otherwise, we have

$$\left\langle \partial_t S\Lambda^s\mathbf{u}^\varepsilon, \Lambda^s\mathbf{u}^\varepsilon\right\rangle \leqslant \left\| \frac{1}{f'}\partial_t\left(f'\left(|a_1^\varepsilon|^2 + |a_2^\varepsilon|^2\right)\right)\right\|_{L^\infty} \left\langle S\Lambda^s\mathbf{u}^\varepsilon, \Lambda^s\mathbf{u}^\varepsilon\right\rangle.$$

Since a_0^ε is bounded in $H^s(\mathbb{R}^n) \subset L^\infty(\mathbb{R}^n)$, there exists C_0 independent of $\varepsilon \in]0, 1]$ such that

$$\|a_0^\varepsilon\|_{L^\infty} \leqslant C_0.$$

So long as $\|\mathbf{u}^\varepsilon\|_{L^\infty} \leqslant 2C_0$, we have:

$$f'\left(|a_1^\varepsilon|^2 + |a_2^\varepsilon|^2\right) \geqslant \inf\left\{f'(y)\ ;\ 0 \leqslant y \leqslant 4C_0^2\right\} = \delta_n > 0,$$

where δ_n is now fixed, since f' is continuous with $f' > 0$. We infer,

$$\left\| \frac{1}{f'}\partial_t\left(f'\left(|a_1^\varepsilon|^2 + |a_2^\varepsilon|^2\right)\right)\right\|_{L^\infty} \leqslant C\left\|\partial_t\left(|a_1^\varepsilon|^2 + |a_2^\varepsilon|^2\right)\right\|_{L^\infty}.$$

Using the equation for a^ε, we see that to estimate $\partial_t a^\varepsilon$, new terms appear, compared to the proof of Theorem 4.1: $\nabla\phi_{\text{eik}} \cdot \nabla a^\varepsilon$ and $a^\varepsilon\Delta\phi_{\text{eik}}$. Since ϕ_{eik} is subquadratic, we have, thanks to Sobolev embeddings:

$$\left\| \frac{1}{f'}\partial_t\left(f'\left(|a_1^\varepsilon|^2 + |a_2^\varepsilon|^2\right)\right)\right\|_{L^\infty} \leqslant C\left(\|\mathbf{u}^\varepsilon\|_{H^s} + \|x\mathbf{u}^\varepsilon\|_{H^{s-1}}\right).$$

For the clarity of the proof, we distinguish the cases for the rest of the computations.

First case: exactly cubic nonlinearity. In the first case of Assumption 4.9, we have noticed that the symmetrizer S is constant. For the quasi-linear estimates involving the matrices A_j, we can mimic the proof of Theorem 4.1:

$$\frac{d}{dt} \langle S\Lambda^s \mathbf{u}^\varepsilon, \Lambda^s \mathbf{u}^\varepsilon \rangle \leqslant C \left(\|\mathbf{u}^\varepsilon\|_{H^s} \right) \langle S\Lambda^s \mathbf{u}^\varepsilon, \Lambda^s \mathbf{u}^\varepsilon \rangle$$
$$+ \left| \left\langle S\Lambda^s \left(\nabla \phi_{\text{eik}} \cdot \nabla \mathbf{u}^\varepsilon \right), \Lambda^s \mathbf{u}^\varepsilon \right\rangle \right|$$
$$+ \left| \left\langle S\Lambda^s \left(M(\nabla^2 \phi_{\text{eik}}) \mathbf{u}^\varepsilon \right), \Lambda^s \mathbf{u}^\varepsilon \right\rangle \right|.$$

For the second term of the right hand side, write:

$$\left\langle S\Lambda^s \left(\partial_j \phi_{\text{eik}} \partial_j \mathbf{u}^\varepsilon \right), \Lambda^s \mathbf{u}^\varepsilon \right\rangle = \left\langle S\partial_j \phi_{\text{eik}} \partial_j \Lambda^s \mathbf{u}^\varepsilon, \Lambda^s \mathbf{u}^\varepsilon \right\rangle$$
$$+ \left\langle S \left[\Lambda^s, \partial_j \phi_{\text{eik}} \partial_j \right] \mathbf{u}^\varepsilon, \Lambda^s \mathbf{u}^\varepsilon \right\rangle.$$

For the first term of the right hand side, an integration by parts yields:

$$\left| \left\langle S\partial_j \phi_{\text{eik}} \partial_j \Lambda^s \mathbf{u}^\varepsilon, \Lambda^s \mathbf{u}^\varepsilon \right\rangle \right| \leqslant \left\| \partial_j \left(S\partial_j \phi_{\text{eik}} \right) \right\|_{L^\infty} \|\mathbf{u}^\varepsilon\|_{H^s}^2 \tag{4.16}$$
$$\leqslant C \|\mathbf{u}^\varepsilon\|_{H^s}^2,$$

where we have used the fact that S is constant and ϕ_{eik} is subquadratic. This also shows that the commutator

$$[\Lambda^s, \partial_j \phi_{\text{eik}} \partial_j]$$

is a pseudo-differential operator of degree $\leqslant s$, with bounded coefficients. We infer:

$$\left| \left\langle S\Lambda^s \left(\nabla \phi_{\text{eik}} \cdot \nabla \mathbf{u}^\varepsilon \right), \Lambda^s \mathbf{u}^\varepsilon \right\rangle \right| \leqslant C \|\mathbf{u}^\varepsilon\|_{H^s}^2.$$

We have obviously

$$\left| \left\langle S\Lambda^s \left(M(\nabla^2 \phi_{\text{eik}}) \mathbf{u}^\varepsilon \right), \Lambda^s \mathbf{u}^\varepsilon \right\rangle \right| \leqslant C \|\mathbf{u}^\varepsilon\|_{H^s}^2.$$

This yields:

$$\frac{d}{dt} \langle S\Lambda^s \mathbf{u}^\varepsilon, \Lambda^s \mathbf{u}^\varepsilon \rangle \leqslant C \left(\|\mathbf{u}^\varepsilon\|_{H^s} \right) \langle S\Lambda^s \mathbf{u}^\varepsilon, \Lambda^s \mathbf{u}^\varepsilon \rangle, \tag{4.17}$$

and we conclude like in Theorem 4.1.

Second case. If we assume only $f' > 0$, the assumption $xa_0^\varepsilon \in H^s$ makes it possible to conclude in a similar fashion. We have:

$$\frac{d}{dt} \langle S\Lambda^s \mathbf{u}^\varepsilon, \Lambda^s \mathbf{u}^\varepsilon \rangle \leqslant C \left(\|\mathbf{u}^\varepsilon\|_{H^s} + \|x\mathbf{u}^\varepsilon\|_{H^{s-1}} \right) \langle S\Lambda^s \mathbf{u}^\varepsilon, \Lambda^s \mathbf{u}^\varepsilon \rangle$$
$$+ \left| \left\langle S\Lambda^s \left(\nabla \phi_{\text{eik}} \cdot \nabla \mathbf{u}^\varepsilon \right), \Lambda^s \mathbf{u}^\varepsilon \right\rangle \right|$$
$$+ \left| \left\langle S\Lambda^s \left(M(\nabla^2 \phi_{\text{eik}}) \mathbf{u}^\varepsilon \right), \Lambda^s \mathbf{u}^\varepsilon \right\rangle \right|.$$

The last term is obviously controlled by $\|\mathbf{u}^\varepsilon\|_{H^s}^2$. For the second term, resume the estimate (4.16). Using the definition of S, we find

$$\|\partial_j \left(S\partial_j \phi_{\text{eik}}\right)\|_{L^\infty} \leqslant C \left(\|\mathbf{u}^\varepsilon\|_{L^\infty}\right) \|\langle x \rangle\, a^\varepsilon\|_{L^\infty} \|\partial_j a^\varepsilon\|_{L^\infty}$$

$$\leqslant C \left(\|\mathbf{u}^\varepsilon\|_{L^\infty}\right) \|\langle x \rangle\, a^\varepsilon\|_{H^{s-1}} \|a^\varepsilon\|_{H^s}.$$

We infer:

$$\frac{d}{dt} \left\langle S\Lambda^s \mathbf{u}^\varepsilon, \Lambda^s \mathbf{u}^\varepsilon \right\rangle \leqslant F\big(\|\mathbf{u}^\varepsilon\|_{H^s} + \|x \mathbf{u}^\varepsilon\|_{H^{s-1}}\big) \left\langle S\Lambda^s \mathbf{u}^\varepsilon, \Lambda^s \mathbf{u}^\varepsilon \right\rangle.$$

To close the family of estimates, we show that

$$\frac{d}{dt} \left\langle S\Lambda^{s-1} \left(x\mathbf{u}^\varepsilon\right), \Lambda^{s-1} \left(x\mathbf{u}^\varepsilon\right) \right\rangle$$

can be bounded in a similar fashion. Let $1 \leqslant k \leqslant n$:

$$\partial_t(x_k \mathbf{u}^\varepsilon) + \sum_{j=1}^n A_j(\mathbf{u}^\varepsilon)\partial_j(x_k \mathbf{u}^\varepsilon) + \nabla\phi_{\text{eik}} \cdot \nabla(x_k \mathbf{u}^\varepsilon) + M \left(\nabla^2 \phi_{\text{eik}}\right) x_k \mathbf{u}^\varepsilon$$

$$= \frac{\varepsilon}{2} L(x_k \mathbf{u}^\varepsilon) + A_k(\mathbf{u}^\varepsilon)\mathbf{u}^\varepsilon + \partial_k \phi_{\text{eik}} \mathbf{u}^\varepsilon + \frac{\varepsilon}{2}[x_k, L]\mathbf{u}^\varepsilon.$$

The quasi-linear part and the term $\nabla\phi_{\text{eik}} \cdot \nabla(x_k \mathbf{u}^\varepsilon)$ are estimated like before. The terms $M \left(\nabla^2 \phi_{\text{eik}}\right) x_k \mathbf{u}^\varepsilon$ and $A_k(\mathbf{u}^\varepsilon)\mathbf{u}^\varepsilon$ are controlled in an obvious way. The term $\partial_k \phi_{\text{eik}} \mathbf{u}^\varepsilon$ is controlled by $\langle x \rangle\, \mathbf{u}^\varepsilon$, since ϕ_{eik} is subquadratic: this is a linear perturbation. Finally,

$$[x_k, L] = \begin{pmatrix} 0 & 2\partial_k & 0 & \dots & 0 \\ -2\partial_k & 0 & 0 & \dots & 0 \\ 0 & 0 & & 0_{n \times n} & \end{pmatrix}.$$

Therefore,

$$\left\|\Lambda^{s-1}[x_k, L]\mathbf{u}^\varepsilon\right\|_{L^2} \leqslant 2\|\mathbf{u}^\varepsilon\|_{H^s}.$$

This shows that we obtain a closed family of estimates. Gronwall lemma and a continuity argument yield existence and uniqueness for $\mathbf{u}^\varepsilon$, like for Theorem 4.1. Note that thanks to tame estimates, T_* does not depend on $s > n/2 + 2$. Finally, we define ϕ^ε by

$$\phi^\varepsilon(t) = -\int_0^t \left(\nabla\phi_{\text{eik}} \cdot v^\varepsilon + \frac{1}{2}|v^\varepsilon|^2 + f\left(|a_0^\varepsilon|^2\right)\right)(\tau)d\tau.$$

We check that $\phi^\varepsilon \in C([-T_*, T_*]; L^2)$ and $\partial_t \left(\nabla\phi^\varepsilon - v^\varepsilon\right) = 0$, and the proof of the theorem is complete. $\square$

Note that the phase $\phi_{\text{eik}}+\phi^\varepsilon$ belongs to a somewhat non-standard space: it is the sum of a subquadratic function and an H^∞ function. Even if ϕ^ε goes to zero at infinity, it should not be considered as a small perturbation of ϕ_{eik} in L^∞. Indeed, like for the case $V \equiv 0$, the coupling between the amplitude a^ε and ϕ^ε is so strong that even though $\phi^\varepsilon_{|t=0} = 0$,

$$\partial_t \phi^\varepsilon_{|t=0} = -f\left(|a_0^\varepsilon|^2\right),$$

so ϕ^ε is of order $\mathcal{O}(1)$ in L^∞ for $t > 0$.

The next step in the semi-classical analysis is to study the asymptotic expansion of $(\phi^\varepsilon, a^\varepsilon)$ as $\varepsilon \to 0$. It proceeds along the same lines as in the case $V = 0$ (Sec. 4.2.1), up to the adaptations pointed out in the above proof. We leave out the discussion at this stage, since all the tools have been given, and the conclusion is essentially the same as in Sec. 4.2.1.

4.2.3 *The case $0 < \alpha < 1$*

So far, we have addressed the case of weakly nonlinear geometric optics (Chap. 2), and the supercritical case (4.1). In this paragraph, we discuss the intermediary case of

$$i\varepsilon\partial_t u^\varepsilon + \frac{\varepsilon^2}{2}\Delta u^\varepsilon = Vu^\varepsilon + \varepsilon^\alpha f\left(|u^\varepsilon|^2\right)u^\varepsilon \quad ; \quad u^\varepsilon_{|t=0} = a_0^\varepsilon e^{i\phi_0/\varepsilon}, \quad (4.18)$$

in the case $0 < \alpha < 1$. We have seen in Chap. 2 that the nonlinear term is so strong that we should not expect $u^\varepsilon e^{-i\phi_{\text{eik}}/\varepsilon}$ to be bounded in H^s ($s > 0$) as $\varepsilon \to 0$. So we adapt the point of view of the previous paragraph, that is, Eq. (4.18) with $\alpha = 0$. Again, we write the exact solution as

$$u^\varepsilon = a^\varepsilon e^{i\Phi^\varepsilon/\varepsilon}, \quad \text{with } \Phi^\varepsilon = \phi_{\text{eik}} + \phi^\varepsilon.$$

Let Assumptions 4.7 and 4.9 be satisfied. To simplify the discussion, suppose that we are in the second case of Assumption 4.9 (which is not incompatible with the first case!). The unknown function is the pair $(a^\varepsilon, \phi^\varepsilon)$. We have two unknown functions to solve a single equation, (1.1). We can choose how to balance the terms: we resume the approach followed when $\alpha = 0$. Note that this approach would also be efficient for the case $\alpha \geqslant 1$, with the serious drawback that we still assume $f' > 0$, an assumption proven to be unnecessary when $\alpha \geqslant 1$ (see Chap. 2). We impose:

$$\begin{cases} \partial_t\phi^\varepsilon + \dfrac{1}{2}|\nabla\phi^\varepsilon|^2 + \nabla\phi_{\text{eik}} \cdot \nabla\phi^\varepsilon + \varepsilon^\alpha f\left(|a^\varepsilon|^2\right) = 0, \\[2mm] \partial_t a^\varepsilon + \nabla\phi^\varepsilon \cdot \nabla a^\varepsilon + \nabla\phi_{\text{eik}} \cdot \nabla a^\varepsilon + \dfrac{1}{2}a^\varepsilon\Delta\phi^\varepsilon + \dfrac{1}{2}a^\varepsilon\Delta\phi_{\text{eik}} = i\dfrac{\varepsilon}{2}\Delta a^\varepsilon, \\[2mm] \phi^\varepsilon_{|t=0} = 0 \quad ; \quad a^\varepsilon_{|t=0} = a_0^\varepsilon. \end{cases}$$

This is the same system as before, with only f replaced by $\varepsilon^\alpha f$. Mimicking the analysis of the previous paragraph, we work with the unknown $\mathbf{u}^\varepsilon$ given by the same definition: it solves the system (4.15), where only the matrices A_j have changed, and now depend on ε. The symmetrizer is the same as before, with f' replaced by $\varepsilon^\alpha f'$: the matrix $S = S^\varepsilon$ is not bounded as $\varepsilon \to 0$, but its inverse is. We claim that we can still proceed as before, thanks to this remark and the following reasons:

- The matrix S^ε is diagonal.
- The matrix M is block diagonal.
- The matrices $S^\varepsilon A_j^\varepsilon$ are independent of $\varepsilon \in \,]0,1]$.
- The inverse of S^ε is uniformly bounded on compact sets, as $\varepsilon \to 0$.

Gronwall lemma then implies the analogue of Theorem 4.10: in particular, $\mathbf{u}^\varepsilon$ exists locally in time, with H^s-norm uniformly bounded as $\varepsilon \to 0$. Note that since $\phi^\varepsilon\big|_{t=0} = 0$, we have:

$$(S^\varepsilon \Lambda^s \mathbf{u}^\varepsilon, \Lambda^s \mathbf{u}^\varepsilon)\big|_{t=0} = \mathcal{O}(1),$$

and we infer more precisely:

$$\|a^\varepsilon\|_{L^\infty([-T_*,T_*];H^s)} + \|xa^\varepsilon\|_{L^\infty([-T_*,T_*];H^s)} = \mathcal{O}(1),$$
$$\|\phi^\varepsilon\|_{L^\infty([-T_*,T_*];H^s)} + \|x\nabla\phi^\varepsilon\|_{L^\infty([-T_*,T_*];H^s)} = \mathcal{O}\left(\varepsilon^\alpha\right).$$

It seems natural to change unknown functions, and work with $\widetilde\phi^\varepsilon = \varepsilon^{-\alpha}\phi^\varepsilon$ instead of ϕ^ε. With this, we somehow correct the shift in the cascade of equations caused by the factor ε^α in front of the nonlinearity. We find

$$\begin{cases} \partial_t\widetilde\phi^\varepsilon + \dfrac{\varepsilon^\alpha}{2}\left|\nabla\widetilde\phi^\varepsilon\right|^2 + \nabla\phi_{\text{eik}}\cdot\nabla\widetilde\phi^\varepsilon + f\left(|a^\varepsilon|^2\right) = 0, \\[2mm] \partial_t a^\varepsilon + \varepsilon^\alpha\nabla\widetilde\phi^\varepsilon\cdot\nabla a^\varepsilon + \nabla\phi_{\text{eik}}\cdot\nabla a^\varepsilon + \dfrac{\varepsilon^\alpha}{2}a^\varepsilon\Delta\widetilde\phi^\varepsilon + \dfrac{1}{2}a^\varepsilon\Delta\phi_{\text{eik}} = i\dfrac{\varepsilon}{2}\Delta a^\varepsilon, \\[2mm] \widetilde\phi^\varepsilon_{|t=0} = 0 \quad ; \quad a^\varepsilon_{|t=0} = a_0^\varepsilon. \end{cases}$$

The pairs $(\widetilde\phi^\varepsilon, a^\varepsilon)$ and $(\partial_t\widetilde\phi^\varepsilon, \partial_t a^\varepsilon)$ are bounded in $C([-T_*,T_*];H^s)$. Therefore, Arzela–Ascoli's theorem shows that a subsequence is convergent, and the limit is given by:

$$\begin{cases} \partial_t\widetilde\phi + \nabla\phi_{\text{eik}}\cdot\nabla\widetilde\phi + f\left(|a|^2\right) = 0 \,; \quad \widetilde\phi_{|t=0} = 0, \\[2mm] \partial_t a + \nabla\phi_{\text{eik}}\cdot\nabla a + \dfrac{1}{2}a\Delta\phi_{\text{eik}} = 0 \,; \quad a_{|t=0} = a_0. \end{cases}$$

We see that a solves the same transport equation as in the linear case, Eq. (1.19); $\widetilde\phi$ is given by an ordinary differential equation along the rays associated to ϕ_{eik}, with a source term showing nonlinear effect: $f\left(|a|^2\right)$. By uniqueness, the whole sequence is convergent. Roughly speaking, we see that if

$$\mathbf{w}^\varepsilon = {}^t\left(\nabla\left(\widetilde{\phi}^\varepsilon - \widetilde{\phi}\right), a^\varepsilon - a\right),$$

then Gronwall lemma yields:

$$(S^\varepsilon \partial_x^\alpha \mathbf{w}^\varepsilon, \partial_x^\alpha \mathbf{w}^\varepsilon) \leqslant C\left(\varepsilon + \varepsilon^\alpha\right) \leqslant 2C\varepsilon^\alpha.$$

We infer, for $|t| \leqslant T_*$:

$$\|a^\varepsilon(t) - a(t)\|_{H^s} \leqslant C_s \varepsilon^\alpha \quad ; \quad \|\phi^\varepsilon(t) - \varepsilon^\alpha\widetilde{\phi}\|_{H^s} \leqslant C_s \varepsilon^{2\alpha} t.$$

Three cases must be distinguished:

- If $1/2 < \alpha < 1$, then we can infer
$$\left\|u^\varepsilon - ae^{i\phi_{\text{eik}}/\varepsilon}e^{i\widetilde{\phi}/\varepsilon^{1-\alpha}}\right\|_{L^\infty([-T_*,T_*];L^2\cap L^\infty)} \xrightarrow[\varepsilon\to 0]{} 0.$$

- If $\alpha = 1/2$, then we can infer a similar result for small time only:
$$\left\|u^\varepsilon - ae^{i\phi_{\text{eik}}/\varepsilon}e^{i\widetilde{\phi}/\varepsilon^{1-\alpha}}\right\|_{L^\infty([-t,t];L^2\cap L^\infty)} \to 0 \quad \text{as } \varepsilon \text{ and } t \to 0.$$

- If $0 < \alpha < 1/2$, then we must pursue the analysis, and compute a corrector of order $\varepsilon^{2\alpha}$.

We shall not go further into detailed computations, but instead, discuss the whole analysis in a rather loose fashion. However, we note that all the ingredients have been given for a complete justification.

Let $N = [1/\alpha]$, where $[r]$ is the largest integer not larger than $r > 0$. We construct $a^{(1)}, \ldots, a^{(N)}$ and $\widetilde{\phi}^{(1)}, \ldots, \widetilde{\phi}^{(N)}$ such that:

$$\left\|a^\varepsilon - a - \varepsilon^\alpha a^{(1)} - \ldots - \varepsilon^{N\alpha}a^{(N)}\right\|_{L^\infty([-T_*,T_*];H^s)} +$$
$$+ \left\|\widetilde{\phi}^\varepsilon - \widetilde{\phi} - \varepsilon^\alpha\widetilde{\phi}^{(1)} - \ldots - \varepsilon^{N\alpha}\widetilde{\phi}^{(N)}\right\|_{L^\infty([-T_*,T_*];H^s)} = o\left(\varepsilon^{N\alpha}\right).$$

But since $N + 1 > 1/\alpha$, we have:

$$\left\|\phi^\varepsilon - \varepsilon^\alpha\widetilde{\phi} - \varepsilon^{2\alpha}\widetilde{\phi}^{(1)} - \ldots - \varepsilon^{N\alpha}\widetilde{\phi}^{(N-1)}\right\|_{L^\infty([-T_*,T_*];H^s)} = \mathcal{O}\left(\varepsilon^{(N+1)\alpha}\right)$$
$$= o(\varepsilon).$$

We infer:

$$\left\|u^\varepsilon - ae^{i\phi_{\text{eik}}/\varepsilon + i\phi_{\text{app}}^\varepsilon}\right\|_{L^\infty([-T_*,T_*];L^2\cap L^\infty)} = o(1),$$

where

$$\phi_{\text{app}}^{\varepsilon} = \frac{\widetilde{\phi}}{\varepsilon^{1-\alpha}} + \frac{\widetilde{\phi}^{(1)}}{\varepsilon^{1-2\alpha}} + \cdots + \frac{\widetilde{\phi}^{(N-1)}}{\varepsilon^{1-N\alpha}}.$$

Remark 4.11. In the case $\alpha = 1$, $N = 1$, and the above analysis shows that one phase shift factor appears: we retrieve the result of Chap. 2, and $\widetilde{\phi}$ coincides with the function G of Eq. (2.7) (under the unnecessary assumption $f' > 0$). If $\alpha > 1$, then $N = 0$, and we see that $ae^{i\phi_{\text{eik}}/\varepsilon}$ is a good approximation for u^{ε}.

To conclude this paragraph, we consider the convergence of quadratic observables. It follows from the pointwise description of the wave function u^{ε}. Since the nonlinear effects are present at leading order only through $\phi_{\text{app}}^{\varepsilon}$, the quadratic observables converge to the same quantities as in the linear case (just like for the case $\alpha = 1$):

$$|u^{\varepsilon}|^2 = \rho^{\varepsilon} \xrightarrow[\varepsilon \to 0]{} \rho \quad ; \quad \varepsilon \operatorname{Im}\left(\overline{u}^{\varepsilon} \nabla u^{\varepsilon}\right) = J^{\varepsilon} \xrightarrow[\varepsilon \to 0]{} \rho v,$$

where $(\rho, v) = (|a|^2, \nabla\phi_{\text{eik}})$ solves the Euler equation

$$\begin{cases} \partial_t v + v \cdot \nabla v + \nabla V = 0 & ; \quad v_{|t=0} = \nabla\phi_0, \\ \partial_t \rho + \operatorname{div}(\rho v) = 0 & ; \quad \rho_{|t=0} = |a_0|^2. \end{cases}$$

The pressure is given by $p = 0$, and the external force is ∇V. We see that even if $\alpha = 1$ is the critical threshold as far as WKB analysis is concerned, when it turns to quadratic observables, the critical threshold becomes $\alpha = 0$ (when the pressure law in Euler equations depends on f).

4.3   Higher order homogeneous nonlinearities

If we consider a quintic nonlinearity in (4.1), $f(y) = y^2$, the previous approach fails. Essentially, the symmetrizer

$$S = \begin{pmatrix} I_2 & 0 \\ 0 & \frac{1}{4f'(|u^{\varepsilon}|^2)} I_n \end{pmatrix}.$$

becomes singular at the zeroes of u^{ε}. Since we have no control on the zeroes of u^{ε}, the approach must be modified. We present the result of [Alazard and Carles (2007b)], for the case of

$$i\varepsilon\partial_t u^{\varepsilon} + \frac{\varepsilon^2}{2}\Delta u^{\varepsilon} = |u^{\varepsilon}|^{2\sigma} u^{\varepsilon} \quad ; \quad u_{|t=0}^{\varepsilon} = a_0^{\varepsilon} e^{i\phi_0/\varepsilon}. \tag{4.19}$$

We assume $V = 0$: in the spirit of Sec. 4.2.2, inserting an external potential adds no technical difficulty, but makes the presentation heavier. We also assume that the nonlinearity is homogeneous, and smooth: $\sigma \in \mathbb{N} \setminus \{0\}$.

Several ideas are natural in view of the results of Sec. 4.2. First, assuming that we could construct $(\Phi^\varepsilon, a^\varepsilon)$ solution to Eq. (4.4), bounded in Sobolev spaces, then passing to the limit, we expect $(\Phi^\varepsilon, a^\varepsilon)$ to converge to (Φ, a) given by Eq. (4.5). Write

$$u^\varepsilon \underset{\varepsilon \to 0}{\sim} a e^{i\Phi/\varepsilon} e^{i(\Phi^\varepsilon - \Phi)/\varepsilon}.$$

The only ε-oscillatory factor is measured by Φ: set $a^\varepsilon := u^\varepsilon e^{-i\Phi/\varepsilon}$. For a^ε to be bounded in H^s, we need another information: we need a rate of convergence of Φ^ε towards Φ,

$$\Phi^\varepsilon - \Phi = \mathcal{O}(\varepsilon).$$

Otherwise, ∇a^ε may not be bounded as $\varepsilon \to 0$. The study led in Sec. 4.2.1 shows that for this property to be satisfied, we have to know a_0^ε up to an error of order $\mathcal{O}(\varepsilon)$. We refuse to count the derivatives when not necessary:

Assumption 4.12. There exists $a_0 \in H^\infty$ such that for all $s \geqslant 0$,

$$\|a_0^\varepsilon - a_0\|_{H^s} = \mathcal{O}(\varepsilon).$$

Lemma 4.13. *Let* $\sigma \in \mathbb{N}$ *and* $\phi_0, a_0 \in H^\infty$. *There exists* $T^* > 0$ *such that*

$$\begin{cases} \partial_t \phi + \dfrac{1}{2}|\nabla \phi|^2 + |a|^{2\sigma} = 0 & ; \quad \phi_{|t=0} = \phi_0, \\[2mm] \partial_t a + \nabla \phi \cdot \nabla a + \dfrac{1}{2}a\Delta\phi = 0 & ; \quad a_{|t=0} = a_0. \end{cases} \tag{4.20}$$

has a unique solution $(\phi, a) \in C^\infty([-T^*, T^*]; H^\infty(\mathbb{R}^n))^2$.

Proof. In Chap. 3, we have considered the system

$$\begin{cases} \partial_t v + v \cdot \nabla v + \nabla\left(|a|^{2\sigma}\right) = 0 & ; \quad v_{|t=0} = \nabla\phi_0. \\[2mm] \partial_t a + v \cdot \nabla a + \dfrac{1}{2}a\,\mathrm{div}\,v = 0 & ; \quad a_{|t=0} = a_0. \end{cases}$$

Thanks to the idea of [Makino *et al.* (1986)], we have proved that it possesses a unique solution $(v, a) \in C^\infty([-T^*, T^*]; H^\infty(\mathbb{R}^n))^2$. Now set

$$\phi(t, x) = \phi_0(x) - \int_0^t \left(\frac{1}{2}|v(\tau, x)|^2 + |a(\tau, x)|^{2\sigma}\right) d\tau.$$

We check that $\partial_t(\nabla\phi - v) = \nabla\partial_t\phi - \partial_t v = 0$, and $\phi \in C([-T^*, T^*]; L^2)$. Hence the lemma. $\qquad\square$

Define
$$a^\varepsilon := u^\varepsilon e^{-i\phi/\varepsilon}.$$

Equation (4.19) is *equivalent* to

$$\begin{cases} \partial_t a^\varepsilon + \nabla\phi \cdot \nabla a^\varepsilon + \frac{1}{2}a^\varepsilon \Delta\phi = i\frac{\varepsilon}{2}\Delta a^\varepsilon - \frac{i}{\varepsilon}\left(|a^\varepsilon|^{2\sigma} - |a|^{2\sigma}\right)a^\varepsilon, \\ a^\varepsilon_{|t=0} = a^\varepsilon_0. \end{cases} \qquad (4.21)$$

The main problem to prove that a^ε is bounded in Sobolev spaces is the presence of the singular factor $1/\varepsilon$ on the right hand side. On the other hand, the analysis of Sec. 4.2 shows that it is natural to expect

$$a^\varepsilon = a + \mathcal{O}(\varepsilon).$$

In this case, the singular factor $1/\varepsilon$ is compensated. Nevertheless, this argument does not seem closed: apparently, to prove that a^ε is bounded in Sobolev spaces, we need a more precise information. The idea in [Alazard and Carles (2007b)] consists in introducing an extra unknown function in order to obtain a closed system of estimates. The approach of considering more unknown functions that in the initial problem has proven very efficient in several contexts. We can mention the study of blow-up for the nonlinear wave equation, [Alinhac (1995b)] (see also [Alinhac (1995a, 2002)]), low Mach number limit of the full Navier–Stokes equations [Alazard (2006)], or geometric optics for the incompressible Euler or Navier-Stokes equations [Cheverry (2004, 2006); Cheverry and Guès (2007)].

Inspired by the analysis of E. Grenier, the idea is to symmetrize the equations, and to obtain a system for the family of unknown functions which is hyperbolic symmetric. Split the term $|a^\varepsilon|^{2\sigma} - |a|^{2\sigma}$ as a product

$$|a^\varepsilon|^2 - |a|^{2\sigma} = \varepsilon g^\varepsilon q^\varepsilon = \varepsilon(GQ)(|a^\varepsilon|^2, |a|^2)$$
$$= \varepsilon G(r_1, r_2)Q(r_1, r_2)\big|_{(r_1,r_2)=(|a^\varepsilon|^2,|a|^2)},$$

where q^ε satisfies an equation of the form

$$\partial_t q^\varepsilon + L(a, \phi, \partial_x)q^\varepsilon + g^\varepsilon \operatorname{div}\left(\operatorname{Im}(\overline{a}^\varepsilon \nabla a^\varepsilon)\right) = 0, \qquad (4.22)$$

and L is a first order differential operator. Introduce the position densities

$$\rho := |a|^2 \quad ; \quad \rho^\varepsilon := |a^\varepsilon|^2 = |u^\varepsilon|^2.$$

Recall that $v = \nabla\phi$. Elementary computations show that:

$$\partial_t \rho + \operatorname{div}(\rho v) = 0, \qquad (4.23)$$

$$\partial_t \rho^\varepsilon + \operatorname{div}\operatorname{Im}\left(\varepsilon\overline{u}^\varepsilon \nabla u^\varepsilon\right) = 0, \qquad (4.24)$$

$$\partial_t \rho^\varepsilon + \operatorname{div}\left(\operatorname{Im}(\varepsilon\overline{a}^\varepsilon \nabla a^\varepsilon) + \rho^\varepsilon v\right) = 0. \qquad (4.25)$$

Denote
$$\beta^\varepsilon := \varepsilon q^\varepsilon = B(r_1, r_2)\big|_{(r_1, r_2) = (|a^\varepsilon|^2, |a|^2)} \quad ; \quad J^\varepsilon := \varepsilon \operatorname{Im}(\overline{a}^\varepsilon \nabla a^\varepsilon).$$
By writing
$$\partial_t \beta^\varepsilon = (\partial_{r_1} B)(\rho^\varepsilon, \rho)\partial_t \rho^\varepsilon + (\partial_{r_2} B)(\rho^\varepsilon, \rho)\partial_t \rho,$$
we compute, from (4.23) and (4.25):
$$\partial_t \beta^\varepsilon + (\partial_{r_1} B)(\rho^\varepsilon, \rho)\operatorname{div}(J^\varepsilon + \rho^\varepsilon v) + (\partial_{r_2} B)(\rho^\varepsilon, \rho)\operatorname{div}(\rho v) = 0.$$
Hence, in order to have an equation of the desired form (4.22), we impose
$$\partial_{r_1} B(r_1, r_2) = G(r_1, r_2).$$
Since on the other hand,
$$G(r_1, r_2) B(r_1, r_2) = r_1^\sigma - r_2^\sigma,$$
this suggests to choose β^ε such that
$$(\beta^\varepsilon)^2 = \frac{2}{\sigma + 1}(\rho^\varepsilon)^{\sigma+1} - 2\rho^\sigma \rho^\varepsilon + f(\rho). \tag{4.26}$$
To obtain an operator L which is linear with respect to β^ε we choose
$$(\beta^\varepsilon)^2 = \frac{2}{\sigma + 1}(\rho^\varepsilon)^{\sigma+1} - \frac{2}{\sigma + 1}\rho^{\sigma+1} - 2\rho^\sigma(\rho^\varepsilon - \rho). \tag{4.27}$$
With this choice, we formally compute:
$$\partial_t \beta^\varepsilon + \varepsilon g^\varepsilon \operatorname{div}(\operatorname{Im}(\overline{a}^\varepsilon \nabla a^\varepsilon)) + v \cdot \nabla \beta^\varepsilon + \frac{\sigma + 1}{2}\beta^\varepsilon \operatorname{div} v = 0.$$
This equation is derived rigorously in [Alazard and Carles (2007b)], and we refer to the paper for the complete proof. Examine the right hand side of (4.27). Taylor's formula yields
$$\frac{2}{\sigma + 1}(\rho^\varepsilon)^{\sigma+1} - \frac{2}{\sigma + 1}\rho^{\sigma+1} - 2\rho^\sigma(\rho^\varepsilon - \rho) = (\rho^\varepsilon - \rho)^2 Q_\sigma(\rho^\varepsilon, \rho),$$
where Q_σ is given by:
$$Q_\sigma(r_1, r_2) := 2\sigma \int_0^1 (1 - s)\,(r_2 + s(r_1 - r_2))^{\sigma-1}\, ds. \tag{4.28}$$
Note that there exists C_σ such that:
$$Q_\sigma(r_1, r_2) \geqslant C_\sigma \left(r_1^{\sigma-1} + r_2^{\sigma-1}\right). \tag{4.29}$$
This is the same convexity inequality as the one we have used in Sec. 3.2.

Notation 4.14. Let $\sigma \in \mathbb{N}$. Introduce
$$G_\sigma(r_1, r_2) = \frac{P_\sigma(r_1, r_2)}{\sqrt{Q_\sigma(r_1, r_2)}} \quad ; \quad B_\sigma(r_1, r_2) := (r_1 - r_2)\sqrt{Q_\sigma(r_1, r_2)},$$
where Q_σ is given by (4.28) and
$$P_\sigma(r_1, r_2) = \frac{r_1^\sigma - r_2^\sigma}{r_1 - r_2} = \sum_{\ell=0}^{\sigma-1} r_1^{\sigma-1-\ell} r_2^\ell.$$

Example 4.15. For $\sigma = 1, 2, 3$, we compute

$$G_1 = 1, \qquad\qquad\qquad B_1 = r_1 - r_2.$$

$$G_2 = \sqrt{\frac{3}{2}} \frac{r_1 + r_2}{\sqrt{r_1 + 2r_2}}, \qquad B_2 = \sqrt{\frac{2}{3}}(r_1 - r_2)\sqrt{r_1 + 2r_2}.$$

$$G_3 = \sqrt{2} \frac{r_1^2 + r_1 \bar{r}_2 + r_2^2}{\sqrt{(r_1 - r_2)^2 + 2r_2^2}}, \qquad B_3 = \frac{1}{\sqrt{2}}(r_1 - r_2)\sqrt{(r_1 - r_2)^2 + 2r_2^2}.$$

A remarkable fact is that, although the functions G_σ and B_σ are not smooth for $\sigma \geqslant 2$, one can compute an evolution equation for the unknown β^ε, as mentioned above.

Denote $\psi^\varepsilon = \nabla a^\varepsilon$. Noticing that

$$g^\varepsilon \operatorname{div}(\operatorname{Im}(\overline{a}^\varepsilon \psi^\varepsilon)) = \operatorname{Im}(g^\varepsilon \overline{a}^\varepsilon \operatorname{div} \psi^\varepsilon),$$

we find that the unknown $(a^\varepsilon, \psi^\varepsilon, q^\varepsilon)$ solves

$$\begin{cases} \partial_t a^\varepsilon + v \cdot \nabla a^\varepsilon - i\dfrac{\varepsilon}{2}\Delta a^\varepsilon = -\dfrac{1}{2}a^\varepsilon \operatorname{div} v - ig^\varepsilon q^\varepsilon a^\varepsilon. \\[2mm] \partial_t \psi^\varepsilon + v \cdot \nabla \psi^\varepsilon + ia^\varepsilon g^\varepsilon \nabla q^\varepsilon - i\dfrac{\varepsilon}{2}\Delta \psi^\varepsilon \\[2mm] \qquad = -\dfrac{1}{2}\psi^\varepsilon \operatorname{div} v - \psi^\varepsilon \cdot \nabla v - \dfrac{1}{2}a^\varepsilon \nabla \operatorname{div} v - iq^\varepsilon \nabla(a^\varepsilon g^\varepsilon), \\[2mm] \partial_t q^\varepsilon + v \cdot \nabla q^\varepsilon + \operatorname{Im}(g^\varepsilon \overline{a}^\varepsilon \operatorname{div} \psi^\varepsilon) = -\dfrac{\sigma + 1}{2}q^\varepsilon \operatorname{div} v. \end{cases} \qquad (4.30)$$

Note that by assumption,

$$\|a^\varepsilon_{|t=0}\|_{H^s(\mathbb{R}^n)} + \|\psi^\varepsilon_{|t=0}\|_{H^s(\mathbb{R}^n)} = \mathcal{O}(1), \quad \forall s \geqslant 0. \qquad (4.31)$$

A similar estimate for the initial data of q^ε is a more delicate issue, since B_σ is not a smooth function. An important remark is that q^ε, viewed as a function of a^ε and a, is an homogeneous function of degree $\sigma + 1$. We then use the following lemma, which can be proved by induction on m:

Lemma 4.16. *Let $p \geqslant 1$ and $m \geqslant 2$ be integers and consider $F: \mathbb{R}^p \to \mathbb{C}$. Assume that $F \in C^\infty(\mathbb{R}^p \setminus \{0\})$ is homogeneous of degree m, that is:*

$$F(\lambda y) = \lambda^m F(y), \qquad \forall \lambda \geqslant 0, \forall y \in \mathbb{R}^p.$$

Then, for $n \leqslant 3$, there exists $K > 0$ such that, for all $u \in H^m(\mathbb{R}^n)$ with values in $\mathbb{R}^p$, $F(u) \in H^m(\mathbb{R}^n)$ and

$$\|F(u)\|_{H^m} \leqslant K\|u\|_{H^m}^m.$$

The same is true when $m = 1$ and $n \in \mathbb{N}$.

Remark 4.17. Note that the result is false for $n \geqslant 4$ and $m \geqslant 2$. Also, one must not expect $F(u) \in H^{m+1}(\mathbb{R}^n)$, even for $u \in H^\infty(\mathbb{R}^n)$. For instance, if

$$n = 1 = p, \quad m = 2, \quad F(y) = y|y|, \quad u(x) = xe^{-x^2},$$

then $F(u) \in H^2(\mathbb{R})$ and $F(u) \notin H^3(\mathbb{R})$. Similarly, in general, one must not expect $F_\sigma(u, v) \in H^{\sigma+1}(\mathbb{R}^n)$, even for $(u, v) \in H^\infty(\mathbb{R}^n)^2$.

The left hand side of (4.30) is a first order quasi-linear symmetric hyperbolic system, plus a second order skew-symmetric term. The right hand side can be viewed as a semi-linear source term. Denoting $U^\varepsilon := (2q^\varepsilon, a^\varepsilon, \overline{a}^\varepsilon, \psi^\varepsilon, \overline{\psi}^\varepsilon)$, we see that

$$\partial_t U^\varepsilon + \sum_{1 \leqslant j \leqslant n} A_j(v, a^\varepsilon g^\varepsilon, \overline{a}^\varepsilon g^\varepsilon) \partial_j U^\varepsilon + \varepsilon \mathcal{L}(\partial_x) U^\varepsilon = E(\Phi, U^\varepsilon, a^\varepsilon g^\varepsilon, \nabla(a^\varepsilon g^\varepsilon)),$$

where $\Phi = (\nabla\phi, \nabla^2\phi, \nabla^3\phi)$, the A_j's are Hermitian matrices linear in their arguments, $\mathcal{L}(\partial_x) = \sum L_{jk} \partial_j \partial_k$ is a skew-symmetric second order differential operator with constant coefficients, and E is a C^∞ function of its arguments, vanishing at the origin.

Using the above structure, quasi-linear analysis for hyperbolic systems, and estimates for non-smooth homogeneous functions, the main result in [Alazard and Carles (2007b)] follows:

Theorem 4.18. *Let $n \leqslant 3$, $\phi_0 \in H^\infty$ and let Assumption 4.12 be satisfied. There exists $T \in]0, T^*[$, where T^* is given by Lemma 4.13, such that the following holds. For all $\varepsilon \in]0, 1]$, the Cauchy problem (1.1) has a unique solution $u^\varepsilon \in C([-T, T]; H^\infty(\mathbb{R}^n))$. Moreover,*

$$\sup_{\varepsilon \in]0,1]} \left(\left\| a^\varepsilon \right\|_{L^\infty([-T,T]; H^k(\mathbb{R}^n))} + \left\| q^\varepsilon \right\|_{L^\infty([-T,T]; H^{k-1}(\mathbb{R}^n))} \right) < +\infty, \quad (4.32)$$

where the index k is as follows:

- *If $\sigma = 1$, then $k \in \mathbb{N}$ is arbitrary.*
- *If $\sigma = 2$ and $n = 1$, then we can take $k = 2$.*
- *If $\sigma = 2$ and $2 \leqslant n \leqslant 3$, then we can take $k = 1$.*
- *If $\sigma \geqslant 3$, then we can take $k = \sigma$.*

The assumption $n \leqslant 3$ appears when estimating non-smooth homogeneous functions. It could be removed, up to considering sufficiently large σ. The restriction for k when $\sigma \geqslant 2$ follows from Lemma 4.16. This argument is used not only to estimate q^ε at time $t = 0$, but also the factor $a^\varepsilon g^\varepsilon$ in terms of a^ε, in Eq. (4.30).

Note that for $k = 1$, Eq. (4.32) is exactly Theorem 3.4. From this point of view, Eq. (4.30) can be interpreted as a local form for the modulated energy. This is even more explicit when considering

$$e^\varepsilon := |a^\varepsilon|^2 + |\psi^\varepsilon|^2 + |q^\varepsilon|^2.$$

It satisfies an equation of the form $\partial_t e^\varepsilon + \operatorname{div}(\eta^\varepsilon) + \flat^\varepsilon = \mathcal{O}(e^\varepsilon)$, where $\int \flat^\varepsilon = 0$. Indeed, directly from Eq. (4.30), we compute

$$\partial_t e^\varepsilon + \operatorname{div}(v e^\varepsilon) + 2\operatorname{div}\left(\operatorname{Im}(g^\varepsilon q^\varepsilon \overline{a}^\varepsilon \psi^\varepsilon)\right) + \varepsilon \operatorname{Im}\left(\overline{a}^\varepsilon \Delta a^\varepsilon + \overline{\psi^\varepsilon} \Delta \psi^\varepsilon\right)$$
$$= -\sigma |q^\varepsilon|^2 \operatorname{div} v - \operatorname{Re}\left((2\psi^\varepsilon \cdot \nabla v + a^\varepsilon \nabla \operatorname{div} v)\overline{\psi^\varepsilon}\right).$$

We have thus obtained an evolution equation for a local modulated energy. Gronwall lemma yields

$$\|e^\varepsilon(t)\|_{L^1(\mathbb{R}^n)} \leqslant \|e^\varepsilon(0)\|_{L^1(\mathbb{R}^n)} \exp{(Ct)}.$$

Finally, we check that $(e^\varepsilon(0))_\varepsilon$ is bounded in $L^1(\mathbb{R}^n)$. Therefore, this argument yields Eq. (4.32) for $k = 1$.

To obtain the pointwise asymptotics of the wave function u^ε, we can proceed in a similar spirit, guided by the results of Sec. 4.2.1. We make the following assumption on the initial amplitude:

Assumption 4.19. There exist $a_0, a_1 \in H^\infty$ such that for all $s \geqslant 0$,

$$\|a_0^\varepsilon - a_0 - \varepsilon a_1\|_{H^s} = \mathcal{O}\left(\varepsilon^2\right).$$

Introduce the system

$$
\begin{cases}
\partial_t \phi^{(1)} + \nabla\phi \cdot \nabla\phi^{(1)} + 2\sigma \operatorname{Re}\left(\overline{a} a^{(1)}\right) |a|^{2\sigma-2} = 0, \\
\partial_t a^{(1)} + \nabla\phi \cdot \nabla a^{(1)} + \nabla\phi^{(1)} \cdot \nabla a + \frac{1}{2} a^{(1)} \Delta\phi + \frac{1}{2} a \Delta\phi^{(1)} = \frac{i}{2}\Delta a, \\
\phi^{(1)}\big|_{t=0} = 0 \quad ; \quad a^{(1)}\big|_{t=0} = a_1.
\end{cases}
$$
$$(4.33)$$

Again, at the zeroes of a, Eq. (4.33) ceases to be strictly hyperbolic, and we cannot solve the Cauchy problem by a standard argument. Yet, we can prove:

Lemma 4.20. *Let* $n \geqslant 1$, $\phi_0 \in H^\infty$, *and let Assumption 4.19 be satisfied. Then* (4.33) *has a unique solution* $(\phi^{(1)}, a^{(1)})$ *in* $C([-T^*, T^*]; H^\infty)^2$, *where* T^* *is given by Lemma 4.13.*

Proof. [Sketch of the proof] We transform the equations so as to obtain an auxiliary hyperbolic system for $(\nabla\phi^{(1)}, A_1)$ for some unknown A_1, depending linearly upon $a^{(1)}$. The definition of A_1 depends on the parity of σ. This allows to determine a function $\phi^{(1)}$ and, next, to define a function $a^{(1)}$ by solving the second equation in (4.33). We conclude the proof by checking that $(\phi^{(1)}, a^{(1)})$ does solve (4.33). The first change of unknown consists in considering $v_1 := \nabla\phi^{(1)}$. The first equation in (4.33) yields:

$$\partial_t v_1 + v \cdot \nabla v_1 + 2\sigma\nabla \operatorname{Re}\left(|a|^{2\sigma-2}\overline{a}a^{(1)}\right) = -v_1 \cdot \nabla v.$$

First case: $\sigma \geqslant 2$ **is even.** Consider the new unknown

$$A_1 := |a|^{\sigma-2}\operatorname{Re}\left(\overline{a}a^{(1)}\right).$$

We check that, if $(\phi^{(1)}, a^{(1)})$ solves (4.33), then

$$\begin{cases} \partial_t v_1 + v \cdot \nabla v_1 + 2\sigma|a|^\sigma\nabla A_1 = -v_1 \cdot \nabla v - 2\sigma A_1\nabla\left(|a|^\sigma\right), \\ \partial_t A_1 + v \cdot \nabla A_1 + \dfrac{1}{2}|a|^\sigma \operatorname{div} v_1 = -\dfrac{1}{\sigma}\nabla\left(|a|^\sigma\right)\cdot v_1 - \dfrac{\sigma}{2}A_1 \operatorname{div} v \\ \qquad\qquad\qquad\qquad + \dfrac{i}{2}\operatorname{Re}\left(|a|^{\sigma-2}\overline{a}\Delta a\right). \end{cases} \quad (4.34)$$

This linear system is hyperbolic symmetric, and its coefficients are smooth. In particular, uniqueness for (4.33) follows from the uniqueness for (4.34). Equation (4.34) possesses a unique solution in $C^\infty([-T^*, T^*]; H^\infty)$. We next define $\phi^{(1)}$ and $a^{(1)}$ as announced above.

Second case: σ **is odd.** In this case, $\sigma = 2m + 1$, for some $m \in \mathbb{N}$. We consider the new unknown

$$A_1 := |a|^{\sigma-1}a^{(1)} = |a|^{2m}a^{(1)}.$$

We check that (v_1, A_1) must solve

$$\begin{cases} \partial_t v_1 + v \cdot \nabla v_1 + 2\sigma\operatorname{Re}\left(|a|^{2m}\overline{a}\nabla A_1\right) = -v_1 \cdot \nabla v - 2\sigma\operatorname{Re}\left(A_1\nabla\left(|a|^{2m}\overline{a}\right)\right), \\ \partial_t A_1 + v \cdot \nabla A_1 + \dfrac{1}{2}|a|^{2m}a \operatorname{div} v_1 = -\dfrac{\sigma}{2}A_1 \operatorname{div} v - |a|^{2m}\nabla a \cdot v_1 \\ \qquad\qquad\qquad\qquad + \dfrac{i}{2}|a|^{2m}\Delta a. \end{cases}$$

We can then conclude as in the first case. $\qquad\qquad\qquad\qquad\qquad\qquad\square$

We can now describe the wave function u^ε at leading order as $\varepsilon \to 0$:

Proposition 4.21. *Let* $n \leqslant 3$, $\phi_0 \in H^\infty$, *and let Assumption 4.19 be satisfied. Set* $\widetilde{a} := ae^{i\phi^{(1)}}$. *Then there exists* $\varepsilon_0 > 0$ *such that* $a^\varepsilon \in C([-T^*, T^*]; H^\infty)$ *for* $\varepsilon \in]0, \varepsilon_0]$, *and*

$$\|a^\varepsilon - \widetilde{a}\|_{L^\infty([-T^*, T^*]; H^k)} = \mathcal{O}(\varepsilon),$$

where k *is as in Theorem 4.18.*

Proof.   We indicate the main steps of the proof only. Denote

$$r^\varepsilon = a^\varepsilon - \widetilde{a} \quad ; \quad \widetilde{a}^{(1)} = a^{(1)} e^{i\phi^{(1)}}.$$

From (4.20), (4.21) and (4.33), we see that r^ε solves

$$\begin{cases} \partial_t r^\varepsilon + v \cdot \nabla r^\varepsilon + \dfrac{1}{2} r^\varepsilon \operatorname{div} v - i\dfrac{\varepsilon}{2}\Delta r^\varepsilon = i\dfrac{\varepsilon}{2}\Delta \widetilde{a} - iS^\varepsilon, \\ r^\varepsilon_{|t=0} = a_0^\varepsilon - a_0 = \varepsilon a_1 + \mathcal{O}\left(\varepsilon^2\right), \end{cases}$$

where the term S^ε is given by:

$$S^\varepsilon = \frac{1}{\varepsilon}\left(|a^\varepsilon|^{2\sigma} - |\widetilde{a}|^{2\sigma}\right) a^\varepsilon - 2\sigma \widetilde{a}|\widetilde{a}|^{2\sigma-2} \operatorname{Re}\left(\overline{\widetilde{a}}\widetilde{a}^{(1)}\right).$$

We check that for all $s \geqslant 0$, we have, in $H^s(\mathbb{R}^n)$:

$$S^\varepsilon = \frac{1}{\varepsilon}\left(|a^\varepsilon|^{2\sigma} - |\widetilde{a} + \varepsilon \widetilde{a}^{(1)}|^{2\sigma}\right) a^\varepsilon + 2\sigma r^\varepsilon |\widetilde{a}|^{2\sigma-2} \operatorname{Re}\left(\overline{\widetilde{a}}\widetilde{a}^{(1)}\right) + \mathcal{O}(\varepsilon).$$

The last term should be viewed as a small source term. The second one is linear in r^ε, and is suitable in view of an application of the Gronwall lemma. There remains to handle the first term. At this stage, we can mimic the previous approach. Introduce the nonlinear change of unknown:

$$\widetilde{q}^\varepsilon = \frac{1}{\varepsilon} B_\sigma\left(|a^\varepsilon|^2, |\widetilde{a} + \varepsilon\widetilde{a}^{(1)}|^2\right) \quad ; \quad \widetilde{g}^\varepsilon = G_\sigma\left(|a^\varepsilon|^2, |\widetilde{a} + \varepsilon\widetilde{a}^{(1)}|^2\right),$$

where B_σ and G_σ are defined in Notation 4.14. We check that $(r^\varepsilon, \nabla r^\varepsilon, \widetilde{q}^\varepsilon)$ solves a system of the form (4.30), plus some extra source terms of order $\mathcal{O}(\varepsilon)$ in $H^s(\mathbb{R}^n)$. We also note that the initial data are of order $\mathcal{O}(\varepsilon)$, from Assumption 4.19:

$$\left\|(r^\varepsilon, \nabla r^\varepsilon)|_{t=0}\right\|_{H^s} = \mathcal{O}(\varepsilon), \ \forall s \geqslant 0.$$

We also have

$$\left\|\widetilde{q}^\varepsilon|_{t=0}\right\|_{H^{k-1}} = \mathcal{O}(\varepsilon),$$

where k is as Theorem 4.18.

The proposition then stems from Gronwall lemma and a standard continuity argument, which we sketch for the convenience of the reader (see also [Rauch and Keel (1999)]). Proceeding like in the proof of Theorem 4.18, Gronwall lemma shows that there exists $T > 0$ and $C > 0$ independent of ε, such that

$$\|r^\varepsilon\|_{L^\infty([-T,T];H^k)} + \|\widetilde{q}^\varepsilon\|_{L^\infty([-T,T];H^{k-1})} \leqslant C\varepsilon.$$

The rate $\mathcal{O}(\varepsilon)$ follows from the fact that the initial data are of order $\mathcal{O}(\varepsilon)$, as well as the source term in the system for $(r^\varepsilon, \nabla r^\varepsilon, \widetilde{q}^\varepsilon)$. If a^ε were not smooth on $[0, T^*]$, then there would exist $t^\varepsilon > 0$ such that

$$\|r^\varepsilon(t^\varepsilon)\|_{H^k} + \|\widetilde{q}^\varepsilon(t^\varepsilon)\|_{H^{k-1}} = 1.$$

Let $\underline{t}^{\varepsilon}$ the smallest such $t^{\varepsilon} > 0$:

$$\|r^{\varepsilon}\|_{L^{\infty}([0,\underline{t}^{\varepsilon}];H^{k})} + \|\widetilde{q}^{\varepsilon}\|_{L^{\infty}([0,\underline{t}^{\varepsilon}];H^{k-1})} \leqslant 1.$$

Using this estimate, and Gronwall lemma, we infer

$$\|r^{\varepsilon}\|_{L^{\infty}([0,\underline{t}^{\varepsilon}];H^{k})} + \|\widetilde{q}^{\varepsilon}\|_{L^{\infty}([0,\underline{t}^{\varepsilon}];H^{k-1})} \leqslant \underline{C}\varepsilon,$$

for some $\underline{C} > 0$ independent of ε. For $\varepsilon > 0$ sufficiently small, $\underline{C}\varepsilon \leqslant 1/2$. This contradicts the definition of t^{ε}. Therefore, a^{ε} is smooth on $[0, T^{*}]$ (and on $[-T^{*}, 0]$ by the same argument) provided $\varepsilon \leqslant \varepsilon_{0}$ for some $\varepsilon_{0} > 0$ sufficiently small, and the error estimate follows. $\qquad\square$

4.4 On conservation laws

Recall some important evolution laws for (4.19):

$$\text{Mass:} \quad \frac{d}{dt}\|u^{\varepsilon}(t)\|_{L^{2}} = 0\,.$$

$$\text{Energy:} \quad \frac{d}{dt}\left(\frac{1}{2}\|\varepsilon\nabla_{x}u^{\varepsilon}\|_{L^{2}}^{2} + \frac{1}{\sigma+1}\|u^{\varepsilon}\|_{L^{2\sigma+2}}^{2\sigma+2}\right) = 0\,.$$

$$\text{Momentum:} \quad \frac{d}{dt}\,\text{Im}\int \overline{u}^{\varepsilon}(t,x)\varepsilon\nabla_{x}u^{\varepsilon}(t,x)dx = 0\,.$$

$$\text{Pseudo-conformal law:} \quad \frac{d}{dt}\left(\frac{1}{2}\|J^{\varepsilon}(t)u^{\varepsilon}\|_{L^{2}}^{2} + \frac{t^{2}}{\sigma+1}\|u^{\varepsilon}\|_{L^{2\sigma+2}}^{2\sigma+2}\right)$$
$$= \frac{t}{\sigma+1}(2-n\sigma)\|u^{\varepsilon}\|_{L^{2\sigma+2}}^{2\sigma+2}\,,$$

where $J^{\varepsilon}(t) = x + i\varepsilon t\nabla$. These evolutions are deduced from the usual ones ($\varepsilon = 1$, see e.g. [Cazenave (2003); Sulem and Sulem (1999)]) via the scaling $\psi(t,x) = u(\varepsilon t, \varepsilon x)$. Writing $u^{\varepsilon} = a^{\varepsilon}e^{i\phi/\varepsilon}$, and passing to the limit formally in the above formulae yields:

$$\frac{d}{dt}\|a(t)\|_{L^{2}} = 0\,.$$

$$\frac{d}{dt}\int\left(\frac{1}{2}|a(t,x)|^{2}|\nabla\phi(t,x)|^{2} + \frac{1}{\sigma+1}|a(t,x)|^{2\sigma+2}\right)dx = 0\,.$$

$$\frac{d}{dt}\int |a(t,x)|^{2}\nabla\phi(t,x)dx = 0\,.$$

$$\frac{d}{dt}\int\left(\frac{1}{2}|(x-t\nabla\phi(t,x))\,a(t,x)|^{2} + \frac{t^{2}}{\sigma+1}|a(t,x)|^{2\sigma+2}\right)dx$$
$$= \frac{t}{\sigma+1}(2-n\sigma)\int |a(t,x)|^{2\sigma+2}dx\,.$$

Note that we also have the conservation ([Carles and Nakamura (2004)]):

$$\frac{d}{dt}\operatorname{Re}\int \overline{u}^\varepsilon(t,x)J^\varepsilon(t)u^\varepsilon(t,x)dx = 0\,,$$

which yields:

$$\frac{d}{dt}\int (x - t\nabla\phi(t,x))\,|a(t,x)|^2 dx = 0\,.$$

All these expressions involve only $(|a|^2,\nabla\phi) = (|\widetilde{a}|^2,\nabla\phi)$. Recall that if we set $(\rho,v) = (|a|^2,\nabla\phi)$,

$$\begin{cases} \partial_t v + v\cdot\nabla v + \nabla\left(\rho^\sigma\right) = 0 &\quad;\quad v_{|t=0} = \nabla\phi_0, \\ \partial_t\rho + \operatorname{div}\left(\rho v\right) = 0 &\quad;\quad \rho_{|t=0} = |a_0|^2. \end{cases} \quad (4.35)$$

Rewriting the above evolution laws, we get:

$$\frac{d}{dt}\int_{R^n}\rho(t,x)dx = 0.$$

$$\frac{d}{dt}\int \left(\frac{1}{2}\rho(t,x)|v(t,x)|^2 + \frac{1}{\sigma+1}\rho(t,x)^{\sigma+1}\right)dx = 0.$$

$$\frac{d}{dt}\int \rho(t,x)v(t,x)dx = 0.$$

$$\frac{d}{dt}\int \left(\frac{1}{2}\left|(x - tv(t,x))\right|^2\rho(t,x) + \frac{t^2}{\sigma+1}\rho(t,x)^{\sigma+1}\right)dx$$

$$= \frac{t}{\sigma+1}(2 - n\sigma)\int \rho(t,x)^{\sigma+1}dx.$$

$$\frac{d}{dt}\int (x - tv(t,x))\,\rho(t,x)dx = 0.$$

We thus retrieve formally some evolution laws for the compressible Euler equation (4.35) (see e.g. [Serre (1997); Xin (1998)]), with the pressure law $p(\rho) = c\rho^{\sigma+1}$.

4.5   Focusing nonlinearities

The main feature of the limit system we used is that it enters, up to a change of unknowns, into the framework of quasi-linear hyperbolic systems. This comes from the fact that we consider the defocusing case. Had we worked instead with the focusing case, where $+|u|^{2\sigma}u$ is replaced with $-|u|^{2\sigma}u$, the corresponding limit system would have been ill-posed. We refer to [Métivier (2005)], in which G. Métivier establishes Hadamard's instabilities for non-hyperbolic nonlinear equations.

As an example, consider the Cauchy problem in space dimension $n = 1$

$$\begin{cases} \partial_t \phi + \dfrac{1}{2}|\partial_x \phi|^2 - |a|^{2\sigma} = 0 & ; \quad \phi_{|t=0} = \phi_0, \\ \partial_t a + \partial_x \phi \partial_x a + \dfrac{1}{2}a\partial_x^2 \phi = 0 & ; \quad a_{|t=0} = a_0. \end{cases} \tag{4.36}$$

As opposed to the previous analysis, this system is not hyperbolic, but elliptic. The following result follows from Hadamard's argument (see [Métivier (2005)]).

Proposition 4.22. *Suppose that* (ϕ, a) *in* $C^2([0,T] \times \mathbb{R})^2$ *solves* (4.36). *If* $\phi_0(x)$ *is real analytic near* $\underline{x}$ *and if* $a_0(\underline{x}) > 0$, *then* $a_0(x)$ *is real analytic near* $\underline{x}$. *Consequently, there are smooth initial data for which the Cauchy problem has no solution.*

This shows that to study the semi-classical limit for the focusing analogue of (1.1), working with analytic data, is not only convenient: it is necessary.

On the other hand, data and solutions with analytic regularity seem appropriate. In [Gérard (1993)], P. Gérard works with the analytic regularity, when the space variable x belongs to the torus $\mathbb{T}^n$, without external potential ($V \equiv 0$). Note that the only assumption needed on the nonlinearity f is analyticity near the range of $\left|a_0^{(0)}\right|^2$ (see below for the definition of $a_0^{(0)}$). This includes the case $f(y) = \lambda y^\sigma$, $\lambda \in \mathbb{R}$, $\sigma \in \mathbb{N}$. This result was extended to the case of the whole Euclidean space $\mathbb{R}^n$ in [Thomann (2007)].

The initial phase ϕ_0 is supposed real analytic, and the initial amplitude is analytic in the following sense (see [Sjöstrand (1982)]):

$$a_0^\varepsilon(z) = \sum_{j \geqslant 0} \varepsilon^j a_0^{(j)}(z),$$

where the functions $a_0^{(j)}$ satisfy the following properties. There exist $\ell > 0$, $A > 0$, $B > 0$ such that, for all $j \geqslant 0$, $a_0^{(j)}$ is holomorphic in $\{|\operatorname{Im} z| < \ell\}$ ($z = (z_1, \ldots, z_n) \in \mathbb{C}^n$), and

$$\left|a_0^{(j)}(z)\right| \leqslant AB^j j! \quad \text{on } \{|\operatorname{Im} z| < \ell\}.$$

To consider functions which decay sufficiently at infinity to be in $H^s(\mathbb{R}^n)$, L. Thomann introduces the weight

$$W(z) = \exp\left(1 + z^2\right)^{1/2}, \quad z^2 = z_1^2 + \ldots + z_n^2.$$

The condition on the coefficients $a_0^{(j)}$ becomes:

$$\left| W(z) a_0^{(j)}(z) \right| \leqslant A B^j j! \quad \text{on } \{ |\operatorname{Im} z| < \ell \}.$$

Denoting $\overline{a}(t, z)$ the complex conjugate of $a(\overline{t}, \overline{z})$, P. Gérard constructs a formal solution of the form

$$u^\varepsilon = a^\varepsilon e^{i\phi/\varepsilon}, \quad a^\varepsilon(t, z) = \sum_{j \geqslant 0} \varepsilon^j a^{(j)}(t, z),$$

which satisfies:

$$\begin{cases} \partial_t v = -v \cdot \nabla v - \nabla f \left(a^{(0)} \overline{a}^{(0)} \right), \\ \partial_t a^\varepsilon = -v \cdot \nabla a^\varepsilon - \dfrac{1}{2} a^\varepsilon \operatorname{div} v + i \dfrac{\varepsilon}{2} \Delta a^\varepsilon - \dfrac{i a^\varepsilon}{\varepsilon} \left(f \left(a^\varepsilon \overline{a}^\varepsilon \right) - f \left(a^{(0)} \overline{a}^{(0)} \right) \right). \end{cases}$$

The sum is defined in the sense of J. Sjöstrand: there exists $t_0 > 0$ such that, for all $j \geqslant 0$, $a^{(j)}$ is holomorphic in $\{ |t| < t_0 \} \times \{ |\operatorname{Im} z| < \ell \}$, and

$$\left| W(z) a^{(j)}(t, z) \right| \leqslant A B^j j! \quad \text{on } \{ |t| \leqslant t_0 \} \times \{ |\operatorname{Im} z| < \ell \},$$

with the convention $W \equiv 1$ in the periodic case. To make the argument complete, one has to truncate the formal series, in such a way that the resulting function solves the nonlinear Schrödinger, up to an error term as small as possible. Setting

$$v^\varepsilon = e^{i\phi/\varepsilon} \sum_{j \leqslant 1/(C_0 \varepsilon)} \varepsilon^j a^{(j)}$$

for C_0 sufficiently large, the approximate solution v^ε satisfies:

$$i\varepsilon \partial_t v^\varepsilon + \frac{\varepsilon^2}{2} \Delta v^\varepsilon = f \left(|v^\varepsilon|^2 \right) v^\varepsilon + \mathcal{O} \left(e^{-\delta/\varepsilon} \right),$$

for some $\delta > 0$. Essentially, this source term is sufficiently small to overcome the difficulty pointed out at the end of Sec. 1.2: for small time independent of ε, the exponential growth provided by Gronwall lemma is more than compensated by the term $e^{-\delta/\varepsilon}$, so it is possible to justify nonlinear geometric optics by semi-linear arguments. We refer to [Gérard (1993)] and [Thomann (2007)] for precise statements and complete proofs.

Chapter 5

Some Instability Phenomena

In this chapter, we show some instability phenomena which are related to the supercritical régime in WKB analysis. We first present some ill-posedness results for the "usual" nonlinear Schrödinger equation, that is, with $\varepsilon = 1$ in Eq. (4.19). Roughly speaking, justification of nonlinear geometric optics on very small time intervals yields ill-posedness, and a justification on a longer time interval yields a worse phenomenon, indicating a loss of regularity for the flow map associated to the nonlinear Schrödinger equation. In the last paragraph of this chapter, we show that even at the semi-classical level, some instabilities occur. These are strong instabilities in L^2, but which do not affect the quadratic observables, or the Wigner measures.

5.1 Ill-posedness for nonlinear Schrödinger equations

In this paragraph, we consider the Cauchy problem
$$i\partial_t \psi + \frac{1}{2}\Delta\psi = \omega|\psi|^{2\sigma}\psi \quad ; \quad \psi_{|t=0} = \varphi, \tag{5.1}$$
where $x \in \mathbb{R}^n$, $\sigma \in \mathbb{N} \setminus \{0\}$ and $\omega \in \mathbb{R} \setminus \{0\}$. There is only one β such that the map
$$\psi(t,x) \mapsto \lambda^\beta \psi(\lambda^2 t, \lambda x) \tag{5.2}$$
leaves the equation in (5.1) (but not the initial data) unchanged for all $\lambda > 0$. It is given by
$$\beta = \frac{1}{\sigma}.$$
Recall that for $0 < s < n/2$, the homogeneous Sobolev space $\dot{H}^s$ is defined as
$$\dot{H}^s(\mathbb{R}^n) = \left\{ u \in L^{\frac{2n}{n-2s}}(\mathbb{R}^n) \quad ; \quad \exists v \in L^2(\mathbb{R}^n), \ \widehat{u}(\xi) = \frac{\widehat{v}(\xi)}{|\xi|^s} \right\}.$$

It is equipped with the norm

$$\|u\|_{\dot{H}^s} = \left(\int_{\mathbb{R}^n} |\xi|^{2s} |\widehat{u}(\xi)|^2 \, d\xi \right)^{1/2}.$$

There is only one s such that the $\dot{H}^s$-norm remains unchanged by the spatial scaling of Eq. (5.2) associated to Eq. (5.1), that is

$$f(x) \mapsto \lambda^{1/\sigma} f(\lambda x).$$

It is given by

$$s_c = \frac{n}{2} - \frac{1}{\sigma}.$$

To simplify the discussion, we assume $s_c > 0$.

Definition 5.1. Let $s \geqslant k \geqslant 0$. The Cauchy problem for (5.1) is locally well-posed from $H^s(\mathbb{R}^n)$ to $H^k(\mathbb{R}^n)$ if, for all bounded subset $B \subset H^s(\mathbb{R}^n)$, there exist $T > 0$ and a Banach space $X_T \hookrightarrow C([-T, T]; H^k(\mathbb{R}^n))$ such that:
(1) For all $\varphi \in B \cap H^\infty$, (1.32) has a unique solution $\psi \in C([-T, T]; H^\infty)$.
(2) The mapping $\varphi \in (H^\infty, \|\cdot\|_B) \mapsto \psi \in X_T$ is continuous.

It has been established in [Cazenave and Weissler (1990)] that the Cauchy problem for (5.1) is locally well-posed from H^s to H^s as soon as $s \geqslant s_c$. The situation is different when $0 < s < s_c$ (recall that the notations below are defined at the beginning of the book):

Theorem 5.2 ([Christ *et al.*]). *Let $n \geqslant 1$, $\sigma \in \mathbb{N} \setminus \{0\}$ and $\omega \in \mathbb{R} \setminus \{0\}$. Assume that $0 < s < s_c$.*
(1) *Ill-posedness. The Cauchy problem for (5.1) is not locally well-posed from H^s to H^s: for any $\delta > 0$, we can find families $(\varphi_1^h)_{0 < h \leqslant 1}$ and $(\varphi_2^h)_{0 < h \leqslant 1}$ with*

$$\varphi_1^h, \varphi_2^h \in \mathcal{S}(\mathbb{R}^n) \; ; \; \|\varphi_1^h\|_{H^s}, \|\varphi_2^h\|_{H^s} \leqslant \delta, \; \|\varphi_1^h - \varphi_2^h\|_{H^s} \xrightarrow[h \to 0]{} 0,$$

such that if ψ_1^h and ψ_2^h denote the solutions to (5.1) with these initial data, there exists $0 < t^h \ll 1$ such that

$$\liminf_{h \to 0} \left\| \psi_1^h(t^h) - \psi_2^h(t^h) \right\|_{H^s} > 0.$$

(2) *Norm inflation. We can find $(\psi^h)_{0 < h \leqslant 1}$ solving (5.1), such that*

$$\varphi^h \in \mathcal{S}(\mathbb{R}^n), \quad \|\varphi^h\|_{H^s} \ll 1 \; ; \; \exists t^h \ll 1, \; \left\| \psi^h(t^h) \right\|_{H^s} \gg 1.$$

The proof we give below relies on WKB analysis for very small time. Such an approach to prove ill-posedness results is due to G. Lebeau for the case of the nonlinear wave equation [Lebeau (2001)] (see also [Métivier (2004a)]). For the case of nonlinear Schrödinger equations, a proof in this spirit appears in the appendix of [Burq *et al.* (2005)]. We present the proof given in Appendix B of [Carles (2007b)]. The main idea of the proof is that in a semi-classical régime, if the initial data do not oscillate rapidly, then for very small time, the Laplacian is negligible, and the nonlinear Schrödinger equation can be approximated by an explicitly solvable ordinary differential equation (recall that the semi-classical limit is sometimes referred to as dispersionless limit). Such an idea appears in [Kuksin (1995)], in the context of the nonlinear wave equation.

Proposition 5.3. *Let $n \geqslant 1$, $\sigma \in \mathbb{N} \setminus \{0\}$, $\omega \in \mathbb{R} \setminus \{0\}$ and $a_0^\varepsilon \in \mathcal{S}(\mathbb{R}^n)$. Fix $k > n/2$, and assume that $(a_0^\varepsilon)_{0 < \varepsilon \leqslant 1}$ is bounded in H^k. Consider the initial value problems:*

$$
\begin{cases}
i\varepsilon \partial_t u^\varepsilon + \dfrac{\varepsilon^2}{2} \Delta u^\varepsilon = \omega |u^\varepsilon|^{2\sigma} u^\varepsilon, \\[2mm]
i\varepsilon \partial_t v^\varepsilon = \omega |v^\varepsilon|^{2\sigma} v^\varepsilon, \\[2mm]
u^\varepsilon_{|t=0} = v^\varepsilon_{|t=0} = a_0^\varepsilon.
\end{cases}
$$

We can find $c_0, c_1, \theta > 0$ independent of $\varepsilon \in]0,1]$ such that

$$
\|u^\varepsilon - v^\varepsilon\|_{L^\infty([-c_0\varepsilon|\log\varepsilon|^\theta, c_0\varepsilon|\log\varepsilon|^\theta];H^k)} \lesssim \varepsilon \langle \log \varepsilon \rangle^{c_1}.
$$

Before proving the above estimate, we show that neglecting the Laplacian for very small time is rather natural with WKB analysis in mind. Suppose for instance that $a_0^\varepsilon \to a_0$ as $\varepsilon \to 0$. We have seen in Chap. 4 that, at least in the case $\omega > 0$,

$$
\left\| u^\varepsilon(t) - a(t)e^{i\phi(t)/\varepsilon} \right\|_{L^\infty} = o_{\varepsilon \to 0}(1) + \mathcal{O}(t),
$$

where (ϕ, a) is given by

$$
\begin{cases}
\partial_t \phi + \dfrac{1}{2}|\nabla \phi|^2 + \omega |a|^{2\sigma} = 0 \quad ; \quad \phi_{|t=0} = 0, \\[2mm]
\partial_t a + \nabla \phi \cdot \nabla a + \dfrac{1}{2} a \Delta \phi = 0 \quad ; \quad a_{|t=0} = a_0.
\end{cases}
$$

Since the above approximation is interesting for very small time only, examine the Taylor expansion for ϕ and a as $t \to 0$. For instance, plug formal series of the form

$$
a(t,x) \sim \sum_{j \geqslant 0} t^j a_j(x) \quad ; \quad \phi(t,x) \sim \sum_{j \geqslant 0} t^j \phi_j(x)
$$

into the above system, and identify the powers of t. Formally, we find

$$a(t, x) = a_0(x) + \mathcal{O}\left(t^2\right) \quad ; \quad \phi(t, x) = -t\omega |a_0(x)|^{2\sigma} + \mathcal{O}\left(t^3\right).$$

Since the phase ϕ is divided by ε, the function

$$a_0(x) \exp\left(-i\omega \frac{t}{\varepsilon} |a_0(x)|^{2\sigma}\right)$$

is expected to be a decent approximation of u^ε for $|t|^3 \ll \varepsilon$. This issue is discussed more precisely in Sec. 5.3. Note that the above approximation was derived without assessing any spatial derivative: it is not surprising that, when $a_0^\varepsilon = a_0$, it coincides with the solution v^ε of the above ordinary differential equation. Note also that in the statement of Proposition 5.3, the error estimate is described for $|t| \lesssim \varepsilon |\log \varepsilon|^\theta$, which is still very far from the borderline $|t| \ll \varepsilon^{1/3}$. The technical reason is that for $|t| \lesssim \varepsilon |\log \varepsilon|^\theta$, a semi-linear analysis is sufficient (see below), while for larger time (even for $\varepsilon |\log \varepsilon|^\theta \ll |t| \ll \varepsilon^{1/3}$), it seems that a quasi-linear analysis (in the spirit of Chapters 3 and 4) is needed.

Proof. [Proof of Proposition 5.3] Let $w^\varepsilon = u^\varepsilon - v^\varepsilon$. It solves:

$$i\varepsilon \partial_t w^\varepsilon + \frac{\varepsilon^2}{2} \Delta w^\varepsilon = (g(w^\varepsilon + v^\varepsilon) - g(v^\varepsilon)) - \frac{\varepsilon^2}{2} \Delta v^\varepsilon, \tag{5.3}$$

with $w^\varepsilon_{|t=0} = 0$, where we have set $g(z) = \omega |z|^{2\sigma} z$. For $k \geqslant 0$, we have, from Lemma 1.2:

$$\begin{aligned}
\|w^\varepsilon\|_{L^\infty([-t,t];H^k)} &\lesssim \frac{1}{\varepsilon} \|g(w^\varepsilon + v^\varepsilon) - g(v^\varepsilon)\|_{L^1([-t,t];H^k)} \\
&\quad + \varepsilon \|\Delta v^\varepsilon\|_{L^1([-t,t];H^k)}.
\end{aligned}$$

Using Taylor formula, and the fact that H^k is an algebra since $k > n/2$, we have:

$$\|g(w^\varepsilon(t) + v^\varepsilon(t)) - g(v^\varepsilon(t))\|_{H^k} \lesssim \left(\|w^\varepsilon(t)\|_{H^k}^{2\sigma} + \|v^\varepsilon(t)\|_{H^k}^{2\sigma}\right) \|w^\varepsilon(t)\|_{H^k}.$$

Since we have

$$v^\varepsilon(t, x) = a_0^\varepsilon(x) \exp\left(-i\omega \frac{t}{\varepsilon} |a_0^\varepsilon|^{2\sigma}\right),$$

we check, for all $s \geqslant 0$:

$$\|v^\varepsilon(t)\|_{H^s} \lesssim \left\langle \frac{t}{\varepsilon} \right\rangle^s.$$

On any time interval where we have, say, $\|w^\varepsilon\|_{H^k} \leqslant 1$, we infer:

$$\|w^\varepsilon\|_{L^\infty([-t,t];H^k)} \leqslant \frac{C}{\varepsilon} \int_{-t}^t \left\langle \frac{\tau}{\varepsilon} \right\rangle^{2\sigma k} \|w^\varepsilon(\tau)\|_{H^k} d\tau + C_1 \varepsilon \int_{-t}^t \left\langle \frac{\tau}{\varepsilon} \right\rangle^{k+2} d\tau.$$

Gronwall lemma yields:

$$\|w^\varepsilon\|_{L^\infty([-t,t];H^k)} \lesssim \varepsilon \int_{-t}^t \left\langle \frac{\tau}{\varepsilon} \right\rangle^{k+2} \exp\left(\frac{C}{\varepsilon} \int_\tau^t \left\langle \frac{\tau'}{\varepsilon} \right\rangle^{2\sigma k} d\tau'\right) d\tau.$$

Let $t^\varepsilon = c_0 \varepsilon |\log \varepsilon|^\theta$:

$$\|w^\varepsilon\|_{L^\infty([-t^\varepsilon,t^\varepsilon];H^k)} \lesssim \varepsilon \exp\left(2Cc_0 |\log \varepsilon|^\theta \langle c_0 \log \varepsilon \rangle^{2\sigma k\theta}\right) \int_{-t^\varepsilon}^{t^\varepsilon} \left\langle \frac{\tau}{\varepsilon} \right\rangle^{k+2} d\tau$$

$$\lesssim \exp\left(2Cc_0 |\log \varepsilon|^\theta \langle c_0 \log \varepsilon \rangle^{2\sigma k\theta}\right) \varepsilon^2 \langle \log \varepsilon \rangle^{(k+3)\theta}.$$

For $\theta = (1 + 2\sigma k)^{-1}$ and c_0 sufficiently small, this yields:

$$\|w^\varepsilon\|_{L^\infty([-t^\varepsilon,t^\varepsilon];H^k)} \lesssim \varepsilon \langle \log \varepsilon \rangle^{\frac{3+k}{1+2\sigma k}}.$$

Up to choosing ε sufficiently small, a continuity argument shows that $\|w^\varepsilon\|_{H^k}$ remains bounded by 1 on this time interval, and the proposition follows. □

As mentioned already, the above proof relies on a perturbative analysis (semi-linear approach). The main remark is that the exponential amplification factor $e^{Ct/\varepsilon}$, already pointed out at the end of Sec. 1.2, can be controlled on very small time intervals, of the form given above.

Proof. [Proof of Theorem 5.2] The result is a straightforward consequence of Proposition 5.3 and explicit computations on ordinary differential equations.

For $a_0 \in \mathcal{S}(\mathbb{R}^n)$ with $\|a_0\|_{H^s} \leqslant \delta/2$, and $h > 0$, consider ψ_1^h solving (5.1) with:

$$\varphi_1^h(x) = h^{-\frac{n}{2}+s} a_0\left(\frac{x}{h}\right).$$

Using the parabolic scaling and the scaling of $\dot{H}^s$, define $\mathbf{u}^h$ by:

$$\mathbf{u}^h(t,x) = h^{\frac{n}{2}-s} \psi_1^h\left(h^2 t, hx\right).$$

It solves:

$$i\partial_t \mathbf{u}^h + \frac{1}{2}\Delta \mathbf{u}^h = \omega h^{2-n\sigma+2\sigma s}|\mathbf{u}^h|^{2\sigma}\mathbf{u}^h \quad ; \quad \mathbf{u}^h_{|t=0} = a_0.$$

Let $\varepsilon = h^{\frac{n\sigma}{2}-1-s\sigma} = h^{\sigma(s_c - s)}$: ε and h go to zero simultaneously since $s < s_c$. With this relation between ε and h, we denote the dependence upon one or the other according to the more natural context. Define

$$u_1^\varepsilon(t,x) = \mathbf{u}^h(ht,x) = h^{\frac{n}{2}-s} \psi_1^h\left(h^{\frac{n\sigma}{2}+1-s\sigma} t, hx\right).$$

It solves:

$$i\varepsilon \partial_t u_1^\varepsilon + \frac{\varepsilon^2}{2} \Delta u_1^\varepsilon = \omega |u_1^\varepsilon|^{2\sigma} u_1^\varepsilon \quad ; \quad u_{1|t=0}^\varepsilon = a_0.$$

We go back to ψ_1^h via the formula:

$$\psi_1^h(t,x) = h^{-\frac{n}{2}+s} u_1^\varepsilon \left(\frac{t}{h^{\frac{n\sigma}{2}+1-s\sigma}}, \frac{x}{h} \right).$$

In particular, ψ_1^h and u_1^ε have the same $\dot{H}^s$ norm.

(1) *Ill-posedness.* Define u_2^ε solution to the same equation, but with initial datum

$$u_{2|t=0}^\varepsilon = (1 + \delta^\varepsilon) a_0,$$

with $\delta^\varepsilon = |\log \varepsilon|^{-\theta} \ll 1$, where $\theta > 0$ stems from Proposition 5.3. We define ψ_2^h by the same scaling as above, and it is straightforward to check, for h sufficiently small:

$$\varphi_1^h, \varphi_2^h \in \mathcal{S}(\mathbb{R}^n) \; ; \quad \|\varphi_1^h\|_{H^s}, \|\varphi_2^h\|_{H^s} \leqslant \delta, \quad \|\varphi_1^h - \varphi_2^h\|_{H^s} \underset{h \to 0}{\longrightarrow} 0.$$

From Proposition 5.3, we have, for $k > n/2$ and $t^\varepsilon = c\varepsilon |\log \varepsilon|^\theta$, with $0 < c \leqslant c_0$,

$$\left\| u_j^\varepsilon (t^\varepsilon) - v_j^\varepsilon (t^\varepsilon) \right\|_{H^k} \ll 1, \quad j = 1, 2,$$

where

$$v_1^\varepsilon(t,x) = a_0(x) \exp \left(-i\omega \frac{t}{\varepsilon} |a_0(x)|^{2\sigma} \right),$$

$$v_2^\varepsilon(t,x) = (1 + \delta^\varepsilon) a_0(x) \exp \left(-i\omega \frac{t}{\varepsilon} |(1 + \delta^\varepsilon) a_0(x)|^{2\sigma} \right).$$

We infer:

$$\left\| u_1^\varepsilon (t^\varepsilon) - u_2^\varepsilon (t^\varepsilon) \right\|_{\dot{H}^s} \gtrsim \left\| v_1^\varepsilon (t^\varepsilon) - v_2^\varepsilon (t^\varepsilon) \right\|_{\dot{H}^s}.$$

On the other hand, we have, from the explicit form of v_j^ε:

$$|v_1^\varepsilon(t,x) - v_2^\varepsilon(t,x)| \underset{\varepsilon \to 0}{\sim} |a_0(x)| \left| e^{-i\omega t |a_0(x)|^{2\sigma}/\varepsilon} - e^{-i\omega t |(1+\delta^\varepsilon) a_0(x)|^{2\sigma}/\varepsilon} \right|$$

$$\underset{\varepsilon \to 0}{\sim} 2 |a_0(x)| \left| \sin \left(\frac{\omega t \left((1+\delta^\varepsilon)^{2\sigma} - 1 \right) |a_0(x)|^{2\sigma}}{2\varepsilon} \right) \right|$$

$$\underset{\varepsilon \to 0}{\sim} 2 |a_0(x)| \left| \sin \left(\omega \sigma \frac{t}{\varepsilon} \delta^\varepsilon |a_0(x)|^{2\sigma} \right) \right|.$$

Hence, from the definition of t^ε and δ^ε:

$$|v_1^\varepsilon(t^\varepsilon, x) - v_2^\varepsilon(t^\varepsilon, x)| \underset{\varepsilon \to 0}{\sim} 2|a_0(x)| \left|\sin\left(\omega \sigma c |a_0(x)|^{2\sigma}\right)\right|.$$

Up to adjusting $c \in]0, c_0]$, we infer:

$$\liminf_{\varepsilon \to 0} \|u_1^\varepsilon(t^\varepsilon) - u_2^\varepsilon(t^\varepsilon)\|_{\dot{H}^s} > 0.$$

Since u_j^ε and ψ_j^h have the same $\dot{H}^s$ norm, the first part of Theorem 5.2 follows.

(2) *Norm inflation.* In [Christ *et al.*], this phenomenon appears as a transfer of energy from low to high Fourier modes. It corresponds to the appearance of rapid oscillations in a supercritical WKB régime: even though u^ε is not ε-oscillatory initially, rapid oscillations appear instantaneously.

Still from Proposition 5.3, with $t^\varepsilon = c\varepsilon |\log \varepsilon|^\theta$, we have

$$u_1^\varepsilon(t^\varepsilon, x) \underset{\varepsilon \to 0}{\sim} a_0(x) e^{-i\omega \frac{t^\varepsilon}{\varepsilon}|a_0(\tau)|^{2\sigma}} = a_0(x) e^{-i\omega |a_0(x)|^{2\sigma} \left(\log \frac{1}{\varepsilon}\right)^\theta}.$$

Even though u^ε is not yet ε-oscillatory, "rapid" oscillations have appeared already. If we replace a_0 with $|\log h|^{-\theta'} a_0$, this proves the second part of the theorem. $\qquad\square$

5.2 Loss of regularity for nonlinear Schrödinger equations

The end of the proof of Theorem 5.2 suggests that there is some room left: if we can show that u^ε becomes exactly ε-oscillatory, then we should obtain a stronger result. On the other hand, to prove that u^ε becomes exactly ε-oscillatory, we have to justify some asymptotics on a time interval which is independent of ε. This was achieved in Chap. 4. However, note that it is not necessary to know the pointwise asymptotic behavior of u^ε to know that it is ε-oscillatory: it turns out that the analysis developed in Chap. 3 suffices to do so (for defocusing nonlinearities, $\omega > 0$), and we have:

Theorem 5.4 ([Alazard and Carles (2007a)]). *Let $\omega > 0$, $\sigma \in \mathbb{N} \setminus \{0\}$, and $0 < s < s_c = n/2 - 1/\sigma$. There exists a family $(\varphi^h)_{0 < h \leqslant 1}$ in $\mathcal{S}(\mathbb{R}^n)$ with*

$$\|\varphi^h\|_{H^s(\mathbb{R}^n)} \to 0 \text{ as } h \to 0,$$

a solution ψ^h to (5.1) and $0 < t^h \to 0$, such that:

$$\|\psi^h(t^h)\|_{H^k(\mathbb{R}^n)} \to +\infty \text{ as } h \to 0, \, \forall k > \frac{s}{1 + \sigma(s_c - s)}.$$

In particular, (5.1) is not locally well-posed from H^s to H^k.

Note that in general, the solutions of the above theorem must be understood as the weak solutions given by Proposition 1.28.

From now on, we assume $\omega > 0$, and fix $\omega = +1$ for simplicity.

This result is to be compared with the main result in [Lebeau (2005)], which we recall with notations adapted to make the comparison with the Schrödinger case easier. For $n \geqslant 3$ and energy-supercritical wave equations

$$\left(\partial_t^2 - \Delta\right)u + u^{2\sigma+1} = 0, \quad \sigma \in \mathbb{N}, \ \sigma > \frac{2}{n-2},$$

G. Lebeau shows that one can find a *fixed* initial datum in H^s, $s > 1$, and a sequence of times $0 < t^h \to 0$, such that the H^k norms of the solution are unbounded along the sequence t^h, for $k \in]I(s), s]$. The expression for $I(s)$ is related to the critical Sobolev exponent

$$s_{\text{sob}} = \frac{n}{2}\frac{\sigma}{\sigma+1},$$

which corresponds to the embedding $H^{s_{\text{sob}}}(\mathbb{R}^n) \subset L^{2\sigma+2}(\mathbb{R}^n)$. In [Lebeau (2005)], we find:

$$I(s) = 1 \text{ if } 1 < s \leqslant s_{\text{sob}} \quad ; \quad I(s) = \frac{s}{1+\sigma(s_c-s)} \text{ if } s_{\text{sob}} \leqslant s < s_c. \quad (5.4)$$

Note that we have

$$\frac{s_{\text{sob}}}{1+\sigma(s_c-s_{\text{sob}})} = 1. \quad (5.5)$$

The approach in [Lebeau (2005)] consists in using an *anisotropic* scaling, as opposed to the isotropic scaling used in [Lebeau (2001); Christ *et al.*]. Compare Theorem 5.4 with the approach of [Lebeau (2005)]. Recall that (5.1) has two important (formally) conserved quantities: mass and energy,

$$M(t) = \int_{\mathbb{R}^n} |\psi(t,x)|^2 dx \equiv M(0),$$

$$E(\psi(t)) = \frac{1}{2}\int_{\mathbb{R}^n} |\nabla\psi(t,x)|^2 dx + \frac{1}{\sigma+1}\int_{\mathbb{R}^n} |\psi(t,x)|^{2\sigma+2}dx \equiv E(\varphi). \quad (5.6)$$

In view of (5.5), we obtain, for H^1-supercritical nonlinearities:

Corollary 5.5. *Let $n \geqslant 3$ and $\sigma > \frac{2}{n-2}$. There exists a family $(\varphi^h)_{0<h\leqslant1}$ in $\mathcal{S}(\mathbb{R}^n)$ with*

$$\|\varphi^h\|_{H^1} + \|\varphi^h\|_{L^{2\sigma+2}} \to 0 \text{ as } h \to 0,$$

a solution ψ^h to

$$i\partial_t\psi^h + \frac{1}{2}\Delta\psi^h = |\psi^h|^{2\sigma}\psi^h \quad ; \quad \psi^h_{|t=0} = \varphi^h,$$

and $0 < t^h \to 0$, such that:

$$\|\psi^h(t^h)\|_{H^k(\mathbb{R}^n)} \to +\infty \text{ as } h \to 0, \ \forall k > 1.$$

We thus get the analogue of the result of G. Lebeau when $I(s) = 1$, with the drawback that we consider a *sequence* of initial data only. The information that we don't have for Schrödinger equations, and which is available for wave equations, is the finite speed of propagation, that is used in [Lebeau (2005)] to construct a fixed initial datum. On the other hand, the range for k in Theorem 5.4 is broader when $1 < s < s_{\text{sob}}$, and also, we allow the range $0 < s \leqslant 1$, for which no analogous result is available for the wave equation. However, we choose to present the proof of Theorem 5.4 for $k \geqslant 1$ only, and refer to [Alazard and Carles (2007a)] for the remaining cases. Note that $k \geqslant 1$ suffices to establish Corollary 5.5.

The proof starts like the proof of Theorem 5.2: for $a_0 \in \mathcal{S}(\mathbb{R}^n)$ and $h > 0$, consider ψ^h solving (5.1) with:

$$\varphi^h(x) = h^{-\frac{n}{2}+s} a_0 \left(\frac{x}{h}\right).$$

Let $\varepsilon = h^{\sigma(s_c - s)}$: ε and h go to zero simultaneously since $s < s_c$. Define

$$u^\varepsilon(t, x) = h^{\frac{n}{2}-s} \psi^h \left(h^{\frac{n\sigma}{2}+1-s\sigma} t, hx\right).$$

It solves:

$$i\varepsilon \partial_t u^\varepsilon + \frac{\varepsilon^2}{2} \Delta u^\varepsilon = \omega |u^\varepsilon|^{2\sigma} u^\varepsilon \quad ; \quad u^\varepsilon_{|t=0} = a_0.$$

The approach consists in showing that for some $\tau > 0$ independent of ε,

$$\liminf_{\varepsilon \to 0} \varepsilon^k \|u^\varepsilon(\tau)\|_{\dot{H}^k} > 0, \quad \forall k \geqslant 1. \tag{5.7}$$

Back to ψ, this will yield $t^h = \tau h^2 \varepsilon$ and

$$\|\psi^h(t^h)\|_{\dot{H}^k} \gtrsim h^{s-k} \varepsilon^{-k} = h^{s-k(1+\sigma(s_c-s))}.$$

To complete the above reduction, note that we only have to prove (5.7) for $k = 1$. Indeed, for $k > 1$, there exists $C_k > 0$ such that

$$\|f\|_{\dot{H}^1} \leqslant C_k \|f\|_{L^2}^{1-1/k} \|f\|_{\dot{H}^k}^{1/k}, \quad \forall f \in H^k(\mathbb{R}^n).$$

This inequality is straightforward thanks to Fourier analysis. Note also that thanks to the conservation of mass for u^ε, we have:

$$\|u^\varepsilon(t)\|_{\dot{H}^1} \leqslant C_k \|a_0\|_{L^2}^{1-1/k} \|u^\varepsilon(t)\|_{\dot{H}^k}^{1/k}.$$

Up to replacing a_0 with $|\log h|^{-1} a_0$, the result then follows from:

Proposition 5.6. *Let* $n \geqslant 1$, $a_0 \in \mathcal{S}(\mathbb{R}^n)$ *be non-trivial, and* $\sigma \geqslant 1$. *There exists* $\tau > 0$ *such that*

$$\liminf_{\varepsilon \to 0} \|\varepsilon \nabla u^\varepsilon(\tau)\|_{L^2} > 0.$$

Proof. The result is a consequence of Theorem 3.4, whose proof is the core of [Alazard and Carles (2007a)]. Indeed, Theorem 3.4 shows that for all $\tau \in [-T, T]$,

$$\liminf_{\varepsilon \to 0} \|\varepsilon \nabla u^\varepsilon(\tau)\|_{L^2} \geqslant \liminf_{\varepsilon \to 0} \|v(\tau) u^\varepsilon(\tau)\|_{L^2},$$

where $(v, a) \in C^\infty([-T, T]; H^\infty)^2$ solves

$$\begin{cases} \partial_t v + v \cdot \nabla v + \nabla \left(|a|^{2\sigma} \right) = 0 & ; \ v_{|t=0} = 0. \\ \partial_t a + v \cdot \nabla a + \dfrac{1}{2} a \operatorname{div} v = 0 & ; \ a_{|t=0} = a_0. \end{cases}$$

Theorem 3.4 also yields:

$$\lim_{\varepsilon \to 0} \left\| |u^\varepsilon|^2 - |a|^2 \right\|_{L^\infty([-T,T]; L^{\sigma+1})} = 0.$$

Using Hölder's inequality, we find

$$\|v(\tau) a(\tau)\|_{L^2}^2 = \left\| |v(\tau)|^2 |a(\tau)|^2 \right\|_{L^1}$$
$$\leqslant \left\| |v(\tau)|^2 |u^\varepsilon(\tau)|^2 \right\|_{L^1} + \left\| |v(\tau)|^2 \left(|a(\tau)|^2 - |u^\varepsilon(\tau)|^2 \right) \right\|_{L^1}$$
$$\leqslant \left\| |v(\tau)|^2 |u^\varepsilon(\tau)|^2 \right\|_{L^1} + \left\| |v(\tau)|^2 \right\|_{L^{\frac{\sigma+1}{\sigma}}} \left\| |a(\tau)|^2 - |u^\varepsilon(\tau)|^2 \right\|_{L^{\sigma+1}}.$$

Therefore, for all $\tau \in [-T, T]$,

$$\liminf_{\varepsilon \to 0} \|\varepsilon \nabla u^\varepsilon(\tau)\|_{L^2} \geqslant \|v(\tau) a(\tau)\|_{L^2}.$$

To show that there exists $\tau > 0$ such that $\|v(\tau) a(\tau)\|_{L^2} > 0$, we argue by continuity. We use the identities, in H^s for all $s \geqslant 0$, and as $t \to 0$:

$$v(t, x) = -t \nabla \left(|a_0(x)|^{2\sigma} \right) + \mathcal{O}\left(t^3 \right) \quad ; \quad a(t, x) = a_0(x) + \mathcal{O}\left(t^2 \right).$$

These identities follow directly from the equation satisfied by (v, a).   □

As discussed in Sec. 4.5, for focusing nonlinearities ($\omega < 0$), the above analysis fails. On the other hand, working in an analytic setting is possible, and we have:

Theorem 5.7 ([Thomann (2007)]). *The conclusions of Theorem 5.4 still hold if we assume $\omega < 0$.*

5.3   Instability at the semi-classical level

We have seen in Chap. 4 that a perturbation of the initial amplitude at order ε (that is, the presence of a non-trivial corrector a_1) alters the leading order amplitude for times of order $\mathcal{O}(1)$; see Proposition 4.5 and the discussion below, as well as Proposition 4.21. It seems natural to guess that a perturbation of the initial amplitude at order $\varepsilon^{1-\delta}$ for some $\delta \in]0,1[$ will alter the leading order amplitude for times of order $o(1)$, which can be understood as "instantaneously". In this section, we show that this is the case, and that the above mentioned perturbation becomes visible for times of order $\mathcal{O}(\varepsilon^\delta)$.

For the sake of readability, we detail the approach in the case of a cubic nonlinearity, in the absence of external potential:

$$i\varepsilon\partial_t u^\varepsilon + \frac{\varepsilon^2}{2}\Delta u^\varepsilon = |u^\varepsilon|^2 u^\varepsilon. \tag{5.8}$$

We present only formal computations here, which can be justified thanks to the method detailed in Chap. 4. In particular, we could repeat the argument in the presence of a sub-quadratic external potential, or for higher order defocusing, smooth, nonlinearity in space dimension $n \leqslant 3$. Recall that the notations in the following result have been defined in the beginning of these notes.

Theorem 5.8. *Let $n \geqslant 1$, $a_0, \widetilde{a}_0^\varepsilon \in \mathcal{S}(\mathbb{R}^n)$, $\phi_0 \in C^\infty(\mathbb{R}^n; \mathbb{R})$, where a_0 and ϕ_0 are independent of ε, and $\nabla\phi_0 \in H^\infty$. Let u^ε and v^ε solve the initial value problems:*

$$i\varepsilon\partial_t u^\varepsilon + \frac{\varepsilon^2}{2}\Delta u^\varepsilon = |u^\varepsilon|^2 u^\varepsilon \ ; \ u^\varepsilon\big|_{t=0} = a_0 e^{i\phi_0/\varepsilon}.$$

$$i\varepsilon\partial_t v^\varepsilon + \frac{\varepsilon^2}{2}\Delta v^\varepsilon = |v^\varepsilon|^2 v^\varepsilon \ ; \ v^\varepsilon\big|_{t=0} = \widetilde{a}_0^\varepsilon e^{i\phi_0/\varepsilon}.$$

Assume that there exists $N \in \mathbb{N}$ and $\varepsilon^{1-\frac{1}{N}} \ll \delta^\varepsilon \ll 1$ such that:

$$\|a_0 - \widetilde{a}_0^\varepsilon\|_{H^s} \approx \delta^\varepsilon, \ \forall s \geqslant 0 \ ; \ \limsup_{\varepsilon \to 0} \left\| \frac{\mathrm{Re}(a_0 - \widetilde{a}_0^\varepsilon)\bar{a}_0}{\delta^\varepsilon} \right\|_{L^\infty(\mathbb{R}^n)} \neq 0. \tag{5.9}$$

Then we can find $0 < t^\varepsilon \ll 1$ such that: $\|u^\varepsilon(t^\varepsilon) - v^\varepsilon(t^\varepsilon)\|_{L^2} \gtrsim 1$, and $\|u^\varepsilon(t^\varepsilon) - v^\varepsilon(t^\varepsilon)\|_{L^\infty} \gtrsim 1$. More precisely, this occurs for $t^\varepsilon \delta^\varepsilon \approx \varepsilon$. In particular,

$$\frac{\|u^\varepsilon - v^\varepsilon\|_{L^\infty([0,t^\varepsilon];L^2 \cup L^\infty)}}{\left\|u^\varepsilon_{|t=0} - v^\varepsilon_{|t=0}\right\|_{L^2 \cap L^\infty}} \to +\infty \quad as \ \varepsilon \to 0.$$

Remark 5.9. From the conservation of the L^2 norm for u^ε and v^ε (Eq. (1.24)), the instability cannot be much stronger, at least in L^2, since

$$\|u^\varepsilon(t) - v^\varepsilon(t)\|_{L^2} \leqslant \|u^\varepsilon(t)\|_{L^2} + \|v^\varepsilon(t)\|_{L^2} \lesssim 1.$$

Remark 5.10. The second part of the assumption (5.9) can be viewed as a polarization condition. We could remove it with essentially the same approach as below, up to demanding $\varepsilon^{1/2-1/N} \ll \delta^\varepsilon \ll 1$.

Example 5.11. Consider $a_0, b_0 \in \mathcal{S}(\mathbb{R}^n)$ independent of h, such that $\mathrm{Re}(\overline{a}_0 b_0) \not\equiv 0$, and take $\widetilde{a}_0^\varepsilon = a_0 + \delta^\varepsilon b_0$.

Example 5.12. Consider $a_0 \in \mathcal{S}(\mathbb{R}^n)$ independent of ε and $x^\varepsilon \in \mathbb{R}^n$. We can take $\widetilde{a}_0^\varepsilon(x) = a_0(x - x^\varepsilon)$, provided that $|x^\varepsilon| = \delta^\varepsilon$ and

$$\limsup_{\varepsilon \to 0} \left\| \frac{x^\varepsilon}{|x^\varepsilon|} \cdot \nabla \left(|a_0|^2\right) \right\|_{L^\infty} \neq 0.$$

Typically, we can think of $x^\varepsilon = \delta^\varepsilon e_j$, for $1 \leqslant j \leqslant n$, where $(e_j)_{1 \leqslant j \leqslant n}$ denotes the canonical basis of $\mathbb{R}^n$, and $\partial_j a_0 \not\equiv 0$ (the latter is merely an assumption, since $a_0 \in \mathcal{S}(\mathbb{R}^n)$).

Remark 5.13. The last example is motivated by the result of [Burq and Zworski (2005)]. There, instability is established for

$$\begin{cases} i\varepsilon \partial_t u^\varepsilon + \dfrac{\varepsilon^2}{2} \Delta u^\varepsilon = \dfrac{|x|^2}{2} u^\varepsilon + \varepsilon^2 |u^\varepsilon|^2 u^\varepsilon, \quad x \in \mathbb{R}^3, \\[2mm] u^\varepsilon(0, x) = \dfrac{1}{\varepsilon^{3/2}} \Phi \left(\dfrac{x - x_0}{\varepsilon} \right). \end{cases} \tag{5.10}$$

As above, a small perturbation of x_0 yields an instability phenomenon in the limit $\varepsilon \to 0$. We will come back to this framework at the end of this paragraph.

The above result addresses perturbations which satisfy in particular $\delta^\varepsilon \gg \varepsilon$. This excludes the standard WKB data of the form

$$u^\varepsilon(0, x) = a_0^\varepsilon(x) e^{i\phi_0(x)/\varepsilon}, \quad \text{where } a_0^\varepsilon \sim a_0 + \varepsilon a_1 + \varepsilon^2 a_2 + \dots$$

In that case, a perturbation of a_1 is relevant at time $t^\varepsilon \approx 1$, and the previous result is essentially sharp:

Proposition 5.14. *Let* $n \geqslant 1$, $a_0, a_1 \in \mathcal{S}(\mathbb{R}^n)$, $\phi_0 \in C^\infty(\mathbb{R}^n; \mathbb{R})$ *independent of* ε, *with* $\nabla \phi_0 \in H^\infty$. *Assume that* $\mathrm{Re}(a_0 \overline{a_1}) \not\equiv 0$. *Let* u^ε *and* v^ε *solve the initial value problems:*

$$i\varepsilon \partial_t u^\varepsilon + \frac{\varepsilon^2}{2} \Delta u^\varepsilon = |u^\varepsilon|^2 u^\varepsilon \; ; \; u^\varepsilon\big|_{t=0} = a_0 e^{i\phi_0/\varepsilon}.$$

$$i\varepsilon \partial_t v^\varepsilon + \frac{\varepsilon^2}{2} \Delta v^\varepsilon = |v^\varepsilon|^2 v^\varepsilon \; ; \; v^\varepsilon\big|_{t=0} = (a_0 + \varepsilon a_1) \, e^{i\phi_0/\varepsilon}.$$

Then for any $\tau^\varepsilon \ll 1$, $\|u^\varepsilon - v^\varepsilon\|_{L^\infty([0,\tau^\varepsilon]; L^2)} \ll 1$, *and for* $t > 0$ *independent of* ε, *and arbitrarily small:* $\|u^\varepsilon(t) - v^\varepsilon(t)\|_{L^2} \gtrsim 1$.

This result is a direct consequence of the analysis presented in Chap. 4. Essentially, the reason why u^ε and v^ε diverge from each other for $t > 0$ is the presence of the phase corrector $\Phi^{(1)}$ for v^ε; we have already seen that under the above assumptions, there is no such phase corrector for u^ε. We then remark that $\Phi^{(1)}\big|_{t=0} = 0$ and $\partial_t \Phi^{(1)}\big|_{t=0} \neq 0$, so $\|\Phi^{(1)}(t)\|_{L^\infty} \approx t$ as $t \to 0$. Hence the proposition.

As announced above, we give only the formal aspect of the proof of Theorem 5.8 here, which originates from [Carles (2007b)]. Since the statement of the result involves times which go to zero as $\varepsilon \to 0$, we can use the approximation, see (4.13):

$$\limsup_{\varepsilon \to 0} \left\| u^\varepsilon(t, \cdot) - a(t, \cdot) e^{i\phi(t,\cdot)/\varepsilon} \right\|_{L^2 \cap L^\infty} = \mathcal{O}(t),$$

where (ϕ, a) is given by:

$$\begin{cases} \partial_t \phi + \dfrac{1}{2}|\nabla \phi|^2 + |a|^2 = 0 & ; \ \phi_{|t=0} = \phi_0, \\[2mm] \partial_t a + \nabla \phi \cdot \nabla a + \dfrac{1}{2} a \Delta \phi = 0 & ; \ a_{|t=0} = a_0. \end{cases}$$

To approximate v^ε, we use the same system. However, we do not pass to the limit in the initial data. We have

$$\limsup_{\varepsilon \to 0} \left\| v^\varepsilon(t, \cdot) - \widetilde{a}^\varepsilon(t, \cdot) e^{i\widetilde{\phi}^\varepsilon(t,\cdot)/\varepsilon} \right\|_{L^2 \cap L^\infty} = \mathcal{O}(t),$$

where $(\widetilde{\phi}^\varepsilon, \widetilde{a}^\varepsilon)$ is given by:

$$\begin{cases} \partial_t \widetilde{\phi}^\varepsilon + \dfrac{1}{2}\left|\nabla \widetilde{\phi}^\varepsilon\right|^2 + |\widetilde{a}^\varepsilon|^2 = 0 & ; \ \widetilde{\phi}^\varepsilon_{|t=0} = \phi_0, \\[2mm] \partial_t \widetilde{a}^\varepsilon + \nabla \widetilde{\phi}^\varepsilon \cdot \nabla \widetilde{a}^\varepsilon + \dfrac{1}{2}\widetilde{a}^\varepsilon \Delta \widetilde{\phi}^\varepsilon = 0 & ; \ \widetilde{a}^\varepsilon_{|t=0} = \widetilde{a}^\varepsilon_0. \end{cases}$$

In particular,

$$u^\varepsilon(t, x) - v^\varepsilon(t, x) \approx a(t, x) e^{i\phi(t,x)/\varepsilon} - \widetilde{a}^\varepsilon(t, x) e^{i\widetilde{\phi}^\varepsilon(t,x)/\varepsilon}.$$

The stability analysis of Chap. 4 shows that

$$\|a - \widetilde{a}^\varepsilon\|_{L^\infty([0,T];H^s)} = \mathcal{O}(\delta^\varepsilon),$$

for some time $T > 0$ independent of ε. We infer:

$$u^\varepsilon(t, x) - v^\varepsilon(t, x) \approx a(t, x) \left(e^{i\phi(t,x)/\varepsilon} - e^{i\widetilde{\phi}^\varepsilon(t,x)/\varepsilon} \right),$$

and

$$|u^\varepsilon(t, x) - v^\varepsilon(t, x)| \approx 2|a(t, x)| \left| \sin\left(\frac{\phi(t, x) - \widetilde{\phi}^\varepsilon(t, x)}{2\varepsilon} \right) \right|.$$

The idea is then to consider the Taylor expansion for the phases with respect to the time variable:

$$\phi(t,x) \approx \sum_{j \geqslant 0} t^j \phi_j(x) \quad ; \quad \widetilde{\phi}^\varepsilon(t,x) \approx \sum_{j \geqslant 0} t^j \widetilde{\phi}_j^\varepsilon(x).$$

The notations are consistent when $j = 0$: $\phi_0 = \widetilde{\phi}_0^\varepsilon$. We have already computed:

$$\phi_1(x) = -|a_0(x)|^2 \quad ; \quad \widetilde{\phi}_1^\varepsilon(x) = -|\widetilde{a}_0^\varepsilon(x)|^2.$$

To compute the higher order terms, we see that we must also consider the Taylor expansion in time for a and $\widetilde{a}^\varepsilon$:

$$a(t,x) \approx \sum_{j \geqslant 0} t^j a_j(x) \quad ; \quad \widetilde{a}^\varepsilon(t,x) \approx \sum_{j \geqslant 0} t^j \widetilde{a}_j^\varepsilon(x).$$

Here again, the notations are consistent when $j = 0$. We have

$$a_1 = -\nabla\phi_0 \cdot \nabla a_0 - \frac{1}{2} a_0 \Delta\phi_0 \quad ; \quad \widetilde{a}_1^\varepsilon = -\nabla\phi_0 \cdot \nabla\widetilde{a}_0^\varepsilon - \frac{1}{2}\widetilde{a}_0^\varepsilon \Delta\phi_0.$$

We check that for $j \geqslant 1$, (ϕ_j, a_j) is determined by $(\phi_k, a_k)_{0 \leqslant k \leqslant j-1}$. Also, since ϕ and $\widetilde{\phi}^\varepsilon$ have the same initial datum, we have

$$\phi(t,x) - \widetilde{\phi}^\varepsilon(t,x) \approx \sum_{j \geqslant 1} t^j \left(\phi_j(x) - \widetilde{\phi}_j^\varepsilon(x) \right).$$

The first term of this series is given by

$$\phi_1 - \widetilde{\phi}_1^\varepsilon = |\widetilde{a}_0^\varepsilon|^2 - |a_0|^2 = |a_0 + \widetilde{a}_0^\varepsilon - a_0|^2 - |a_0|^2$$
$$= 2\operatorname{Re}\left(\overline{a}_0 \left(\widetilde{a}_0^\varepsilon - a_0 \right) \right) + |\widetilde{a}_0^\varepsilon - a_0|^2.$$

Note that by assumption, the first term on the last line is *exactly* of order δ^ε (on the support of a_0). The last term is smaller, controlled by $(\delta^\varepsilon)^2$:

$$\phi_1 - \widetilde{\phi}_1^\varepsilon = 2\operatorname{Re}\left(\overline{a}_0 \left(\widetilde{a}_0^\varepsilon - a_0 \right) \right) + \mathcal{O}\left((\delta^\varepsilon)^2 \right).$$

By induction, it is easy to check

$$\phi_j - \widetilde{\phi}_j^\varepsilon = \mathcal{O}(\delta^\varepsilon), \quad \forall j \geqslant 2.$$

Therefore, for any $K \in \mathbb{N} \setminus \{0\}$,

$$\phi(t,x) - \widetilde{\phi}^\varepsilon(t,x) = \sum_{j=1}^{K} t^j \left(\phi_j(x) - \widetilde{\phi}_j^\varepsilon(x) \right) + \mathcal{O}\left(t^{K+1}\delta^\varepsilon \right) \approx t\left(\phi_1(x) - \widetilde{\phi}_1^\varepsilon(x) \right).$$

We infer as $\varepsilon \to 0$ and $t \to 0$:

$$|u^\varepsilon(t,x) - v^\varepsilon(t,x)| \approx 2|a(t,x)| \left| \sin\left(\frac{\phi(t,x) - \widetilde{\phi}^\varepsilon(t,x)}{2\varepsilon} \right) \right|$$

$$\approx 2|a_0(x)| \left| \sin\left(\frac{t}{2\varepsilon} \left(\phi_1(x) - \widetilde{\phi}_1^\varepsilon(x) \right) \right) \right|.$$

The argument of the sine function is of order exactly $t\delta^\varepsilon/\varepsilon$. Therefore, we can find $t = t^\varepsilon$ of order $\varepsilon/\delta^\varepsilon$, with:

- $t^\varepsilon \to 0$ as $\varepsilon \to 0$.
- $\liminf\limits_{\varepsilon \to 0} \|u^\varepsilon(t^\varepsilon) - v^\varepsilon(t^\varepsilon)\|_{L^2 \cap L^\infty} > 0$.

This completes the proof of Theorem 5.8.

Remark 5.15. In view of [Burq and Zworski (2005)], introduce the complex projective distance:

$$u_j \in L^2(\mathbb{R}^n), \quad d_{\mathrm{pr}}(u_1, u_2) := \arccos\left(\frac{|\langle u_1, u_2 \rangle|}{\|u_1\|_{L^2} \|u_2\|_{L^2}}\right).$$

Then we can check that, up to demanding $\varepsilon/\delta^\varepsilon \ll t^\varepsilon \ll 1$,

$$\frac{d_{\mathrm{pr}}\left(u^\varepsilon(t^\varepsilon), v^\varepsilon(t^\varepsilon)\right)}{d_{\mathrm{pr}}\left(u^\varepsilon(0), v^\varepsilon(0)\right)} \to +\infty \quad \text{as } \varepsilon \to 0.$$

Essentially, Theorem 5.8 uses the fact that oscillations of order $\mathcal{O}(1)$ appear for time of order $\varepsilon/\delta^\varepsilon$. The above result uses the fact that for larger time, these oscillations are rapid as $\varepsilon \to 0$ (but not of order $\mathcal{O}(1/\varepsilon)$).

We now turn our attention to the special case where the initial data contain no rapid oscillation: $\phi_0 = 0$. An easy induction shows that in the Taylor expansion for ϕ (resp. a), all the even (resp. odd) powers of t vanish:

$$\phi(t, x) \approx \sum_{j \geqslant 0} t^{2j+1} \phi_{2j+1}(x) \quad ; \quad a(t, x) \approx \sum_{j \geqslant 0} t^{2j} a_{2j}(x).$$

The same holds for $\widetilde{\phi}^\varepsilon$ and $\widetilde{a}^\varepsilon$, since at time $t = 0$, $\widetilde{\phi}^\varepsilon = \phi_0 = 0$. In particular,

$$\phi(t, x) = t\phi_1(x) + \mathcal{O}\left(t^3\right) = -t|a_0(x)|^2 + \mathcal{O}\left(t^3\right).$$

As $\varepsilon \to 0$ and $t \to 0$, we infer

$$u^\varepsilon(t, x) \approx a(t, x)e^{i\phi(t,x)/\varepsilon} \approx a_0(x)e^{-it|a_0(x)|^2/\varepsilon}e^{i\mathcal{O}(t^3)/\varepsilon}.$$

We have therefore:

$$u^\varepsilon(t, x) \approx a_0(x)e^{-it|a_0(x)|^2/\varepsilon} \quad \text{for } |t| \ll \varepsilon^{1/3}.$$

Note that the function of the right hand side involves no spatial derivative/integration. In other words, it is constructed without considering the Laplacian in the nonlinear Schrödinger equation: we recover the approximate solution considered in Proposition 5.3. The above approximation for $|t| \ll \varepsilon^{1/3}$ is expected to remain valid for any smooth nonlinearity, that is even in cases where the rigorous justification of WKB analysis is not known. An advantage of replacing the assumption $|t| \leqslant c\varepsilon|\log\varepsilon|^\theta$ with $|t| \ll \varepsilon^{1/3}$ would be to infer a loss of regularity phenomenon, as in Sec. 5.2, even though the range for the Sobolev index k should be decreased compared to Theorem 5.4. In particular, this would not suffice to prove Corollary 5.5.

Finally, we discuss more precisely the link between this approach and
the result of [Burq and Zworski (2005)] mentioned in Remark 5.13. The
initial data in Eq. (5.10) are not of WKB type. As we will see in the
second part of this book, they correspond to a focusing phenomenon (a
caustic reduced to a point). The important aspect is that the nonlinearity
is cubic, and in space dimension three, the size of the coupling constant, ε^2
in Eq. (5.10), is *supercritical* as far as nonlinear effects near the caustic are
concerned; see Sec. 6.3 and Chap. 7. Therefore, supercriticality is the main
feature shared by Eq. (5.8) and Eq. (5.10), and is the reason why instability
occurs in the limit $\varepsilon \to 0$. Without entering into the details of the proof
of [Burq and Zworski (2005)], we mention that the main idea consists in
neglecting the Laplacian for small times, in order to reduce the problem to
an ordinary differential equation mechanism, like in §5.1.

In the particular case of Eq. (5.10) (where the harmonic potential is
isotropic), this link can be made more explicit. Consider u^ε solution to a
generalization of Eq. (5.10):

$$i\varepsilon\partial_t u^\varepsilon + \frac{\varepsilon^2}{2}\Delta u^\varepsilon = \frac{|x|^2}{2}u^\varepsilon + \varepsilon^k|u^\varepsilon|^2 u^\varepsilon, \quad x \in \mathbb{R}^n.$$

Assume that $1 < k < n$ (hence $n \geqslant 2$). First, introduce U^ε given by a lens
transform:

$$U^\varepsilon(t,x) = \frac{1}{(1+t^2)^{n/4}}e^{i\frac{t}{1+t^2}\frac{|x|^2}{2\varepsilon}}u^\varepsilon\left(\arctan t, \frac{x}{\sqrt{1+t^2}}\right).$$

It solves:

$$\begin{cases} i\varepsilon\partial_t U^\varepsilon + \dfrac{\varepsilon^2}{2}\Delta U^\varepsilon = \varepsilon^k\left(1+t^2\right)^{n/2-1}|U^\varepsilon|^2 U^\varepsilon, \\ U^\varepsilon(0,x) = u^\varepsilon(0,x). \end{cases}$$

Introduce $\gamma = k/n < 1$, $h = \varepsilon^{1-\gamma}$, $t_0^h = h^{\gamma/(1-\gamma)}$ and $\psi = \psi^h$ given by

$$\psi^h(t,x) = U^\varepsilon\left(\frac{t}{\varepsilon^\gamma} - 1, x\right).$$

It solves:

$$\begin{cases} ih\partial_t\psi^h + \dfrac{h^2}{2}\Delta\psi^h = \left(\left(t_0^h\right)^2 + \left(t - t_0^h\right)^2\right)^{n/2-1}|\psi^h|^2\psi^h, \\ \psi^h\left(t_0^h, x\right) = u^\varepsilon(0,x). \end{cases}$$

The above equation is closely akin to the cubic nonlinear Schrödinger equation in a supercritical WKB régime (think of the coupling constant in front
of the nonlinearity as t^{n-2}, and recall that $n \geqslant 2$). Consider the case where

$u^\varepsilon(0,x) = a_0(x)$ is independent of ε. Instabilities as $h \to 0$ can be established in the same spirit as above (with a slightly different scaling, because of the factor in front of the cubic nonlinearity for ψ^h), yielding instabilities for u^ε. More precisely, we can prove

$$\left\| \psi^h(t) - a(t)e^{i\phi(t)/h} \right\|_{L^2 \cap L^\infty} = \mathcal{O}\left(t + h^{\frac{n(k-1)}{n-k}} \right),$$

where (ϕ, a) is given by:

$$\begin{cases} \partial_t \phi + \dfrac{1}{2}|\nabla \phi|^2 + t^{n-2}|a|^2 = 0 & ; \quad \phi_{|t=0} = 0, \\[2mm] \partial_t a + \nabla \phi \cdot \nabla a + \dfrac{1}{2}a\Delta\phi = 0 & ; \quad a_{|t=0} = a_0. \end{cases}$$

The Taylor expansion for ϕ yields:

$$\phi(t,x) = -\frac{t^{n-1}}{n-1}|a_0(x)|^2 + \mathcal{O}\left(t^{2n-1} \right).$$

Therefore, a perturbation of a_0 of order δ^h, with $h^{1-1/N} \ll \delta^h \ll 1$ (as in Theorem 5.8) becomes visible on ψ^h for times t^h such that $(t^h)^{n-1}\delta^h \approx h$. In the case of Eq. (5.10), $1 < k = 2 < n = 3$, and this yields $t^h \approx (h/\delta^h)^{1/2}$. This corresponds to a time

$$\frac{1}{\sqrt{\varepsilon\delta}} - 1$$

for U^ε, that is,

$$\arctan\left(\frac{1}{\sqrt{\varepsilon\delta}} - 1 \right) = \frac{\pi}{2} - \arctan\left(\frac{\sqrt{\varepsilon\delta}}{1 - \sqrt{\varepsilon\delta}} \right) \approx \frac{\pi}{2} - \arctan\left(\sqrt{\varepsilon\delta} \right)$$

for u^ε. Note that instabilities occur for small time for ψ^h, corresponding to time of order $\pi/2$ for u^ε. This is due to the fact that the harmonic potential causes focusing at time $\pi/2$ (see Examples 1.12 and 1.18, and §8.1 below). Note also that U^ε is of order $\mathcal{O}(1)$ in L^∞ when the instability occurs, which implies that u^ε is of order

$$|u^\varepsilon| \approx (\varepsilon\delta)^{-3/4} \ll \varepsilon^{-1+1/(4N)} \ll \varepsilon^{-3/2},$$

since $\varepsilon^{1/3 - 1/(3N)} = h^{1-1/N} \ll \delta \ll 1$. When the instability occurs, the wave function is not as concentrated as in [Burq and Zworski (2005)]; see Eq. (5.10). Heuristically, it is not surprising that instability occurs also for more concentrated data.

Finally, we point out that the instabilities at the semi-classical level affect the wave function, but not the usual quadratic observables. The instability mechanism is due to a phase modulation, and the creation of oscillations whose period is of order $\varepsilon/\delta^\varepsilon$. By assumption,

$$\varepsilon \ll \varepsilon/\delta^\varepsilon \ll 1.$$

Therefore, this scale of oscillation is not detected by the Wigner measure, which accounts for phenomena at scales 1 and ε only. Similarly, the instability does not affect the convergence of the position and current densities for small time. Indeed, these quadratic observables, as well as the Wigner measure, are described by an Euler equation, which is stable on $[-T, T]$, for some $T > 0$. On the other hand, nothing seems to be known as for what happens when the solution to the Euler equation develops singularities. The example developped in Sec. 7.4.3 suggests that the instabilities at the semi-classical level may affect also the Wigner measures after the solution to the Euler equation has become singular.

PART II

Caustic Crossing:
The Case of Focal Points

Chapter 6

Caustic Crossing: Formal Analysis

6.1 Presentation

In this second part, we consider the régime where the WKB analysis breaks down. This analysis is essentially independent of the first part, which may be considered as a motivation only. Roughly speaking, in the linear case, many results are available, while in the nonlinear case, very few phenomena have been identified.

Resume one of the examples given in Chap. 1:

$$i\varepsilon\partial_t u_{\text{lin}}^\varepsilon + \frac{\varepsilon^2}{2}\Delta u_{\text{lin}}^\varepsilon = 0 \quad ; \quad u_{\text{lin}}^\varepsilon(0,x) = a_0(x)e^{-i|x|^2/(2\varepsilon)}. \tag{6.1}$$

We have seen that for $t < 1$, the solution to the eikonal equation, and to the leading order transport equation respectively, are given by:

$$\phi_{\text{eik}}(t,x) = \frac{|x|^2}{2(t-1)} \quad ; \quad a(t,x) = \frac{1}{(1-t)^{n/2}}a_0\left(\frac{x}{1-t}\right).$$

As already noticed, these terms (as well as all the others involved in the WKB analysis) become singular as $t \to 1$. On the other hand, if, say, $a_0 \in \mathcal{S}(\mathbb{R}^n)$, then for every fixed ε, Eq. (6.1) has a unique global solution $u_{\text{lin}}^\varepsilon \in C^\infty(\mathbb{R};\mathcal{S}(\mathbb{R}^n))$. Since we consider the free Schrödinger equation (by "free", we mean linear, and without external potential), $u_{\text{lin}}^\varepsilon$ is given explicitly by the integral

$$u_{\text{lin}}^\varepsilon(t,x) = \frac{1}{(2i\pi t)^{n/2}}\int_{\mathbb{R}^n} e^{i\frac{|x-y|^2}{2\varepsilon t}} u_{\text{lin}}^\varepsilon(0,y)dy$$

$$= \frac{1}{(2i\pi t)^{n/2}}\int_{\mathbb{R}^n} e^{i\frac{|x-y|^2}{2\varepsilon t} - i\frac{|y|^2}{2\varepsilon}} a_0(y)dy.$$

This formula also yields the expression for ϕ_{eik} and a very easily, thanks to the stationary phase formula (which can be found for instance in [Alinhac and Gérard (1991); Grigis and Sjöstrand (1994); Hörmander (1994)]).

Indeed, the last expression can be viewed as an oscillatory integral, with phase

$$\varphi(t, x, y) = \frac{|x - y|^2}{2t} - \frac{|y|^2}{2}.$$

For $t \neq 1$, the critical points of φ, considered as a function of y, are given by:

$$\nabla_y \varphi(t, x, y_c) = 0 \iff \frac{y_c - x}{t} = y_c \iff y_c = \frac{x}{1 - t}.$$

The Hessian of φ is

$$\nabla_y^2 \varphi(t, x, y) = \left(\frac{1}{t} - 1 \right) I_n.$$

Therefore, if $t \neq 1$, φ has a unique critical point, which is non-degenerate. Stationary phase formula then yields:

$$u_{\text{lin}}^\varepsilon(t, x) \underset{\varepsilon \to 0}{\sim} \begin{cases} \dfrac{1}{(1 - t)^{n/2}} a_0 \left(\dfrac{x}{1 - t} \right) e^{i \frac{|x|^2}{2\varepsilon(t-1)}} & \text{if } t < 1, \\[3mm] \dfrac{e^{-in\pi/2}}{(t - 1)^{n/2}} a_0 \left(\dfrac{x}{1 - t} \right) e^{i \frac{|x|^2}{2\varepsilon(t-1)}} & \text{if } t > 1. \end{cases}$$

From this point of view, the main difference between the asymptotics for $t < 1$ and for $t > 1$ is the phase shift $e^{-in\pi/2}$, which is due to the sign change of the Hessian of φ as $t - 1$ changes signs.

For $t = 1$, we can no longer apply stationary phase formula, but we have directly from the integral expression of $u_{\text{lin}}^\varepsilon$:

$$u_{\text{lin}}^\varepsilon(1, x) = \frac{1}{(2i\pi)^{n/2}} \int_{\mathbb{R}^n} e^{i \frac{|x-y|^2}{2\varepsilon} - i \frac{|y|^2}{2\varepsilon}} a_0(y) dy = \frac{e^{i \frac{|x|^2}{2\varepsilon}}}{(i\varepsilon)^{n/2}} \widehat{a}_0 \left(\frac{x}{\varepsilon} \right).$$

By writing

$$u_{\text{lin}}^\varepsilon(1, x) = \frac{1}{(i\varepsilon)^{n/2}} \widehat{a}_0 \left(\frac{x}{\varepsilon} \right) \exp \left(i \frac{\varepsilon}{2} \left| \frac{x}{\varepsilon} \right|^2 \right),$$

we see that

$$u_{\text{lin}}^\varepsilon(1, x) = \frac{1}{(i\varepsilon)^{n/2}} \widehat{a}_0 \left(\frac{x}{\varepsilon} \right) + \mathcal{O}(\varepsilon) \quad \text{in } L^2(\mathbb{R}^n).$$

To summarize, we have:

$$u_{\text{lin}}^\varepsilon(t, x) \underset{\varepsilon \to 0}{\sim} \begin{cases} \dfrac{1}{(1 - t)^{n/2}} a_0 \left(\dfrac{x}{1 - t} \right) e^{i \frac{|x|^2}{2\varepsilon(t-1)}} & \text{if } t < 1, \\[3mm] \dfrac{e^{-in\pi/4}}{\varepsilon^{n/2}} \widehat{a}_0 \left(\dfrac{x}{\varepsilon} \right) & \text{if } t = 1, \\[3mm] \dfrac{e^{-in\pi/2}}{(t - 1)^{n/2}} a_0 \left(\dfrac{x}{1 - t} \right) e^{i \frac{|x|^2}{2\varepsilon(t-1)}} & \text{if } t > 1. \end{cases} \qquad (6.2)$$

We see that with initial data of order $\mathcal{O}(1)$ as $\varepsilon \to 0$, the size of $u_{\text{lin}}^{\varepsilon}$ grows as t approaches one. As $t = 1$, the amplitude is saturated, of order $\mathcal{O}(\varepsilon^{-n/2})$. After that critical time, the amplitude decreases, and is of order $\mathcal{O}(1)$ again. In particular, the family $(\|u^{\varepsilon}(1, \cdot)\|_{L^{\infty}})_{0 < \varepsilon \leqslant 1}$ is unbounded: if $|u^{\varepsilon}|^2$ is viewed as the intensity of a laser beam, it becomes extremely high for $t \approx 1$. This is why the phenomenon is called *caustic* (from the Greek *kaustikos*, after the verb *katein*=to burn).

Recall that in Chap. 1, the caustic was defined as the set where the solution to the eikonal equation becomes singular. This is a geometrical definition, which is equivalent to saying that the caustic is the locus where rays of geometric optics form an envelope. This may seem different from the etymological definition, which refers to an analytical phenomenon (growth of the amplitude). Yet, since in the semi-classical limit, the energy is carried by the rays of geometric optics, these points of view agree. On the other hand, we will see in Sec. 9.2 that in some nonlinear cases, it might be sensible to distinguish these two notions.

It turns out that the result outlined above is fairly general. To study the high frequency limit of linear equations, the use of oscillatory integrals has proven very efficient to describe the wave function itself. Working in the phase space (that is, $(t, x, \tau, \xi) \in T^{*}\mathbb{R}^{1+n}$) instead of the physical space ($(t, x) \in \mathbb{R}^{1+n}$), the singularity corresponding to the caustic disappears. In other words, the caustic phenomenon appears when projecting from the phase space to the physical space. In the case of Eq. (6.1), this approach leads to the following Lagrangian integral:

$$u_{\text{lin}}^{\varepsilon}(t, x) = \frac{1}{(2\pi\varepsilon)^{n/2}} \int_{\mathbb{R}^n} e^{-i\frac{t-1}{2\varepsilon}|\xi|^2 + i\frac{x \cdot \xi}{\varepsilon}} A^{\varepsilon}(\xi) d\xi. \tag{6.3}$$

Obviously, the Lagrangian symbol A^{ε} is given by the formula

$$A^{\varepsilon}(\xi) = \frac{1}{\varepsilon^{n/2}} e^{i\frac{t-1}{2\varepsilon}|\xi|^2} \widehat{u}_{\text{lin}}^{\varepsilon}\left(t, \frac{\xi}{\varepsilon}\right).$$

Note that the right hand side is indeed independent of time, since $u_{\text{lin}}^{\varepsilon}$ solves the free Schrödinger equation. We check that A^{ε} converges as $\varepsilon \to 0$:

$$A^{\varepsilon}(\xi) = \frac{1}{(2\pi\varepsilon)^{n/2}} e^{-i\frac{|\xi|^2}{2\varepsilon}} \int_{\mathbb{R}^n} e^{-i\frac{x \cdot \xi}{\varepsilon}} u_{\text{lin}}^{\varepsilon}(0, x) dx$$

$$= \frac{1}{(2\pi\varepsilon)^{n/2}} \int_{\mathbb{R}^n} e^{-i\frac{|x+\xi|^2}{2\varepsilon}} a_0(x) dx = e^{-in\frac{\pi}{4}} a_0(-\xi) + \mathcal{O}(\varepsilon).$$

Roughly speaking, in general two phenomena occur at leading order in the high frequency limit for linear equations. First, a phase shift appears after the caustic crossing, in the same fashion as for $t > 1$ in Eq. (6.2). This phase shift is called the Maslov index, and is related to the geometry of the caustic; see e.g. [Duistermaat (1974); Maslov and Fedoriuk (1981); Yajima (1979)]. From a technical point of view, it corresponds to a signature change of the Hessian of the phase involved in the oscillatory integral approach: in Eq. (6.3), the Hessian of the phase is also $(1 - t)I_n$, and its signature changes from $+n$ to $-n$ as $t - 1$ changes signs, hence a phase shift of $(-n - (+n))\pi/4 = -n\pi/2$. The second phenomenon, which is not present in the example of Eq. (6.1), is the creation of new phases. Typically, when applying a stationary phase argument beyond the caustic, it may happen that the phase in the oscillatory integral has several non-degenerate critical points, even if it has only one before the caustic. In that case, we need a description of the form:

$$u_{\text{lin}}^{\varepsilon}(t, x) \approx \sum_{\text{critical points}} a_j(t, x)e^{i\phi_j(t,x)/\varepsilon}.$$

All the phases ϕ_j solve the eikonal equation (Eq. (1.10) in the case of Schrödinger equations). Note that in order to describe the wave function, it is not sufficient to consider only the viscosity solution to this Hamilton–Jacobi equation: by selecting only one of the above phases, the error we make is of the same order as the wave function $u_{\text{lin}}^{\varepsilon}$ itself, in L^{∞}; see [Kossioris (1993)]. We refer to [Duistermaat (1974); Maslov and Fedoriuk (1981)] and references therein for a more precise description of the phenomena mentioned above.

Note that if one is interested in quadratic observables rather than in the wave function itself, the Wigner measures share an important feature with the Lagrangian integral approach mentioned above: they unfold the singularities. In other words, the caustic phenomenon does not exist on Wigner measures; see e.g. [Gérard *et al.* (1997); Lions and Paul (1993); Sparber *et al.* (2003)]. On the other hand, the Maslov index is lost when working with Wigner measures.

For nonlinear equations, far less is known. As pointed out in Chap. 1, a new parameter appears for nonlinear problems: the size of the data. Moreover, the nature of the nonlinearity (dissipative, accretive, conservative, Lipschitzean...) is crucial, as discussed below. So potentially, a huge variety of phenomena is likely to happen. Let us mention the most important results on the subject, concerning hyperbolic equations: [Joly *et al.* (1995, 1996a, 2000)] (see also [Joly *et al.* (1997a)]). As a typical striking result of this work, consider the following semi-linear wave equation, with highly oscillatory initial data:

$$\begin{cases} \left(\partial_t^2 - \Delta\right) u^\varepsilon + |\partial_t u^\varepsilon|^{p-1} \partial_t u^\varepsilon = 0. \\ u^\varepsilon(0,x) = \varepsilon U_0\left(x, \frac{\phi_0(x)}{\varepsilon}\right) \quad ; \quad \partial_t u^\varepsilon(0,x) = U_1\left(x, \frac{\phi_0(x)}{\varepsilon}\right), \end{cases} \quad (6.4)$$

where the functions $U_j(x, \cdot)$ are 2π-periodic for all $x \in \mathbb{R}^n$, and $p > 1$ (not necessarily an integer). One can check that in a WKB régime, the above equation is of weakly nonlinear type: the eikonal equation is the same as in the linear case, and the nonlinearity is present in the leading order transport equation. Consider the case where the initial phase ϕ_0 is spherically symmetric, with

$$\phi_0(x) = |x| =: r.$$

As mentioned in Chap. 1, the eikonal equation associated to the wave equation is

$$\left(\partial_t \phi\right)^2 = |\nabla \phi|^2.$$

In the particular case under consideration, we find the two solutions

$$\phi_\pm = \phi_\pm(t,r) = r \pm t.$$

We see that the rays associated to ϕ_- meet at the origin for positive time: as in the case of Eq. (6.1), a caustic reduced to a point is formed. Note that Eq. (6.4) is a dissipative equation: the energy associated to the linear equation is a non-increasing function of time,

$$\frac{d}{dt} \int_{\mathbb{R}^n} \left(\left(\partial_t u^\varepsilon(t,x)\right)^2 + |\nabla u^\varepsilon(t,x)|^2 \right) dx \leqslant 0.$$

To give a flavor of the results of J.-L. Joly, G. Métivier and J. Rauch, let us describe qualitatively the main result of [Joly et al. (1995)], generalized in [Joly et al. (2000)]. Suppose that $p > 2$. As the wave focuses at the origin, its amplitude grows, and the dissipative mechanism becomes very strong. The part of the wave carrying the most energy is highly dissipated. Since the highest energy terms are those which are highly oscillatory, the rapid oscillations of u^ε are absorbed at the focus: past the focus, u^ε is no longer ε-oscillatory at leading order.

Of course, the above mechanism is due to the nature of the nonlinearity: it is dissipative, and the phenomenon sketched here is due to this aspect. On the other hand, the nonlinear Schrödinger equations which we consider here, of the form

$$i\varepsilon \partial_t u^\varepsilon + \frac{\varepsilon^2}{2} \Delta u^\varepsilon = |u^\varepsilon|^{2\sigma} u^\varepsilon,$$

are conservative and Hamiltonian: the L^2-norm of u^ε is conserved, and the L^2-norm of $\varepsilon \nabla u^\varepsilon$ is bounded; see Eq. (1.24) and Eq. (1.25). So the above dissipation mechanism is less likely to occur. Before describing some caustic phenomena for this nonlinear Schrödinger equation, we present an idea due to J. Hunter and J. Keller, which suggests that the discussion presented in §1.2 for relevant phenomena in a WKB régime, should be repeated, with some differences though, near a caustic.

6.2   The idea of J. Hunter and J. Keller

The idea presented in [Hunter and Keller (1987)] for conservation laws is essentially the following. In a WKB régime, according to the amplitude of a wave, the nonlinearity can influence the geometry of the propagation (supercritical case, where the nonlinearity is present in the eikonal equation), or simply the leading order amplitude (weakly nonlinear régime), or can even be negligible at leading order. In the linear case, near a caustic, the amplitude of the wave is altered, as we have seen on the example of Eq. (6.1). The amplification factor depends on the geometry of the caustic, and the maximal amplification corresponds to a focal point: the energy concentrates on a single point rather than on, say, a curve or a surface. Therefore, we can resume a similar discussion: according to the amplitude of the wave near the caustic, the nonlinearity should have several possible effects at leading order. Note that the amplification is localized near the caustic, in some boundary layer. The idea in [Hunter and Keller (1987)] consists in saying that the important quantity to consider is the *average* nonlinear effect near the caustic.

For instance, in the case of Eq. (6.1), Eq. (6.2) shows that the amplification near the caustic is of order $\varepsilon^{-n/2}$, and that the boundary layer associated to the focal point is of order ε.

Before describing more precisely this idea for solutions to the nonlinear Schrödinger equation, resume the case of the semi-linear wave equation mentioned in the previous paragraph, and consider spherically symmetric initial data:

$$\begin{cases} \left(\partial_t^2 - \Delta\right) u^\varepsilon + \left|\partial_t u^\varepsilon\right|^{p-1} \partial_t u^\varepsilon = 0, & (t, x) \in [0, T] \times \mathbb{R}^3, \\ u^\varepsilon(0, x) = \varepsilon^{J+1} U_0 \left(r, \dfrac{r - r_0}{\varepsilon}\right) \ ; \ \partial_t u^\varepsilon(0, x) = \varepsilon^J U_1 \left(r, \dfrac{r - r_0}{\varepsilon}\right). \end{cases} \quad (6.5)$$

The new parameter $J \geqslant 0$ measures the size of the initial data. For $r_0 = 0$, and functions $U_j(r, \cdot)$ which are 2π-periodic for all $r \geqslant 0$, we recover the framework of [Joly *et al.* (1995)]. We consider rather the case $r_0 > 0$ and $\operatorname{supp} U_j(r, .) \subset [-z_0, z_0]$ for some $z_0 > 0$ independent of $r \geqslant 0$ (the initial data are *short pulses*, as opposed to the previous wave trains). In the linear case, the solution to

$$\begin{cases} \left(\partial_t^2 - \Delta\right) v^\varepsilon = 0, & (t, x) \in [0, T] \times \mathbb{R}^3, \\ v^\varepsilon(0, x) = \varepsilon^{J+1} U_0\left(r, \dfrac{r - r_0}{\varepsilon}\right) & ; \quad \partial_t v^\varepsilon(0, x) = \varepsilon^J U_1\left(r, \dfrac{r - r_0}{\varepsilon}\right), \end{cases}$$

is given by, since we consider a three-dimensional setting:

$$v^\varepsilon(t, r) = \begin{cases} \varepsilon^{J+1} \dfrac{g^\varepsilon(t + r) - g^\varepsilon(t - r)}{2r} & \text{for } r > 0, \\ \varepsilon^{J+1} \left(g^\varepsilon\right)'(t) & \text{for } r = 0, \end{cases}$$

where $g^\varepsilon(r)$ is essentially $r\left(U_0 + V_1\right)\left(r, \frac{r - r_0}{\varepsilon}\right)$, where V_1 is an anti-derivative of U_1 with respect to its second variable. We see that $\partial_t v^\varepsilon$ is of order ε^J outside the origin, and of order ε^{J-1} near the origin. The characteristic boundary layer about the origin is of order ε, since the pulses are supported on a domain of thickness of order ε, and propagate along the characteristic directions, $r \pm t = \text{Const.}$, see Fig. 6.1. Plugging this estimate

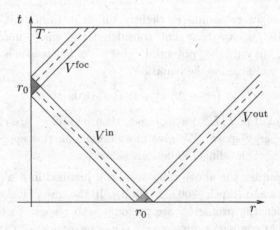

Fig. 6.1 Propagation of short pulses.

in the nonlinear potential, we have, for $r \approx 0$:

$$\left|\partial_t v^\varepsilon\right|^{p-1} \approx \varepsilon^{(p-1)(J-1)}.$$

Integrated on the focusing region, we find:

$$\int_{|r|\lesssim\varepsilon} |\partial_t v^\varepsilon|^{p-1} \approx \varepsilon^{(p-1)(J-1)+1}.$$

This suggests that if $(p-1)(J-1)+1 > 0$, then in the nonlinear case (6.5), nonlinear effects are negligible at the focus, while they become relevant for $(p-1)(J-1)+1 = 0$. To resume the terminology of [Hunter and Keller (1987)], the former case is referred to as "linear caustic" while the latter is called "nonlinear caustic". We can also check that the WKB régime is linear if $J > 0$ ("linear geometric optics", or "linear propagation"), weakly nonlinear if $J = 0$, as in the previous paragraph ("nonlinear propagation"). This leads to the following distinctions:

	$J > 0$	$J = 0$
$J > \frac{p-2}{p-1}$	linear propagation linear caustic	nonlinear propagation linear caustic
$J = \frac{p-2}{p-1}$	linear propagation nonlinear caustic	nonlinear propagation nonlinear caustic
$J < \frac{p-2}{p-1}$	linear propagation supercritical caustic	nonlinear propagation supercritical caustic

The above line of reasoning is slightly different from that of [Hunter and Keller (1987)]: we have not considered the whole nonlinear term $|\partial_t v^\varepsilon|^{p-1}\partial_t v^\varepsilon$, but only the "potential" $|\partial_t v^\varepsilon|^{p-1}$. The reason is that energy estimates are available for the equation

$$\left(\partial_t^2 - \Delta\right)u^\varepsilon + V^\varepsilon \partial_t u^\varepsilon = 0,$$

and if V^ε is small in $L_t^1 L_x^\infty$ for instance, then u^ε and v^ε are close to each other in the energy space. We give more details on this approach in the case of nonlinear Schrödinger equations below.

For short pulses, the above table has been justified in a series of three articles, [Carles and Rauch (2002, 2004a,b)]. In the case of a slowly varying envelopes (when the profile U_j are periodic with respect to their second variable, and not compactly supported), the same distinctions are expected, but a justification for all the cases is not available so far.

Notice that assuming that the above table remains valid for slowly varying envelopes, the result of [Joly *et al.* (1995)] corresponds to $J = 0 < \frac{p-2}{p-1}$: the supercritical effect near the caustic is the absorption of oscillations. On the other hand, the analogous phenomenon for the pulses is the absorption of the pulse at the focal point [Carles and Rauch (2004b)]. Indeed, formally, a pulse has zero mean value:

$$\frac{1}{2M} \int_{-M}^{M} f(z)dz \xrightarrow[M \to \infty]{} 0 \quad \text{if } f \text{ is compactly supported.}$$

So in a way, like for the case of wave trains, only the mean value of the wave remains past the focus. We leave out the discussion for the semi-linear wave equation at this stage.

Back to the case of nonlinear Schrödinger equations, consider the initial value problem

$$i\varepsilon\partial_t u^\varepsilon + \frac{\varepsilon^2}{2}\Delta u^\varepsilon = \varepsilon^\alpha |u^\varepsilon|^{2\sigma} u^\varepsilon \quad ; \quad u^\varepsilon(0, x) = a_0(x)e^{i\phi_0(x)/\varepsilon}. \tag{6.6}$$

Recall that the factor ε^α can be viewed as a scaling factor for the size of the initial data; see the end of §1.1. We have seen in the first part of these notes that the value $\alpha = 1$ is critical for the WKB methods: if $\alpha > 1$, then the nonlinearity does not affect the transport equation (1.19), while if $\alpha = 1$, then the nonlinearity appears in the right hand side of (1.19). To resume the terminology of [Hunter and Keller (1987)], we say that $\alpha > 1$ corresponds to a "linear geometric optics", or "linear propagation", while $\alpha = 1$ corresponds to a "nonlinear geometric optics", or "nonlinear propagation". The idea presented in [Hunter and Keller (1987)] consists in saying that according to the geometry of the caustic $\mathcal{C}$, different notions of criticality exist, as far as α is concerned, near the caustic. In the linear setting, the influence of the caustic is relevant only in a neighborhood of this set (essentially, in a boundary layer whose size depends on ε and the geometry of $\mathcal{C}$). View the nonlinearity in Eq. (6.6) as a potential, and assume that the nonlinear effects are negligible near the caustic: then $u^\varepsilon \sim v^\varepsilon$ near $\mathcal{C}$, where v^ε solves the free Schrödinger equation. Consider the term $\varepsilon^\alpha |u^\varepsilon|^{2\sigma}$ as a (nonlinear) potential. The average nonlinear effect near $\mathcal{C}$ is expected to be:

$$\varepsilon^{-1} \int_{\mathcal{C}(\varepsilon)} \varepsilon^\alpha |u^\varepsilon|^{2\sigma} \sim \varepsilon^{-1} \int_{\mathcal{C}(\varepsilon)} \varepsilon^\alpha |v^\varepsilon|^{2\sigma},$$

where $\mathcal{C}(\varepsilon)$ is the region where caustic effects are relevant, and the factor ε^{-1} is due to the integration in time (recall that there is an ε in front of the time derivative in Eq. (6.6)). The idea of this heuristic argument is that when the nonlinear effects are negligible near $\mathcal{C}$ (in the sense that the uniform norm of $u^\varepsilon - v^\varepsilon$ is small compared to that of v^ε near $\mathcal{C}$), the above approximation should be valid. On the other hand, it is expected that it ceases to be valid precisely when nonlinear effects can no longer be neglected near the caustic: $u^\varepsilon - v^\varepsilon$ is (at least) of the same order of magnitude as v^ε in $L^\infty(\mathcal{C}(\varepsilon))$. Like for the case of the semi-linear wave equation (6.5), we have not considered the whole nonlinearity $|u^\varepsilon|^{2\sigma} u^\varepsilon$, but only the nonlinear potential $|u^\varepsilon|^{2\sigma}$. Like for the wave equation, this is due to the fact that energy estimates are available for Schrödinger equations, showing that the norm of $|u^\varepsilon|^{2\sigma}$ in $L_t^1 L_x^\infty$ measures the influence of the whole nonlinear term in Eq. (6.6). Indeed, write

$$\left(i\varepsilon\partial_t + \frac{\varepsilon^2}{2}\Delta\right)(u^\varepsilon - v^\varepsilon) = \varepsilon^\alpha |u^\varepsilon|^{2\sigma} u^\varepsilon = \varepsilon^\alpha |u^\varepsilon|^{2\sigma}(u^\varepsilon - v^\varepsilon) + \varepsilon^\alpha |u^\varepsilon|^{2\sigma} v^\varepsilon.$$

Lemma 1.2 yields:

$$\begin{aligned}
\|u^\varepsilon - v^\varepsilon\|_{L^\infty(I;L^2)} &\leqslant \varepsilon^{\alpha-1} \int_I \left\||u^\varepsilon|^{2\sigma} v^\varepsilon\right\|_{L^2} \\
&\leqslant \varepsilon^{\alpha-1} \|u^\varepsilon\|_{L^1(I;L^\infty)}^{2\sigma} \|v^\varepsilon\|_{L^\infty(I;L^2)} \\
&\leqslant C\varepsilon^{\alpha-1} \|u^\varepsilon\|_{L^1(I;L^\infty)}^{2\sigma},
\end{aligned}$$

where C does not depend on ε, since the L^2 norm of v^ε is conserved, and independent of ε. In the case of a linear caustic, we can replace u^ε by v^ε in the last term of the above inequality, hence the above discussion.

Practically, assume that in the linear case, v^ε has an amplitude $\varepsilon^{-\ell}$ in a boundary layer of size ε^k; then the above quantity is

$$\varepsilon^{-1} \int_{\mathcal{C}(\varepsilon)} \varepsilon^\alpha |v^\varepsilon|^{2\sigma} \sim \varepsilon^{-1} \varepsilon^\alpha \left|\varepsilon^{-\ell}\right|^{2\sigma} \varepsilon^k.$$

The value α is then critical when the above cumulated effects are not negligible:

$$\alpha_c = 1 + 2\ell\sigma - k.$$

When $\alpha > \alpha_c$, the nonlinear effects are expected to be negligible near the caustic: "linear caustic". The case $\alpha = \alpha_c$ corresponds to a "nonlinear caustic". We illustrate this discussion in the case of two geometries below:

- When the caustic is reduced to a single point.
- When the caustic is a cusp.

6.3   The case of a focal point

In the case of a focal point, which corresponds to a quadratic initial phase (e.g. $\phi_0(x) = -|x|^2/2$), we have seen that $k = 1$ and $\ell = n/2$. See Eq. (6.2) and Fig. 6.2. This leads us to the value: $\alpha_c(\text{focal point}) = n\sigma$. Therefore, the following distinctions are expected:

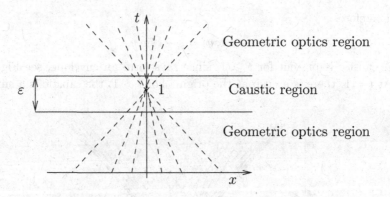

Fig. 6.2   Rays of geometric optics: focal point.

	$\alpha > 1$	$\alpha = 1$
$\alpha > n\sigma$	linear propagation linear caustic	nonlinear propagation linear caustic
$\alpha = n\sigma$	linear propagation, nonlinear caustic	nonlinear propagation nonlinear caustic

Note that as in the case of the semi-linear wave equation, the entries for rows and columns are fairly independent, as suggested in [Hunter and Keller (1987)]. In Chap. 7, we show that the above distinctions are relevant, and we describe the asymptotics of the wave functions in each of the four cases. Unlike for the semi-linear wave equation, we have note mentioned the case of a "supercritical caustic". A complete description in that case is not available yet, and we give a very partial answer in Chap. 9.

6.4   The case of a cusp

It seems that there are no rigorous results concerning nonlinear effects near a cusped caustic for nonlinear Schrödinger equations. Therefore, the discussion in this paragraph is only formal.

In space dimension $n = 1$, a cusped caustic appears for instance if the initial phase is given by

$$\phi_0(x) = \cos x.$$

The rays of geometric optics are given

$$\partial_t x(t,y) = \xi(t,y) \quad ; \quad x(0,y) = y \quad ; \quad \partial_t \xi(t,y) = 0 \quad ; \quad \xi(0,y) = -\sin y.$$

Therefore,

$$x(t, y) = y - t \sin y.$$

A caustic is present for $t \geqslant 1$, since rays form an envelope, see Fig. 6.3. At $t = 1$, there is a cusp at the origin. For $t > 1$, the caustic is a smooth

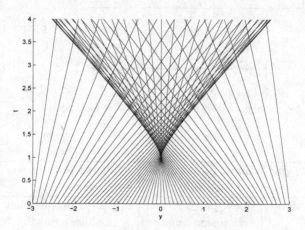

Fig. 6.3   Rays of geometric optics: cusped caustic.

curve. Moreover, for $t > 1$, three phases are necessary to describe the wave function inside the caustic region. To see this, recall that the linear solution is given by

$$u_{\text{lin}}^{\varepsilon}(t, x) = \frac{1}{(2i\pi t)^{1/2}} \int_{\mathbb{R}} e^{i \frac{|x-y|^2}{2\varepsilon t} + i \frac{\cos y}{\varepsilon}} a_0(y) dy.$$

The critical points y_c associated to the phase of this integral are such that

$$\frac{y_c - x}{t} = \sin y_c,$$

that is $x = y_c - t \sin y_c$. To compute the number of critical points, we map $y \mapsto y - t \sin y$ for various values of t. At time $t = 0.5$ (Fig. 6.4), there is only one critical point: the caustic is not formed yet. At time $t = 1$ (Fig. 6.5), we see that the tangent is vertical at $(x, y) = (0, 0)$: this corresponds to the cusp. At time $t = 1.5$ (Fig. 6.6), there are three critical points for $-x_0(1.5) < x < x_0(1.5)$, for some $x_0(1.5) \approx 0.3$. This corresponds to the three phases inside the caustic region. Note that $|x| = x_0(1.5)$ corresponds to the caustic itself. For $t = 4$ (Fig. 6.7), the caustic is broader, and the corresponding $x_0(4)$ is larger: $x_0(4) > 2$.

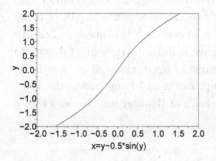

Fig. 6.4   Time $t = 0.5$.

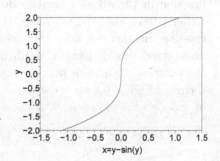

Fig. 6.5   Time $t = 1$.

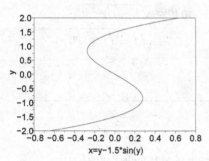

Fig. 6.6   Time $t = 1.5$.

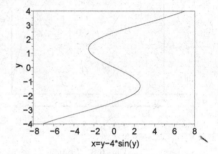

Fig. 6.7   Time $t = 4$.

To apply the argument of J. Hunter and J. Keller, we must distinguish two families of points: the cusp (at $(t, x) = (1, 0)$), and smooth points of the caustic (for $t > 1$ and $|x| = x_0(t)$).

At smooth points, the wave function is described thanks to the Airy function: on the caustic, it is not possible to apply the usual stationary phase formula, because the critical points are degenerate, but the third order derivative is not zero. This was first noticed in [Ludwig (1966)]. See also [Duistermaat (1974); Hörmander (1994); Hunter and Keller (1987)]. Typically, near the caustic, the linear solution can be approximated by

$$\varepsilon^{-1/6} \left(\alpha Ai \left(\frac{\psi(t, x)}{\varepsilon^{2/3}} \right) + \beta \varepsilon^{1/3} Ai' \left(\frac{\psi(t, x)}{\varepsilon^{2/3}} \right) \right) e^{i\rho(t,x)/\varepsilon},$$

where Ai stands for the Airy function, for $\alpha, \beta \in \mathbb{C}$ and some smooth functions ψ and ρ, with $\psi = 0$ on $\mathcal{C}$. This yields $\ell = 1/6$. For the value of k, it is tempting to take $k = 2/3$, in view of the argument of the Airy function in the above formula. However, it is suggested in [Hunter and Keller (1987)] that the relevant quantity to consider is the length of a ray crossing this layer of order $\varepsilon^{2/3}$, which is of order $\varepsilon^{1/3}$, hence $k = 1/3$. To understand this derivation, recall that by definition, rays are tangent to the caustic. At smooth points of the caustic, approximate the caustic by a circle. Figure 6.8 shows why the length of a ray lying inside the layer of order $\varepsilon^{2/3}$ is of order $\varepsilon^{1/3}$. The approach of [Hunter and Keller (1987)]

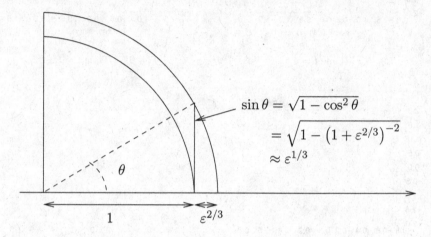

$$\sin\theta = \sqrt{1 - \cos^2\theta}$$
$$= \sqrt{1 - \left(1 + \varepsilon^{2/3}\right)^{-2}}$$
$$\approx \varepsilon^{1/3}$$

Fig. 6.8   Length of a ray inside a typical boundary layer.

therefore suggests:

$$\alpha_c(\text{smooth point in 1D}) = \frac{\sigma + 2}{3}.$$

At the cusp, the wave function grows like $\varepsilon^{-1/4}$, in a region of order $\varepsilon^{1/2}$ (see [Duistermaat (1974); Hunter and Keller (1987)]), hence

$$\alpha_c(\text{cusp in 1D}) = 1 - \frac{2\sigma}{4} - \frac{1}{2} = \frac{\sigma + 1}{2}.$$

We see that we have

$$\alpha_c(\text{cusp in 1D}) > \alpha_c(\text{smooth point in 1D}) \iff \sigma > 1.$$

Therefore, nonlinear effects might be stronger either at the cusp, or at smooth points of the caustic, according to the power of the nonlinearity.

This may seem paradoxical, since the cusp is expected to concentrate more energy than a smooth point on the caustic. However, we invite the reader to consult [Joly *et al.* (2000)] (or [Joly *et al.* (1997b)]): the analytical results (to prove estimates in Lebesgue spaces for oscillatory integrals) differ from the topological results of [Duistermaat (1974)]. In the case studied by J.-L. Joly, G. Métivier and J. Rauch, the critical index for L^p estimates is given by smooth points of the caustic, and the influence of the cusp is negligible. However, the approach of [Joly *et al.* (2000)] does not seem to yield directly estimates which can be used in the case of Schrödinger equations. Even in the apparently simple case considered in this paragraph, the following two questions remain open so far:

- Find α_c such that nonlinear effects are negligible near the caustic for $\alpha > \alpha_c$, but not for $\alpha < \alpha_c$.
- For $\alpha = \alpha_c$, describe the (possible) leading order nonlinear effects at the caustic.

Chapter 7

Focal Point without External Potential

7.1 Presentation

In this chapter, we consider the initial value problem

$$i\varepsilon\partial_t u^\varepsilon + \frac{\varepsilon^2}{2}\Delta u^\varepsilon = \varepsilon^\alpha |u^\varepsilon|^{2\sigma}u^\varepsilon \quad ; \quad u^\varepsilon(0,x) = a_0(x)e^{-i|x|^2/(2\varepsilon)}. \quad (7.1)$$

To simplify the notations and the discussions, we consider only non-negative time: $t \geqslant 0$. Recall that we have derived formally the following distinctions in the previous chapter:

	$\alpha > 1$	$\alpha = 1$
$\alpha > n\sigma$	linear propagation linear caustic	nonlinear propagation linear caustic
$\alpha = n\sigma$	linear propagation, nonlinear caustic	nonlinear propagation nonlinear caustic

Consider the solution to the free Schrödinger equation which coincides with u^ε at some fixed time τ:

$$i\varepsilon\partial_t v^\varepsilon_\tau + \frac{\varepsilon^2}{2}\Delta v^\varepsilon_\tau = 0 \quad ; \quad v^\varepsilon_\tau(\tau,x) = u^\varepsilon(\tau,x). \quad (7.2)$$

We can then give a more precise interpretation of the above table, where the expression $f^\varepsilon - g^\varepsilon = o(h^\varepsilon)$ stands for $\|f^\varepsilon - g^\varepsilon\|_{L^2} = o\left(\|h^\varepsilon\|_{L^2}\right)$ as $\varepsilon \to 0$:

- If $\alpha > \max(1, n\sigma)$, then we expect $u^\varepsilon(t) - v^\varepsilon_0(t) = o(v^\varepsilon_0(t))$ for all t.
- If $\alpha = 1 > n\sigma$, then outside the focal point at $t = 1$, we already know that this is a weakly nonlinear régime: the leading order nonlinear effect consists of a self-phase modulation, and u^ε is not comparable to v^ε_0. On the other hand, the nonlinearity is negligible near the focal point: $u^\varepsilon(t) - v^\varepsilon_1(t) = o(v^\varepsilon_1(t))$ for $|t-1| \leqslant C\varepsilon$. Recall that ε also measures the influence zone of the focal point.

- If $\alpha = n\sigma > 1$, then we expect $u^\varepsilon(t) - v_0^\varepsilon(t) = o(v_0^\varepsilon(t))$ for $t < 1$ and $u^\varepsilon(t) - v_\tau^\varepsilon(t) = o(v_\tau^\varepsilon(t))$ for $t > 1$, for (any) $\tau > 1$ independent of ε. These two relations express the fact that the nonlinearity in Eq. (7.1) is negligible off $t = 1$. On the other hand, it should not be negligible near $t = 1$, and therefore,

$$\liminf_{\varepsilon \to 0} \|v_0^\varepsilon(t) - v_\tau^\varepsilon(t)\|_{L^2} > 0, \forall t \geqslant 0, \ \forall \tau > 1, \quad \text{in general.}$$

  We will see that at leading order, v_0^ε and v_τ^ε differ in terms of a scattering operator associated to the nonlinear Schrödinger equation.

- If $\alpha = 1 = n\sigma$, then we don't expect $u^\varepsilon(t) - v_\tau^\varepsilon(t) = o(v_\tau^\varepsilon(t))$ for any $t \neq \tau$: the solution u^ε never behaves like a free solution.

These four assertions have been justified in [Carles (2000b)], thus proving that the previous table is correct.

For technical reasons, we will always assume that the nonlinearity is H^1-subcritical (see Sec. 1.4), and some lower bounds on σ:

$$0 < \sigma < \infty \text{ if } n = 1; \ \frac{1}{2} < \sigma < \infty \text{ if } n = 2; \ \frac{2}{n+2} < \sigma < \frac{2}{n-2} \text{ if } n \geqslant 3.$$

We also assume that the initial amplitude a_0 is in $\Sigma(1)$, defined in §1.4.2. For the sake of readability, we drop the index 1, and consider therefore:

$$a_0 \in \Sigma := \left\{ f \in H^1(\mathbb{R}^n) \ ; \ x \mapsto \langle x \rangle \, f(x) \in L^2(\mathbb{R}^n) \right\}.$$

The space Σ is equipped with the norm

$$\|f\|_\Sigma = \|f\|_{L^2} + \|\nabla f\|_{L^2} + \|xf\|_{L^2}.$$

Obviously, for fixed ε, $u^\varepsilon(0, \cdot) \in \Sigma$.

There are at least two ways to present the results. First, we may write the asymptotics for the function u^ε itself. Second, we can proceed as in [Joly *et al.* (1996a, 2000)], and write the solution as a modified Lagrangian integral. This approach was followed in [Carles (2000b)]. We will present the results from both points of view here. In our case, the generalization of the representation (6.3) is:

$$u^\varepsilon(t, x) = \frac{1}{(2\pi\varepsilon)^{n/2}} \int_{\mathbb{R}^n} e^{-i\frac{t-1}{2\varepsilon}|\xi|^2 + i\frac{x \cdot \xi}{\varepsilon}} A^\varepsilon(t, \xi) d\xi. \tag{7.3}$$

Note that unlike in the linear case, the Lagrangian symbol A^ε depends on time. It is still given by

$$A^\varepsilon(t, \xi) = \frac{1}{\varepsilon^{n/2}} e^{i\frac{t-1}{2\varepsilon}|\xi|^2} \widehat{u^\varepsilon}\left(t, \frac{\xi}{\varepsilon}\right).$$

We have the following preliminary result:

Lemma 7.1. *Let $n \geqslant 1$ and $a_0 \in \Sigma$. The initial Lagrangian symbol converges in Σ:*
$$A^\varepsilon(0, \xi) = e^{-in\pi/4} a_0(-\xi) + o(1) \quad in \ \Sigma, \ as \ \varepsilon \to 0.$$
Moreover, the Lagrangian symbol satisfies the equation:
$$i\partial_t A^\varepsilon(t, \xi) = \varepsilon^{\alpha - n/2 - 1} e^{i\frac{t-1}{2\varepsilon}|\xi|^2} \mathcal{F}\left(|u^\varepsilon|^{2\sigma} u^\varepsilon\right)\left(t, \frac{\xi}{\varepsilon}\right). \qquad (7.4)$$

Proof.   As in the linear case, we have
$$A^\varepsilon(0, \xi) = \frac{1}{(2\pi\varepsilon)^{n/2}} \int_{\mathbb{R}^n} e^{-i\frac{|x+\xi|^2}{2\varepsilon}} a_0(x) dx$$
$$= \frac{1}{(2\pi\varepsilon)^{n/2}} \int_{\mathbb{R}^n} e^{-i\frac{|x|^2}{2\varepsilon}} a_0(x - \xi) dx.$$
Recalling that
$$\mathcal{F}\left(e^{-i|x|^2/(2\varepsilon)}\right)(\eta) = \left(\frac{\varepsilon}{i}\right)^{n/2} e^{i\varepsilon|\eta|^2/2},$$
$$\mathcal{F}_{x \to \eta}\left(a_0(x - \xi)\right)(\eta) = e^{i\eta \cdot \xi} \widehat{a_0}(\eta),$$
Parseval formula yields:
$$A^\varepsilon(0, \xi) = \frac{1}{(2i\pi)^{n/2}} \int e^{i\eta \cdot \xi} e^{-i\varepsilon|\eta|^2/2} \widehat{a_0}(\eta) d\eta.$$
Then
$$A^\varepsilon(0, \xi) - e^{-in\pi/4} a_0(-\xi) = \frac{1}{(2i\pi)^{n/2}} \int e^{i\eta \cdot \xi} \left(e^{-i\varepsilon|\eta|^2/2} - 1\right) \widehat{a_0}(\eta) d\eta.$$
By Plancherel equality, we infer
$$\left\| A^\varepsilon(0, \cdot) - e^{-in\pi/4} a_0(-\cdot) \right\|_{L^2} = \left\| \left(e^{-i\varepsilon|\cdot|^2/2} - 1\right) \widehat{a_0}(\cdot) \right\|_{L^2}.$$
Now since
$$\left| e^{-i\varepsilon|\eta|^2/2} - 1 \right| = 2 \left| \sin\left(\varepsilon \frac{|\eta|^2}{4}\right) \right|,$$
the Dominated Convergence Theorem yields:
$$\lim_{\varepsilon \to 0} \left\| A^\varepsilon(0, \cdot) - e^{-in\pi/4} a_0(-\cdot) \right\|_{L^2} = 0.$$
Noting that
$$\nabla\left(A^\varepsilon(0, \xi) - e^{-in\pi/4} a_0(-\xi)\right) = \frac{i}{(2i\pi)^{n/2}} \int e^{i\eta \cdot \xi} \left(e^{-i\varepsilon|\eta|^2/2} - 1\right) \eta \widehat{a_0}(\eta) d\eta,$$
$$\xi\left(A^\varepsilon(0, \xi) - e^{-in\pi/4} a_0(-\xi)\right) =$$
$$= \frac{i}{(2i\pi)^{n/2}} \int e^{i\eta \cdot \xi} \partial_\eta \left(\left(e^{-i\varepsilon|\eta|^2/2} - 1\right) \widehat{a_0}(\eta)\right) d\eta,$$
and since $a_0 \in \Sigma$, the estimate of the lemma follows by the same argument. The second part of the lemma is straightforward. $\qquad \square$

For $t \geqslant 0$, we check the following identities:

$$\|A^\varepsilon(t)\|_{L^2} = \|u^\varepsilon(t)\|_{L^2} \quad ; \quad \|\nabla A^\varepsilon(t)\|_{L^2} = \|J^\varepsilon(t)u^\varepsilon(t)\|_{L^2}$$
$$\|\xi A^\varepsilon(t)\|_{L^2} = \|\varepsilon \nabla u^\varepsilon(t)\|_{L^2} , \tag{7.5}$$

where the operator J^ε is given by:

$$J^\varepsilon(t) = \frac{x}{\varepsilon} + i(t-1)\nabla.$$

Note that we have

$$J^\varepsilon(t) = X + iT\nabla_X\big|_{(T,X)=((t-1)/\varepsilon, x/\varepsilon)}.$$

We recover the operator $X + iT\nabla$, introduced in [Ginibre and Velo (1979)], which is classical in the scattering theory for nonlinear Schrödinger equation; see also e.g. [Tsutsumi and Yajima (1984); Hayashi and Tsutsumi (1987); Cazenave and Weissler (1992)].

Let us mention another point of view, from which the introduction of this operator is also very natural. Recall that in Chap. 2 (and also in §4.3), we have performed H^s estimates not directly on u^ε, but on $u^\varepsilon e^{-i\phi_{\mathrm{eik}}/\varepsilon}$. The idea is that the H^s norm of the latter is bounded uniformly in $\varepsilon \in]0,1]$, while the former is not. In the present case, we have

$$\phi_{\mathrm{eik}}(t,x) = \frac{|x|^2}{2(t-1)},$$

so it is natural to consider

$$e^{i\phi_{\mathrm{eik}}(t,x)/\varepsilon} \nabla \left(e^{-i\phi_{\mathrm{eik}}(t,x)/\varepsilon}. \right) = -i\frac{x}{\varepsilon(t-1)} + \nabla.$$

We can be more precise by recalling that the concentration rate for the approximate solution in the linear case is exactly $1-t$, see (6.2). Therefore, up to an irrelevant factor i, we retrieve

$$i(t-1)e^{i\phi_{\mathrm{eik}}(t,x)/\varepsilon} \nabla \left(e^{-i\phi_{\mathrm{eik}}(t,x)/\varepsilon}. \right) = J^\varepsilon(t).$$

This formula has two interesting consequences from a technical point of view:

- The operator acts on gauge invariant nonlinearities like a derivative: if $G(z) = F\left(|z|^2\right) z$ is C^1, then

$$J^\varepsilon(t)G(u) = \partial_z G(u)J^\varepsilon(t)u - \partial_{\bar z}G(u)\overline{J^\varepsilon(t)u}.$$

- Weighted Gagliardo–Nirenberg inequalities are available:

$$\|u\|_{L^r} \leqslant \frac{C_r}{|1-t|^{\delta(r)}} \|u\|_{L^2}^{1-\delta(r)} \|J^\varepsilon(t)u\|_{L^2}^{\delta(r)},$$

$$\text{where } \delta(r) := n\left(\frac{1}{2} - \frac{1}{r}\right) \in [0,1[, \tag{7.6}$$

and C_r depends only on r and n. This is a direct consequence of the standard inequality (without the weight $|1-t|$), where $J^\varepsilon(t)$ is replaced by ∇.

Example 7.2. Apply the operator $J^\varepsilon(t)$ to the approximate solution of the linear case, that is, the right hand side of (6.2). For $t < 1$, we find:

$$J^\varepsilon(t)\left(\frac{1}{(1-t)^{n/2}}a_0\left(\frac{x}{1-t}\right)e^{i\frac{|x|^2}{2\varepsilon(t-1)}}\right) = \frac{i}{(1-t)^{n/2}}\nabla a_0\left(\frac{x}{1-t}\right)e^{i\frac{|x|^2}{2\varepsilon(t-1)}}.$$

When applying the weighted Gagliardo–Nirenberg inequality (7.6) to this function, we see that both the left hand side and the right hand side are of order

$$|1-t|^{-\delta(r)}.$$

This suggests that for $t \neq 1$, in the semi-classical régime, the operator J^ε yields sharp estimates, as far as the parameters t and ε are concerned.

For $t = 1$, the inequality (7.6) becomes singular. Instead, resume the standard inequality, rescaled by ε:

$$\|u\|_{L^r} \leqslant \frac{C_r}{\varepsilon^{\delta(r)}} \|u\|_{L^2}^{1-\delta(r)} \|\varepsilon\nabla u\|_{L^2}^{\delta(r)}.$$

Then again, the power $\varepsilon^{-\delta(r)}$ is such that when applide to

$$\frac{1}{(i\varepsilon)^{n/2}}\widehat{a}_0\left(\frac{x}{\varepsilon}\right),$$

both sides of the above estimates are of order $\varepsilon^{-\delta(r)}$.

The conclusion suggested by this example, and which turns out to be useful in the nonlinear estimates, is that the operator J^ε yields good estimates off $t = 1$, while near $t = 1$, the natural operator is $\varepsilon\nabla$.

Before describing more precisely the results, we mention a third important property of the operator J^ε: it commutes with the linear Schrödinger operator,

$$\left[i\varepsilon\partial_t + \frac{\varepsilon^2}{2}\Delta, J^\varepsilon(t)\right] = 0.$$

This property, classical in the case $\varepsilon = 1$ [Hayashi and Tsutsumi (1987)], stems from the fact that J^ε can be factorized in a different way. Let

$$U^\varepsilon(t) = \exp\left(i\varepsilon\frac{t}{2}\Delta\right)$$

denote the group associated to the free semi-classical Schrödinger equation. We have:

$$J^\varepsilon(t) = U^\varepsilon(t-1)\frac{x}{\varepsilon}U^\varepsilon(1-t).$$

This expression implies the above commutation with the linear Schrödinger operator. We will see in Sec. 8.3 that the existence of such an operator with nice properties both for nonlinear estimates and for linear commutators does not seem to be generic, in the presence of an external potential.

7.2 Linear propagation, linear caustic

In view of Lemma 7.1, denote

$$A_0(\xi) = e^{-in\pi/4}a_0(-\xi).$$

Proposition 7.3. *Assume* $\alpha > \max(1, n\sigma)$. *Then*

$$A^\varepsilon \xrightarrow[\varepsilon\to 0]{} A_0 \quad in \ L^\infty_{\mathrm{loc}}\left(\mathbb{R};\Sigma\right).$$

Equivalently,

$$\|\mathcal{B}^\varepsilon\left(u^\varepsilon - v_0^\varepsilon\right)\|_{L^\infty_{\mathrm{loc}}(\mathbb{R};L^2)} \xrightarrow[\varepsilon\to 0]{} 0, \quad for \ all \ \mathcal{B}^\varepsilon \in \{\mathrm{Id}, \varepsilon\nabla, J^\varepsilon\}.$$

To make the proof more intuitive, we distinguish the special case $n = 1$ from the general case $n \geqslant 1$.

The case $n = 1$. In space dimension $n = 1$, we can use the Sobolev embedding $H^1(\mathbb{R}) \hookrightarrow L^\infty(\mathbb{R})$. More precisely, Gagliardo–Nirenberg inequality shows that there exists C independent of ε and t such that for all $u \in \Sigma$:

$$\|u\|_{L^\infty} \leqslant \frac{C}{(\varepsilon + |t-1|)^{1/2}}\|u\|_{L^2}^{1/2}\left(\|\varepsilon\partial_x u\|_{L^2} + \|J^\varepsilon(t)u\|_{L^2}\right)^{1/2}. \tag{7.7}$$

We also note that (for any $n \geqslant 1$)

$$\begin{aligned}
\|v_0^\varepsilon(t)\|_{L^2} &= \|v_0^\varepsilon(0)\|_{L^2} = \|a_0\|_{L^2}, \\
\|\varepsilon\nabla v_0^\varepsilon(t)\|_{L^2} &= \|\varepsilon\nabla v_0^\varepsilon(0)\|_{L^2} = \mathcal{O}(1), \\
\|J^\varepsilon(t)v_0^\varepsilon(t)\|_{L^2} &= \|J^\varepsilon(0)v_0^\varepsilon(0)\|_{L^2} = \|\nabla a_0\|_{L^2}.
\end{aligned} \tag{7.8}$$

Consider the error term $w^\varepsilon = u^\varepsilon - v_0^\varepsilon$. It solves:

$$i\varepsilon \partial_t w^\varepsilon + \frac{\varepsilon^2}{2} \Delta w^\varepsilon = \varepsilon^\alpha |u^\varepsilon|^{2\sigma} u^\varepsilon \quad ; \quad w_{|t=0}^\varepsilon = 0. \tag{7.9}$$

Lemma 1.2 yields, for $t > 0$:

$$\|w^\varepsilon\|_{L^\infty([0,t];L^2)} \leqslant \varepsilon^{\alpha-1} \int_0^t \left\| |u^\varepsilon(\tau)|^{2\sigma} u^\varepsilon(\tau) \right\|_{L^2} d\tau$$

$$\leqslant \varepsilon^{\alpha-1} \|a_0\|_{L^2} \int_0^t \|u^\varepsilon(\tau)\|_{L^\infty}^{2\sigma} d\tau,$$

where we have used the conservation of the L^2-norm of u^ε. Recalling that $u^\varepsilon = w^\varepsilon + v_0^\varepsilon$, (7.7) and Eq. (7.8) yield:

$$\|u^\varepsilon(\tau)\|_{L^\infty}^{2\sigma} \leqslant C_\sigma \left(\|w^\varepsilon(\tau)\|_{L^\infty}^{2\sigma} + \|v_0^\varepsilon(\tau)\|_{L^\infty}^{2\sigma} \right)$$

$$\leqslant C \left(\|w^\varepsilon(\tau)\|_{L^\infty}^{2\sigma} + \frac{1}{(\varepsilon + |\tau - 1|)^\sigma} \right).$$

Since w^ε is expected to be a relatively small error estimate, it should satisfy at least the same estimates as v_0^ε. From Proposition 1.26, $u^\varepsilon \in C(\mathbb{R}; \Sigma)$, hence $w^\varepsilon \in C(\mathbb{R}; \Sigma)$. Since $w_{|t=0}^\varepsilon = 0$, there exists $t^\varepsilon > 0$ such that

$$\|J^\varepsilon(\tau) w^\varepsilon(\tau)\|_{L^2} \leqslant 1 \tag{7.10}$$

for $\tau \in [0, t^\varepsilon]$. Recall that from the conservation of the energy for u^ε, Eq. (1.25),

$$\frac{d}{dt} \left(\frac{1}{2} \|\varepsilon \nabla u^\varepsilon(t)\|_{L^2}^2 + \frac{\varepsilon^\alpha}{\sigma+1} \|u^\varepsilon(t)\|_{L^{2\sigma+2}}^{2\sigma+2} \right) = 0.$$

Therefore, there exists C independent of ε such that

$$\|\varepsilon \nabla u^\varepsilon(t)\|_{L^2} \leqslant C, \quad \forall t \in \mathbb{R}.$$

So, there exists C' independent of ε, such that

$$\|\varepsilon \nabla w^\varepsilon(t)\|_{L^2} \leqslant C', \quad \forall t \in \mathbb{R}.$$

In view of (7.7), we infer, so long as (7.10) holds,

$$\|w^\varepsilon(\tau)\|_{L^\infty} \leqslant \frac{C}{(\varepsilon + |\tau - 1|)^{1/2}},$$

for some constant C independent of ε. We infer, so long as (7.10) holds:

$$\|w^\varepsilon\|_{L^\infty([0,t];L^2)} \leqslant C\varepsilon^{\alpha-1} \int_0^t \frac{d\tau}{(\varepsilon + |\tau - 1|)^\sigma}.$$

Fix $T > 1$. Distinguishing the regions $\{|\tau - 1| \geqslant \varepsilon\}$ and $\{|\tau - 1| > \varepsilon\}$, for $t \leqslant T$, the latest integral is controlled by:

$$\int_0^t \frac{d\tau}{(\varepsilon + |\tau - 1|)^\sigma} \leqslant \int_0^{1-\varepsilon} \frac{d\tau}{(1-\tau)^\sigma} + \int_{1-\varepsilon}^{1+\varepsilon} \frac{d\tau}{\varepsilon^\sigma} + \int_{1-\varepsilon}^T \frac{d\tau}{(\tau - 1)^\sigma}$$

$$\leqslant C\left(\max\left(\varepsilon^{1-\sigma}, \log\frac{1}{\varepsilon}, 1\right) + \varepsilon^{1-\sigma}\right),$$

where we have distinguished the three cases, $\sigma > 1$, $\sigma = 1$ and $0 < \sigma < 1$. Therefore, if (7.10) holds on $[0, T]$, we infer:

$$\|w^\varepsilon\|_{L^\infty([0,t];L^2)} \leqslant C\max\left(\varepsilon^{\alpha-\sigma}, \varepsilon^{\alpha-1}\log\frac{1}{\varepsilon}\right). \tag{7.11}$$

The strategy is to obtain similar estimates for $\varepsilon\nabla w^\varepsilon$ and $J^\varepsilon w^\varepsilon$. Applying the operator $\varepsilon\nabla$ to Eq. (7.9), we find:

$$\left(i\varepsilon\partial_t + \frac{\varepsilon^2}{2}\Delta\right)\varepsilon\nabla w^\varepsilon = \varepsilon^{1+\alpha}\nabla\left(|u^\varepsilon|^{2\sigma}u^\varepsilon\right) = (\sigma+1)\varepsilon^\alpha |u^\varepsilon|^{2\sigma}\varepsilon\nabla u^\varepsilon$$

$$+ \sigma\varepsilon^\alpha (u^\varepsilon)^{\sigma+1}(\overline{u}^\varepsilon)^{\sigma-1}\varepsilon\nabla\overline{u}^\varepsilon,$$

along with the Cauchy data $\varepsilon\nabla w^\varepsilon_{|t=0} = 0$. From the conservation of the energy for u^ε, we can mimic the previous computations, and find, so long as (7.10) holds:

$$\|\varepsilon\nabla w^\varepsilon\|_{L^\infty([0,t];L^2)} \leqslant C\max\left(\varepsilon^{\alpha-\sigma}, \varepsilon^{\alpha-1}\log\frac{1}{\varepsilon}\right). \tag{7.12}$$

To complete the argument, apply the operator J^ε to Eq. (7.9). We have seen that J^ε behaves like the gradient: it commute with the linear Schrödinger operator, and acts on gauge invariant nonlinearities like a derivatives. Therefore, so long as (7.10) holds:

$$\|J^\varepsilon w^\varepsilon\|_{L^\infty([0,t];L^2)} \leqslant C\max\left(\varepsilon^{\alpha-\sigma}, \varepsilon^{\alpha-1}\log\frac{1}{\varepsilon}\right)\|J^\varepsilon u^\varepsilon\|_{L^\infty([0,t];L^2)}.$$

Since $u^\varepsilon = w^\varepsilon + v_0^\varepsilon$, we have:

$$\|J^\varepsilon u^\varepsilon\|_{L^\infty([0,t];L^2)} \leqslant \|J^\varepsilon w^\varepsilon\|_{L^\infty([0,t];L^2)} + \|J^\varepsilon v^\varepsilon\|_{L^\infty([0,t];L^2)} \leqslant 1 + \|\nabla a_0\|_{L^2},$$

so long as (7.10) holds. We infer:

$$\|J^\varepsilon w^\varepsilon\|_{L^\infty([0,t];L^2)} \leqslant C\max\left(\varepsilon^{\alpha-\sigma}, \varepsilon^{\alpha-1}\log\frac{1}{\varepsilon}\right). \tag{7.13}$$

Therefore, for every $T > 1$, there exists $\varepsilon(T) > 0$ such that (7.10) holds on $[0, T]$ for $0 < \varepsilon \leqslant \varepsilon(T)$. The proposition in the case $n = 1$ then follows from (7.11), (7.12) and (7.13).

The case $n \geqslant 2$. First, notice that since

$$\sigma > \frac{2}{n+2} \quad (\text{even for } n = 2),$$

we always have $n\sigma > 1$, so $\max(1, n\sigma) = n\sigma$.

Since we work at the level of Σ regularity, we cannot expect L^∞ estimates when $n \geqslant 2$. To overcome this issue, we do not use the mere energy estimate provided by Lemma 1.2, but rather Strichartz estimates, which we now recall.

Definition 7.4. A pair (q, r) is **admissible** if $2 \leqslant r < \frac{2n}{n-2}$ ($2 \leqslant r \leqslant \infty$ if $n = 1$, $2 \leqslant r < \infty$ if $n = 2$) and

$$\frac{2}{q} = \delta(r) := n\left(\frac{1}{2} - \frac{1}{r}\right).$$

Notation 7.5. For $f^\varepsilon = f^\varepsilon(t, x)$ and $t > 0$, we write

$$\|f^\varepsilon\|_{L^q_t(L^r)} := \|f^\varepsilon\|_{L^q(0,t;L^r(\mathbb{R}^n))} = \left(\int_0^t \left(\int_{\mathbb{R}^n} |f^\varepsilon(\tau, x)|^r \, dx\right)^{q/r} d\tau\right)^{1/q},$$

with the usual modification when q or r is infinite.

Strichartz estimates are classically given with $\varepsilon = 1$ (see [Ginibre and Velo (1985b); Kato (1987); Yajima (1987); Ginibre and Velo (1992); Keel and Tao (1998)]). Using the scaling

$$u^\varepsilon(t, x) = \frac{1}{\varepsilon^{n/2}} \psi^\varepsilon \left(\frac{t}{\varepsilon}, \frac{x}{\varepsilon}\right),$$

we get the following lemma.

Lemma 7.6 (Strichartz estimates). *Denote $U_0^\varepsilon(t) = e^{i\varepsilon\frac{t}{2}\Delta}$.*
(1) Homogeneous Strichartz estimate. *For any admissible pair (q, r), there exists C_q independent of ε such that*

$$\varepsilon^{1/q} \|U_0^\varepsilon \varphi\|_{L^q(\mathbb{R};L^r)} \leqslant C_q \|\varphi\|_{L^2}, \quad \forall \varphi \in L^2(\mathbb{R}^n).$$

(2) Inhomogeneous Strichartz estimate. *For a time interval I, denote*

$$D_I^\varepsilon(F)(t, x) = \int_{I \cap \{\tau \leqslant t\}} U_0^\varepsilon(t - \tau) F(\tau, x) d\tau.$$

For all admissible pairs (q_1, r_1) and (q_2, r_2), and any interval I, there exists $C = C_{r_1, r_2}$ independent of ε and I such that

$$\varepsilon^{1/q_1 + 1/q_2} \|D_I^\varepsilon(F)\|_{L^{q_1}(I;L^{r_1})} \leqslant C \|F\|_{L^{q_2'}\left(I;L^{r_2'}\right)}, \tag{7.14}$$

for all $F \in L^{q_2'}(I; L^{r_2'})$.

The proof of Proposition 7.3 highly relies on the technical Proposition 7.8 below, which can be understood as an adaptation of the Gronwall lemma. Before stating and proving it, we need some preliminaries:

Lemma 7.7. *Let $n \geqslant 2$, and assume $\frac{2}{n+2} < \sigma < \frac{2}{n-2}$. There exists $\underline{q}$, $\underline{r}$, $\underline{s}$ and $\underline{k}$ satisfying*

$$\frac{1}{\underline{r}'} = \frac{1}{\underline{r}} + \frac{2\sigma}{\underline{s}} \quad ; \quad \frac{1}{\underline{q}'} = \frac{1}{\underline{q}} + \frac{2\sigma}{\underline{k}}, \tag{7.15}$$

and the additional conditions:

- *The pair $(\underline{q}, \underline{r})$ is admissible,*
- *$0 < \frac{1}{\underline{k}} < \delta(\underline{s}) < 1$.*

Proof. With $\delta(\underline{s}) = 1$, the first part of Eq. (7.15) becomes

$$\delta(\underline{r}) = \sigma \left(\frac{n}{2} - 1 \right),$$

and this expression is less than 1 for $\sigma < \frac{2}{n-2}$. Still with $\delta(\underline{s}) = 1$, the second part of Eq. (7.15) yields

$$\frac{2}{\underline{k}} = 1 - \frac{n}{2} + \frac{1}{\sigma},$$

which lies in $]0, 2[$ for $\frac{2}{n+2} < \sigma < \frac{2}{n-2}$. By continuity, these conditions are still satisfied for $\delta(\underline{s})$ close to 1 and $\delta(\underline{s}) < 1$. $\square$

Consider again $w^\varepsilon = u^\varepsilon - v_0^\varepsilon$. It solves Eq. (7.9). We now prove a general estimate for the integral equation,

$$\begin{aligned} w^\varepsilon(t) = U_0^\varepsilon(t - t_0) w_0^\varepsilon &- i\varepsilon^{\alpha-1} \int_{t_0}^t U_0^\varepsilon(t - \tau) F^\varepsilon(w^\varepsilon(\tau)) d\tau \\ &- i\varepsilon^{-1} \int_{t_0}^t U_0^\varepsilon(t - \tau) h^\varepsilon(\tau) d\tau. \end{aligned} \tag{7.16}$$

Writing

$$|u^\varepsilon|^{2\sigma} u^\varepsilon = |u^\varepsilon|^{2\sigma} w^\varepsilon + |u^\varepsilon|^{2\sigma} v_0^\varepsilon,$$

the goal is to consider Eq. (7.16) with

$$F^\varepsilon(w^\varepsilon) = |u^\varepsilon|^{2\sigma} w^\varepsilon \quad ; \quad h^\varepsilon = \varepsilon^\alpha |u^\varepsilon|^{2\sigma} v_0^\varepsilon.$$

Proposition 7.8. *Let $t_1 > t_0$, with $|t_1 - t_0| \leqslant 2$. Assume that there exists a constant C independent of t and ε such that for $t_0 \leqslant t \leqslant t_1$,*

$$\|F^\varepsilon(w^\varepsilon)(t)\|_{L^{\underline{r}'}} \leqslant \frac{C}{(\varepsilon + |1 - t|)^{2\sigma\delta(\underline{s})}} \|w^\varepsilon(t)\|_{L^{\underline{r}}}, \tag{7.17}$$

and define

$$D^\varepsilon(t_0, t_1) := \left(\int_{t_0}^{t_1} \frac{dt}{(\varepsilon + |t - 1|)^{\underline{k}\delta(\underline{s})}} \right)^{2\sigma/\underline{k}}.$$

Then there exist C^* independent of ε, t_0 and t_1, such that for any admissible pair $(\underline{q}, \underline{r})$,

$$\|w^\varepsilon\|_{L^{\underline{q}}(t_0, t_1; L^{\underline{r}})} \leqslant C^* \varepsilon^{-1/\underline{q}} \|w_0^\varepsilon\|_{L^2} + C_{\underline{q}, q} \varepsilon^{-1 - 1/\underline{q} - 1/q} \|h^\varepsilon\|_{L^{q'}(t_0, t_1; L^{r'})}$$
$$+ C^* \varepsilon^{\alpha - n\sigma + 2\sigma(\delta(\underline{s}) - 1/\underline{k})} D^\varepsilon(t_0, t_1) \|w^\varepsilon\|_{L^{\underline{q}}(t_0, t_1; L^{\underline{r}})}.$$

Proof. Apply Strichartz inequalities to Eq. (7.16) with $q_1 = \underline{q}$, $r_1 = \underline{r}$, and $q_2 = q$, $r_2 = r$ for the term with $F^\varepsilon(w^\varepsilon)$, $q_2 = q$, $r_2 = r$ for the term with h^ε, it yields

$$\|w^\varepsilon\|_{L^{\underline{q}}(t_0, t_1; L^{\underline{r}})} \leqslant C \varepsilon^{-1/\underline{q}} \|w_0^\varepsilon\|_{L^2} + C_{\underline{q}, q} \varepsilon^{-1 - 1/\underline{q} - 1/q} \|h^\varepsilon\|_{L^{q'}(t_0, t_1; L^{r'})}$$
$$+ C \varepsilon^{\alpha - 1 - 2/\underline{q}} \|F^\varepsilon(w^\varepsilon)\|_{L^{\underline{q}'}(t_0, t_1; L^{\underline{r}'})}.$$

Then estimate the space norm of the last term by (7.17) and apply Hölder inequality in time, thanks to Eq. (7.15). Using Eq. (7.15) and the fact that $(\underline{q}, \underline{r})$ is admissible, we compute:

$$-1 - \frac{2}{\underline{q}} = -\frac{2}{\underline{q}} - 1 + \frac{2\sigma}{\underline{k}} - \frac{2\sigma}{\underline{k}} = -\frac{4}{\underline{q}} - \frac{2\sigma}{\underline{k}}$$

$$= -2n \left(\frac{1}{2} - \frac{1}{\underline{r}} \right) - \frac{2\sigma}{\underline{k}}$$

$$= -n + n \left(1 - \frac{2\sigma}{\underline{s}} \right) - \frac{2\sigma}{\underline{k}}$$

$$= 2\sigma n \left(\frac{1}{2} - \frac{1}{\underline{s}} \right) - n\sigma - \frac{2\sigma}{\underline{k}}$$

$$= -n\sigma + 2\sigma\delta(\underline{s}) - \frac{2\sigma}{\underline{k}}.$$

The result follows. $\qquad\square$

We will rather use the following corollary:

Corollary 7.9. *Suppose the assumptions of Proposition 7.8 are satisfied. Assume moreover that $\alpha \geqslant n\sigma$ and $C^* \varepsilon^{2\sigma(\delta(\underline{s}) - 1/\underline{k})} D^\varepsilon(t_0, t_1) \leqslant 1/2$. Since $\underline{k}\delta(\underline{s}) > 1$, this holds in either of the two cases:*

- *$0 \leqslant t_0 \leqslant t_1 \leqslant 1 - \Lambda\varepsilon$ or $1 + \Lambda\varepsilon \leqslant t_0 \leqslant t_1$, with $\Lambda \geqslant \Lambda_0$ sufficiently large, or*

- $t_0, t_1 \in [1 - \Lambda\varepsilon, 1 + \Lambda\varepsilon]$, with $|t_1 - t_0|/\varepsilon \leqslant \eta$ *sufficiently small.*

Note that the parameters Λ_0 *and* η *are independent of* ε. *Then for all admissible pair* (q, r), *we have:*

$$\|w^\varepsilon\|_{L^\infty(t_0,t_1;L^2)} \leqslant C\|w_0^\varepsilon\|_{L^2} + C_{q,q}\varepsilon^{-1-1/q}\|h^\varepsilon\|_{L^{q'}(t_0,t_1;L^{r'})}. \qquad (7.18)$$

Proof. The additional assumption implies that the last term in the estimate of Proposition 7.8 can be "absorbed" by the left-hand side, up to doubling the constants,

$$\|w^\varepsilon\|_{L^q(t_0,t_1;L^r)} \leqslant C\varepsilon^{-1/q}\|w_0^\varepsilon\|_{L^2} + C\varepsilon^{-1-1/q-1/q}\|h^\varepsilon\|_{L^{q'}(t_0,t_1;L^{r'})}. \quad (7.19)$$

Now apply Strichartz inequalities to Eq. (7.16) again, but with $q_1 = \infty$, $r_1 = 2$, and $q_2 = \underline{q}$, $r_2 = \underline{r}$ for the term with $F^\varepsilon(w^\varepsilon)$, $q_2 = q$, $r_2 = r$ for the term with h^ε. It yields

$$\begin{aligned}
\|w^\varepsilon\|_{L^\infty(t_0,t_1;L^2)} \leqslant &\, C\|w_0^\varepsilon\|_{L^2} + C\varepsilon^{-1-1/q}\|h^\varepsilon\|_{L^{q'}(t_0,t_1;L^{r'})} \\
&+ C\varepsilon^{\alpha-1-1/\underline{q}}\|F^\varepsilon(w^\varepsilon)\|_{L^{\underline{q}'}(t_0,t_1;L^{\underline{r}'})}.
\end{aligned}$$

Like before,

$$\begin{aligned}
\varepsilon^{n\sigma-1}\|F^\varepsilon(w^\varepsilon)\|_{L^{\underline{q}'}(t_0,t_1;L^{\underline{r}'})} &\lesssim \varepsilon^{2/\underline{q}+2\sigma(\delta(\underline{s})-1/\underline{k})}D^\varepsilon(t_0,t_1)\|w^\varepsilon\|_{L^{\underline{q}}(t_0,t_1;L^{\underline{r}})} \\
&\lesssim \varepsilon^{2/\underline{q}}\|w^\varepsilon\|_{L^{\underline{q}}(t_0,t_1;L^{\underline{r}})},
\end{aligned}$$

and the corollary follows from (7.19), since $\alpha \geqslant n\sigma$. $\square$

We now essentially proceed like in the one-dimensional case. Inequality (7.7) is now replaced by (see (7.6))

$$\|u\|_{L^p} \leqslant \frac{C}{(\varepsilon + |t - 1|)^{\delta(p)}}\|u\|_{L^2}^{1-\delta(p)}\left(\|\varepsilon\nabla u\|_{L^2} + \|J^\varepsilon(t)u\|_{L^2}\right)^{\delta(p)}, \quad (7.20)$$

for all $p \in [2, 2/(n-2)[$. Note that we still have the *a priori* estimates:

$$\begin{aligned}
&\|u^\varepsilon(t)\|_{L^2} = \|v_0^\varepsilon(t)\|_{L^2} = \|a_0\|_{L^2}, \\
&\|\varepsilon\nabla u^\varepsilon(t)\|_{L^2} + \|\varepsilon\nabla v_0^\varepsilon(t)\|_{L^2} = \mathcal{O}(1), \\
&\|J^\varepsilon(t)v_0^\varepsilon(t)\|_{L^2} = \|J^\varepsilon(0)v_0^\varepsilon(0)\|_{L^2} = \|\nabla a_0\|_{L^2}.
\end{aligned}$$

Since $w^\varepsilon \in C(\mathbb{R}; \Sigma)$, there exists $t^\varepsilon > 0$ such that

$$\|J^\varepsilon(\tau)w^\varepsilon(\tau)\|_{L^2} \leqslant 1 \qquad (7.21)$$

for $\tau \in [0, t^\varepsilon]$. So long as (7.21) holds, (7.20) yields, since $\underline{s} \in [2, 2/(n-2)[$,

$$\|u^\varepsilon(\tau)\|_{L^{\underline{s}}} \leqslant \frac{C}{(\varepsilon + |\tau - 1|)^{\delta(\underline{s})}}. \qquad (7.22)$$

Let $T > 1$. Split the time interval

$$[1 - \Lambda_0 \varepsilon, 1 + \Lambda_0 \varepsilon]$$

provided by Corollary 7.9 into $\approx 2\Lambda_0/\eta$ intervals of length $\leqslant \eta$. Applying Corollary 7.9 $\approx 2 + 2\Lambda_0/\eta$ times yields, for $t \leqslant T$ and so long as Eq. (7.21) holds:

$$\|w^\varepsilon\|_{L^\infty(0,t;L^2)} \lesssim \varepsilon^{\alpha - 1 - 1/q} \left\| |u^\varepsilon|^{2\sigma} v_0^\varepsilon \right\|_{L^{q'}(0,t;L^{r'})},$$

for all admissible pair (q, r). Take $(q, r) = (\underline{q}, \underline{r})$. Hölder's inequality yields, in view of Lemma 7.7:

$$\left\| |u^\varepsilon|^{2\sigma} v_0^\varepsilon \right\|_{L^{\underline{q}'}(0,t;L^{\underline{r}'})} \leqslant \|u^\varepsilon\|_{L^{\underline{k}}(0,t;L^{\underline{s}})}^{2\sigma} \|v_0^\varepsilon\|_{L^{\underline{q}}(0,t;L^{\underline{r}})}$$

$$\leqslant C \left(\int_0^T \frac{d\tau}{(\varepsilon + |\tau - 1|)^{\underline{k}\delta(\underline{s})}} \right)^{2\sigma/\underline{k}} \varepsilon^{-1/\underline{q}} \|a_0\|_{L^2},$$

where we have used (7.22) and the homogeneous Strichartz estimate for v_0^ε. Distinguishing the regions $\{|\tau - 1| \geqslant \varepsilon\}$ and $\{|\tau - 1| < \varepsilon\}$, we infer, since $\underline{k}\delta(\underline{s}) > 1$,

$$\|w^\varepsilon\|_{L^\infty(0,t;L^2)} \leqslant C \varepsilon^{\alpha - 1 - 2/\underline{q} - 2\sigma\delta(\underline{s}) + 2\sigma/\underline{k}}$$

Again, notice that

$$-1 - 2/\underline{q} - 2\sigma\delta(\underline{s}) + 2\sigma/\underline{k} = -n\sigma.$$

Therefore, so long as (7.21) holds:

$$\|w^\varepsilon\|_{L^\infty(0,t;L^2)} \leqslant C \varepsilon^{\alpha - n\sigma}. \tag{7.23}$$

For $\mathcal{B}^\varepsilon \in \{\varepsilon\nabla, J^\varepsilon\}$, apply $\mathcal{B}^\varepsilon$ to Eq. (7.9):

$$\left(i\varepsilon\partial_t + \frac{\varepsilon^2}{2}\Delta \right) \mathcal{B}^\varepsilon w^\varepsilon = \varepsilon^\alpha \mathcal{B}^\varepsilon \left(|u^\varepsilon|^{2\sigma} u^\varepsilon \right) \ ; \ \mathcal{B}^\varepsilon w_{|t=0}^\varepsilon = 0.$$

Since $\mathcal{B}^\varepsilon \left(|u^\varepsilon|^{2\sigma} u^\varepsilon \right)$ is a linear combination of terms of the form

$$(u^\varepsilon)^j \, (\overline{u}^\varepsilon)^{2\sigma - j} \, \mathcal{B}^\varepsilon u^\varepsilon \text{ and } (u^\varepsilon)^\ell \, (\overline{u}^\varepsilon)^{2\sigma - \ell} \, \overline{\mathcal{B}^\varepsilon u^\varepsilon},$$

we mimic the above approach. Write $\mathcal{B}^\varepsilon u^\varepsilon = \mathcal{B}^\varepsilon w^\varepsilon + \mathcal{B}^\varepsilon v_0^\varepsilon$. The function F^ε is now chosen in order to contain all the terms of the form

$$(u^\varepsilon)^j \, (\overline{u}^\varepsilon)^{2\sigma - j} \, \mathcal{B}^\varepsilon w^\varepsilon \text{ and } (u^\varepsilon)^\ell \, (\overline{u}^\varepsilon)^{2\sigma - \ell} \, \overline{\mathcal{B}^\varepsilon w^\varepsilon},$$

and h^ε contains all the terms of the form

$$\varepsilon^\alpha \, (u^\varepsilon)^j \, (\overline{u}^\varepsilon)^{2\sigma - j} \, \mathcal{B}^\varepsilon v_0^\varepsilon \text{ and } \varepsilon^\alpha \, (u^\varepsilon)^\ell \, (\overline{u}^\varepsilon)^{2\sigma - \ell} \, \overline{\mathcal{B}^\varepsilon v_0^\varepsilon}.$$

Since $\mathcal{B}^\varepsilon v_0^\varepsilon$ solves the free semi-classical Schrödinger equation, Strichartz inequalities yield:

$$\|\mathcal{B}^\varepsilon v_0^\varepsilon\|_{L^q(\mathbb{R};L^r)} \leqslant C_q \varepsilon^{-1/q} \|a_0\|_\Sigma,$$

where C_q is independent of ε. We can therefore follow the same lines as above, and conclude: so long as (7.21) holds,

$$\|\mathcal{B}^\varepsilon w^\varepsilon\|_{L^\infty(0,t;L^2)} \leqslant C\varepsilon^{\alpha-n\sigma}. \tag{7.24}$$

Like in the one-dimensional case, a continuity argument shows that for any $T > 1$, there exists $\varepsilon(T) > 0$ such that (7.21) holds on $[0,T]$ for all $\varepsilon \in]0, \varepsilon(T)]$. The proposition follows. Note that we have the more precise error estimate:

$$\|\mathcal{B}^\varepsilon w^\varepsilon\|_{L^\infty(0,T;L^2)} \leqslant C(T)\varepsilon^{\alpha-n\sigma}, \quad \forall \mathcal{B}^\varepsilon \in \{\mathrm{Id}, \varepsilon\nabla, J^\varepsilon\}.$$

7.3 Nonlinear propagation, linear caustic

In this paragraph, we assume $\alpha = 1 > n\sigma$. For technical reasons, we will treat the case $n = 1$ only, and we assume in addition $\sigma \geqslant 1/2$. Essentially, notice that Lemma 7.7 cannot be used when $n \geqslant 2$ and $\sigma < 1/n$. For results in the case $n \geqslant 2$ and $\alpha = 1 > n\sigma$, we invite the reader to consult [Carles (2000b)]. Here, we first explain how to derive suitable approximate solutions in the general case $n \geqslant 1$, and then we justify the asymptotics for $n = 1$.

Outside the focal point, we can use the approximate solution studied in Chap. 2. We first make the expressions given in Sec. 2.3 as explicit as possible. The rays of geometric optics are now given by

$$x(t,y) = (1-t)y.$$

Therefore, when $t < 1$, the inverse mapping is given by:

$$y(t,x) = \frac{x}{1-t}.$$

In the general case of the space dimension n, the Jacobi's determinant is given by

$$J_t(y) = \det \nabla_y x(t,y) = (1-t)^n.$$

We infer that for $t < 1$, the leading order approximate solution constructed in Chap. 2 is given by:

$$u_{\mathrm{app}}^\varepsilon(t,x) = \frac{1}{(1-t)^{n/2}} a_0\left(\frac{x}{1-t}\right) e^{i\frac{|x|^2}{2\varepsilon(t-1)}} e^{iG(t,x)},$$

where G is given by:

$$G(t,x) = - \left| a_0 \left(\frac{x}{1-t} \right) \right|^{2\sigma} \int_0^t \frac{d\tau}{|1-\tau|^{n\sigma}}.$$

Loosely speaking, since no nonlinear effect is expected near the focal point, a natural candidate for an approximate solution past the caustic consists in continuing $u_{\text{app}}^\varepsilon$ for $t > 1$ by taking into account linear effects at a focal point. Since the linear effects at a focal point consist of a phase shift at leading order (Maslov index), define:

$$u_{\text{app}}^\varepsilon(t,x) = \begin{cases} \dfrac{1}{(1-t)^{n/2}} a_0 \left(\dfrac{x}{1-t} \right) e^{i\frac{|x|^2}{2\varepsilon(t-1)}} e^{iG(t,x)} & \text{if } t < 1, \\[4mm] \dfrac{e^{-in\pi/2}}{(t-1)^{n/2}} a_0 \left(\dfrac{x}{1-t} \right) e^{i\frac{|x|^2}{2\varepsilon(t-1)}} e^{iG(t,x)} & \text{if } t > 1. \end{cases}$$

Note that the phase shift G is defined globally in time: the map

$$\tau \mapsto \frac{1}{|1-\tau|^{n\sigma}}$$

is locally integrable, since $n\sigma < 1$.

We now explain how to derive a global in time approximate solution from the Lagrangian integral point of view. Suppose that $A^\varepsilon(t,\xi)$ converges to some function $A(t,\xi)$ as $\varepsilon \to 0$, on some time interval $[0,T]$, $T > 0$. Formally, Eq. (7.3) yields

$$u^\varepsilon(t,x) \approx \frac{1}{(2\pi\varepsilon)^{n/2}} \int_{\mathbb{R}^n} e^{-i\frac{t-1}{2\varepsilon}|\xi|^2 + i\frac{x\cdot\xi}{\varepsilon}} A(t,\xi) d\xi.$$

For $t \neq 1$, we can apply stationary phase formula to the right hand side, and find:

$$\frac{1}{(2\pi\varepsilon)^{n/2}} \int_{\mathbb{R}^n} e^{-i\frac{t-1}{2\varepsilon}|\xi|^2 + i\frac{x\cdot\xi}{\varepsilon}} A(t,\xi) d\xi \approx \frac{e^{in\pi/4 \operatorname{sgn}(1-t)}}{|t-1|^{n/2}} A \left(t, \frac{x}{t-1} \right) e^{i\frac{|x|^2}{2\varepsilon(t-1)}}.$$

We infer

$$|u^\varepsilon|^{2\sigma} u^\varepsilon(t,x) \approx \frac{e^{in\pi/4 \operatorname{sgn}(1-t)}}{|t-1|^{n/2+n\sigma}} |A|^{2\sigma} A \left(t, \frac{x}{t-1} \right) e^{i\frac{|x|^2}{2\varepsilon(t-1)}}.$$

Using stationary phase formula again, we obtain

$$\mathcal{F} \left(|u^\varepsilon|^{2\sigma} u^\varepsilon \right) \left(t, \frac{\xi}{\varepsilon} \right) \approx$$

$$\approx \frac{1}{(2\pi)^{n/2}} \frac{e^{in\pi/4 \operatorname{sgn}(1-t)}}{|t-1|^{n/2+n\sigma}} \int e^{i\frac{|x|^2}{2\varepsilon(t-1)} - i\frac{x\cdot\xi}{\varepsilon}} |A|^{2\sigma} A \left(t, \frac{x}{t-1} \right) dx$$

$$\approx \frac{\varepsilon^{n/2}}{|t-1|^{n\sigma}} |A|^{2\sigma} A(t,\xi) e^{-i\frac{t-1}{2\varepsilon}|\xi|^2}.$$

Since the time evolution of A^ε is given by Eq. (7.4), we expect the limiting equation ($\alpha = 1$):

$$i\partial_t A(t,\xi) = \frac{1}{|t-1|^{n\sigma}} |A(t,\xi)|^{2\sigma} A(t,\xi) \quad ; \quad A_{|t=0} = A_0, \qquad (7.25)$$

where we recall that A_0 is defined by

$$A_0(\xi) = e^{-in\pi/4} a_0(-\xi).$$

Notice that the modulus of A is independent of time (since $\partial_t |A|^2 = 0$), so

$$A(t,\xi) = A_0(\xi) e^{ig(t,\xi)},$$

where

$$g(t,\xi) = -|A_0(\xi)|^{2\sigma} \int_0^t \frac{d\tau}{|\tau-1|^{n\sigma}} = -|a_0(-\xi)|^{2\sigma} \int_0^t \frac{d\tau}{|\tau-1|^{n\sigma}}.$$

Note that up to a scaling in time, we recover the previous function G:

$$g(t,\xi) = G(t,(t-1)\xi).$$

Unlike G, g is defined for all t: the Lagrangian integral unfolds the singularity at $t = 1$.

Proposition 7.10. *Assume $n = 1$, and $\alpha = 1 > \sigma \geqslant 1/2$. Then*

$$A^\varepsilon \xrightarrow[\varepsilon \to 0]{} A_0 e^{ig} \quad \text{in } L^\infty_{\text{loc}}(\mathbb{R}; \Sigma).$$

This implies, for all $\mathcal{B}^\varepsilon \in \{\text{Id}, \varepsilon\partial_x, J^\varepsilon\}$, all $\beta \in [0,1[$ and all $T > 0$,

$$\left\| \mathcal{B}^\varepsilon \left(u^\varepsilon - v^\varepsilon_{1-\varepsilon^\beta} \right) \right\|_{L^\infty(1-\varepsilon^\beta, 1+\varepsilon^\beta; L^2)} \lesssim \varepsilon^{\beta-\sigma}$$

$$\sup_{|t-1| \geqslant \varepsilon^\beta, t \leqslant T} \left\| \mathcal{B}^\varepsilon(t) \left(u^\varepsilon(t) - u^\varepsilon_{\text{app}}(t) \right) \right\|_{L^2} \xrightarrow[\varepsilon \to 0]{} 0.$$

Remark 7.11. The statement of the proposition suggests that more information is available in terms of the Lagrangian symbol than in terms of the wave function directly. This aspect is also present in the analysis of the case "nonlinear propagation, nonlinear caustic", see §7.5. Of course, by construction, the Lagrangian integral unfolds the singularity at the focal point, which makes it possible to have a uniform in time statement. Moreover, from a technical point of view, Lagrangian integrals make it easier to consider non-smooth functions (here, $z \mapsto |z|^\sigma$), since they come along with energy estimates which cost one derivative less than the error estimates outside the focal point presented in Chap. 2. This aspect is also crucial in §7.5.

Proof. First, we verify that the first point implies the second assertion. We use the following lemma.

Lemma 7.12. *Let* $n \geqslant 1$ *and* $f \in C(\mathbb{R}; L^2)$. *Denote*

$$\Lambda^\varepsilon(t, x) = \frac{1}{(2\pi\varepsilon)^{n/2}} \int_{\mathbb{R}^n} e^{-i\frac{t-1}{2\varepsilon}|\xi|^2 + i\frac{x \cdot \xi}{\varepsilon}} f(t, \xi) d\xi,$$

and

$$F^\varepsilon(t, x) = e^{i\frac{|x|^2}{2\varepsilon(t-1)}} \left(\frac{i}{1-t}\right)^{n/2} f\left(t, \frac{x}{t-1}\right).$$

Then there exists $h \in C(\mathbb{R}^2; \mathbb{R}_+)$ *with* $h(t, 0) = 0$ *such that for all* $t \neq 1$,

$$\|\Lambda^\varepsilon(t) - F^\varepsilon(t)\|_{L^2} = h\left(t, \frac{\varepsilon}{1-t}\right).$$

If, in addition, $f \in C(\mathbb{R}; H^2)$, *then a little more can be said about* h:

$$\exists C > 0, \quad h(t, \lambda) \leqslant C\lambda \|f(t, \cdot)\|_{H^2}.$$

Remark 7.13. We do not use the last point of this lemma in this section. It will be used in Sec. 7.5.

Proof. [Proof of Lemma 7.12] From Parseval's formula,

$$\Lambda^\varepsilon(t, x) = e^{i\frac{|x|^2}{2\varepsilon(t-1)}} \left(\frac{i}{1-t}\right)^{n/2} \int_{\mathbb{R}^n} e^{i\varepsilon|\eta|^2/(2(t-1))} e^{ix \cdot \eta/(1-t)} \mathcal{F}_{\xi \to \eta}^{-1} f(t, \eta) d\eta.$$

Define, for $\lambda \in \mathbb{R}$,

$$h(t, \lambda) = \left\| \left(e^{i\frac{\lambda}{2}|\cdot|^2} - 1\right) \mathcal{F}^{-1} f(t, \cdot) \right\|_{L^2}.$$

Since $f \in C(\mathbb{R}; L^2)$, $h \in C(\mathbb{R}^2; \mathbb{R})$. The property $h(t, 0) = 0$ then follows from the Dominated Convergence Theorem.

When $f \in C(\mathbb{R}; H^2)$, we use the general inequality $|e^{i\theta} - 1| \leqslant |\theta|$. □

Introduce

$$\tilde{u}_{\text{app}}^\varepsilon(t, x) = \frac{1}{\sqrt{2\pi\varepsilon}} \int_{\mathbb{R}} e^{-i\frac{t-1}{2\varepsilon}|\xi|^2 + i\frac{x \cdot \xi}{\varepsilon}} A_0(\xi) e^{ig(t, \xi)} d\xi.$$

With $f(t, \xi) = A_0(\xi) e^{ig(t, \xi)}$, we check that $f \in C(\mathbb{R}; \Sigma)$. Lemma 7.12 shows that for any $\beta \in [0, 1[$ and any $T > 0$,

$$\sup_{|t-1| \geqslant \varepsilon^\beta, t \leqslant T} \left\| \tilde{u}_{\text{app}}^\varepsilon(t) - u_{\text{app}}^\varepsilon(t) \right\|_{L^2} \xrightarrow[\varepsilon \to 0]{} 0.$$

We check easily that since $\partial_x f, xf \in C(\mathbb{R}; L^2)$, we have moreover

$$\sup_{|t-1| \geqslant \varepsilon^\beta, t \leqslant T} \left\| \mathcal{B}^\varepsilon(t) \left(\widetilde{u}^\varepsilon_{\text{app}}(t) - u^\varepsilon_{\text{app}}(t) \right) \right\|_{L^2} \xrightarrow[\varepsilon \to 0]{} 0, \quad \forall \mathcal{B}^\varepsilon \in \{ \mathrm{Id}, \varepsilon\partial_x, J^\varepsilon \}.$$

Notice, on the other hand,

$$\left\| A^\varepsilon(t) - A_0 e^{ig(t)} \right\|_\Sigma = \sum_{\mathcal{B}^\varepsilon \in \{\mathrm{Id}, \varepsilon\partial_x, J^\varepsilon\}} \left\| \mathcal{B}^\varepsilon(t) \left(u^\varepsilon(t) - \widetilde{u}^\varepsilon_{\text{app}}(t) \right) \right\|_{L^2}.$$

Therefore, the first part of the proposition implies: for any $\beta \in [0, 1[$ and any $T > 0$,

$$\sup_{|t-1| \geqslant \varepsilon^\beta, t \leqslant T} \sum_{\mathcal{B}^\varepsilon \in \{\mathrm{Id}, \varepsilon\partial_x, J^\varepsilon\}} \left\| \mathcal{B}^\varepsilon(t) \left(u^\varepsilon(t) - \widetilde{u}^\varepsilon_{\text{app}}(t) \right) \right\|_{L^2} \xrightarrow[\varepsilon \to 0]{} 0.$$

For the region the region $\{|t - 1| \leqslant \varepsilon^\beta\}$, denote

$$w^\varepsilon_\beta = u^\varepsilon - v^\varepsilon_{1-\varepsilon^\beta}.$$

By definition of $v^\varepsilon_{1-\varepsilon^\beta}$, it solves:

$$i\varepsilon \partial_t w^\varepsilon_\beta + \frac{\varepsilon^2}{2} \partial_x^2 w^\varepsilon_\beta = \varepsilon |u^\varepsilon|^{2\sigma} u^\varepsilon \quad ; \quad w^\varepsilon_{\beta|t=1-\varepsilon^\beta} = 0.$$

For $\mathcal{B}^\varepsilon \in \{\mathrm{Id}, \varepsilon\partial_x, J^\varepsilon\}$, apply the operator $\mathcal{B}^\varepsilon$ to the above equation, and the energy estimate of Lemma 1.2:

$$\sup_{|t-1| \leqslant \varepsilon^\beta} \left\| \mathcal{B}^\varepsilon(t) w^\varepsilon_\beta(t) \right\|_{L^2} \lesssim \int_{|\tau-1| \leqslant \varepsilon^\beta} \| u^\varepsilon(\tau) \|^{2\sigma}_{L^\infty} \, d\tau \sup_{|t-1| \leqslant \varepsilon^\beta} \left\| \mathcal{B}^\varepsilon(t) u^\varepsilon(t) \right\|_{L^2}.$$

From the conservations of mass and energy for u^ε, the L^2 norms of u^ε and $\varepsilon\partial_x u^\varepsilon$ are bounded independent of ε. Gagliardo-Nirenberg inequality yields

$$\sup_{t \in \mathbb{R}} \| u^\varepsilon(t) \|_{L^\infty} \leqslant C\varepsilon^{-1/2},$$

hence

$$\sup_{|t-1| \leqslant \varepsilon^\beta} \left\| \mathcal{B}^\varepsilon(t) w^\varepsilon_\beta(t) \right\|_{L^2} \lesssim \varepsilon^{\beta-\sigma} \sup_{|t-1| \leqslant \varepsilon^\beta} \left\| \mathcal{B}^\varepsilon(t) u^\varepsilon(t) \right\|_{L^2}.$$

Since

$$\sum_{\mathcal{B}^\varepsilon \in \{\mathrm{Id}, \varepsilon\partial_x, J^\varepsilon\}} \left\| \mathcal{B}^\varepsilon(t) u^\varepsilon(t) \right\|_{L^2} = \| A^\varepsilon(t) \|_\Sigma,$$

the first part of the proposition yields:

$$\sup_{|t-1| \leqslant \varepsilon^\beta} \left\| \mathcal{B}^\varepsilon(t) w^\varepsilon_\beta(t) \right\|_{L^2} \lesssim \varepsilon^{\beta-\sigma}.$$

Therefore, we just have to prove the first assertion of the proposition.

Fix $T > 0$ and $\beta \in]0, 1[$. Again, we distinguish the regions $\{t \leqslant 1 - \varepsilon^\beta\}$, $\{|t - 1| < \varepsilon^\beta\}$ and $1 + \varepsilon^\beta \leqslant t \leqslant T\}$. We check that by construction, $\widetilde{u}^\varepsilon_{\text{app}}$ satisfies

$$i\varepsilon \partial_t \widetilde{u}^\varepsilon_{\text{app}} + \frac{\varepsilon^2}{2} \partial_x^2 \widetilde{u}^\varepsilon_{\text{app}} = \varepsilon \left| \widetilde{u}^\varepsilon_{\text{app}} \right|^{2\sigma} \widetilde{u}^\varepsilon_{\text{app}} - \varepsilon r^\varepsilon,$$

where the error term r^ε is given by:

$$r^\varepsilon(t, x) = \left| \widetilde{u}^\varepsilon_{\text{app}}(t, x) \right|^{2\sigma} \widetilde{u}^\varepsilon_{\text{app}}(t, x)$$
$$- \frac{1}{\sqrt{2\pi\varepsilon}} \int_{\mathbb{R}} e^{-i \frac{t-1}{2\varepsilon} |\xi|^2 + i \frac{x \cdot \xi}{\varepsilon}} A_0(\xi) e^{ig(t,\xi)} \times (-\partial_t g(t, \xi)) \, d\xi$$
$$= \left| \widetilde{u}^\varepsilon_{\text{app}}(t, x) \right|^{2\sigma} \widetilde{u}^\varepsilon_{\text{app}}(t, x)$$
$$- \frac{1}{\sqrt{2\pi\varepsilon}} \frac{1}{|1 - t|^\sigma} \int_{\mathbb{R}} e^{-i \frac{t-1}{2\varepsilon} |\xi|^2 + i \frac{x \cdot \xi}{\varepsilon}} |A_0(\xi)|^{2\sigma} A_0(\xi) e^{ig(t,\xi)} \, d\xi.$$

Note that the two terms involved in the definition of r^ε are of the form F^ε and Λ^ε respectively, as in Lemma 7.12, with

$$f(t, \xi) = |A_0(\xi)|^{2\sigma} A_0(\xi) e^{ig(t,\xi)}.$$

We check that $f \in C(\mathbb{R}; \Sigma)$. Therefore, since $\beta < 1$, Lemma 7.12 and Eq. (7.5) yield

$$\sup_{|t-1| \geqslant \varepsilon^\beta, t \leqslant T} \sum_{\mathcal{B}^\varepsilon \in \{\text{Id}, \varepsilon \partial_x, J^\varepsilon\}} \|\mathcal{B}^\varepsilon(t) r^\varepsilon(t)\|_{L^2} \xrightarrow[\varepsilon \to 0]{} 0. \tag{7.26}$$

Let $w^\varepsilon = u^\varepsilon - \widetilde{u}^\varepsilon_{\text{app}}$. It solves

$$i\varepsilon \partial_t w^\varepsilon + \frac{\varepsilon^2}{2} \partial_x^2 w^\varepsilon = \varepsilon \left(|u^\varepsilon|^{2\sigma} u^\varepsilon - \left| \widetilde{u}^\varepsilon_{\text{app}} \right|^{2\sigma} \widetilde{u}^\varepsilon_{\text{app}} \right) + \varepsilon r^\varepsilon. \tag{7.27}$$

From Lemma 7.1,

$$\sum_{\mathcal{B}^\varepsilon \in \{\text{Id}, \varepsilon \partial_x, J^\varepsilon\}} \|\mathcal{B}^\varepsilon(0) w^\varepsilon(0)\|_{L^2} \xrightarrow[\varepsilon \to 0]{} 0.$$

We then proceed as in Sec. 7.2: there exists $\varepsilon_0 > 0$ such that for every $\varepsilon \in]0, \varepsilon_0]$, there exists $t^\varepsilon > 0$ such that

$$\|J^\varepsilon(\tau) w^\varepsilon(\tau)\|_{L^2} \leqslant 1 \tag{7.28}$$

for $\tau \in [0, t^\varepsilon]$. So long as (7.28) holds, the weighted Gagliardo–Nirenberg inequality (7.7) yields, along with the conservations of mass and energy for u^ε:

$$\|w^\varepsilon(t)\|_{L^\infty} \leqslant \frac{C}{\left(\varepsilon + |t - 1|^{1/2} \right)},$$

for some C independent of ε. Since on the other hand $A_0 e^{ig} \in C(\mathbb{R}; \Sigma)$, Eq. (7.5) shows that there exists C independent of ε such that

$$\sum_{\mathcal{B}^\varepsilon \in \{\mathrm{Id}, \varepsilon\partial_x, J^\varepsilon\}} \left\| \mathcal{B}^\varepsilon(t)\widetilde{u}^\varepsilon_{\mathrm{app}}(t) \right\|_{L^2} \leqslant C, \quad \forall t \in \mathbb{R}.$$

Apply $\mathcal{B}^\varepsilon \in \{\mathrm{Id}, \varepsilon\partial_x, J^\varepsilon\}$ to Eq. (7.27), and write $u^\varepsilon = w^\varepsilon + \widetilde{u}^\varepsilon_{\mathrm{app}}$. The energy estimate of Lemma 1.2 shows that so long as (7.28) holds,

$$\left\| \mathcal{B}^\varepsilon w^\varepsilon \right\|_{L^\infty(0,t;L^2)} \lesssim \int_0^t \left\| |u^\varepsilon(\tau)|^{2\sigma} \, \mathcal{B}^\varepsilon(\tau) w^\varepsilon(\tau) \right\|_{L^2} d\tau$$

$$+ \int_0^t \left\| \left(|u^\varepsilon(\tau)|^{2\sigma} - |\widetilde{u}^\varepsilon_{\mathrm{app}}(\tau)|^{2\sigma} \right) \mathcal{B}^\varepsilon(\tau) \widetilde{u}^\varepsilon_{\mathrm{app}}(\tau) \right\|_{L^2} d\tau$$

$$+ \int_0^t \left\| \mathcal{B}^\varepsilon(\tau) r^\varepsilon(\tau) \right\|_{L^2} d\tau + o(1)$$

$$\lesssim \int_0^t \left\| u^\varepsilon(\tau) \right\|_{L^\infty}^{2\sigma} \left\| \mathcal{B}^\varepsilon(\tau) w^\varepsilon(\tau) \right\|_{L^2} d\tau$$

$$+ \int_0^t \left\| \left(|u^\varepsilon(\tau)|^{2\sigma-1} + |\widetilde{u}^\varepsilon_{\mathrm{app}}(\tau)|^{2\sigma-1} \right) |w^\varepsilon(\tau)| \right\|_{L^\infty} d\tau$$

$$+ \int_0^t \left\| \mathcal{B}^\varepsilon(\tau) r^\varepsilon(\tau) \right\|_{L^2} d\tau + o(1),$$

where we have used the assumption $\sigma \geqslant 1/2$ and the uniform boundedness of $\mathcal{B}^\varepsilon \widetilde{u}^\varepsilon_{\mathrm{app}}$ in L^2. We infer

$$\left\| \mathcal{B}^\varepsilon w^\varepsilon \right\|_{L^\infty(0,t;L^2)} \lesssim \int_0^t \frac{1}{(\varepsilon + |\tau - 1|)^\sigma} \left\| \mathcal{B}^\varepsilon(\tau) w^\varepsilon(\tau) \right\|_{L^2} d\tau$$

$$+ \int_0^t \frac{1}{(\varepsilon + |\tau - 1|)^{\sigma - 1/2}} \left\| w^\varepsilon(\tau) \right\|_{L^\infty} d\tau$$

$$+ \int_0^t \left\| \mathcal{B}^\varepsilon(\tau) r^\varepsilon(\tau) \right\|_{L^2} d\tau + o(1)$$

$$\lesssim \int_0^t \frac{1}{(\varepsilon + |\tau - 1|)^\sigma} \left\| \mathcal{B}^\varepsilon(\tau) w^\varepsilon(\tau) \right\|_{L^2} d\tau$$

$$+ \int_0^t \frac{1}{(\varepsilon + |\tau - 1|)^\sigma} \sum_{\mathcal{K}^\varepsilon \in \{\mathrm{Id}, \varepsilon\partial_x, J^\varepsilon\}} \left\| \mathcal{K}^\varepsilon(\tau) w^\varepsilon(\tau) \right\|_{L^2} d\tau$$

$$+ \int_0^t \left\| \mathcal{B}^\varepsilon(\tau) r^\varepsilon(\tau) \right\|_{L^2} d\tau + o(1).$$

Summing over $\mathcal{B}^\varepsilon \in \{\mathrm{Id}, \varepsilon\partial_x, J^\varepsilon\}$, Gronwall lemma and (7.26) yield, so long as (7.28) holds:

$$\sum_{\mathcal{B}^\varepsilon \in \{\mathrm{Id}, \varepsilon\partial_x, J^\varepsilon\}} \left\| \mathcal{B}^\varepsilon w^\varepsilon \right\|_{L^\infty(0,t;L^2)} \xrightarrow[\varepsilon \to 0]{} 0,$$

provided that $t \leqslant 1 - \varepsilon^{\beta}$ and $\beta < 1$. Therefore, for every $\beta \in [0, 1[$, there exists $\varepsilon(\beta) > 0$ such that (7.28) holds on $[0, 1 - \varepsilon^{\beta}]$ for $\varepsilon \in]0, \varepsilon(\beta)]$.

For the region $\{|t-1| \leqslant \varepsilon^{\beta}\}$, assume moreover $\beta > \sigma$, which is consistent with the previous assumption $\beta \in [0, 1[$, since $\sigma < 1$. We go back to the evolution equation for A^{ε}, Eq. (7.4): since $\alpha = 1$, and in view of Eq. (7.5),

$$\partial_t \|A^{\varepsilon}(t)\|_{\Sigma} \leqslant \|\partial_t A^{\varepsilon}(t)\|_{\Sigma} \leqslant \sum_{\mathcal{B}^{\varepsilon} \in \{\mathrm{Id}, \varepsilon\partial_x, J^{\varepsilon}\}} \left\| \mathcal{B}^{\varepsilon}(t) \left(|u^{\varepsilon}(t)|^{2\sigma} u^{\varepsilon}(t) \right) \right\|_{L^2}$$

$$\lesssim \|u^{\varepsilon}(t)\|_{L^{\infty}}^{2\sigma} \sum_{\mathcal{B}^{\varepsilon}(t) \in \{\mathrm{Id}, \varepsilon\partial_x, J^{\varepsilon}\}} \|\mathcal{B}^{\varepsilon}(t) u^{\varepsilon}(t)\|_{L^2}$$

$$\lesssim \|u^{\varepsilon}(t)\|_{L^{\infty}}^{2\sigma} \|A^{\varepsilon}(t)\|_{\Sigma} \lesssim \varepsilon^{-\sigma} \|A^{\varepsilon}(t)\|_{\Sigma},$$

where we have used the conservations of mass and energy for u^{ε}, and Gagliardo–Nirenberg inequality. Since we have seen that A^{ε} is uniformly bounded in Σ at time $\iota = 1 - \varepsilon^{\beta}$, Gronwall lemma yields

$$\sup_{|t-1| \leqslant \varepsilon^{\beta}} \|A^{\varepsilon}(t)\|_{\Sigma} \lesssim e^{C\varepsilon^{\beta-\sigma}} \lesssim 1,$$

since $\beta > \sigma$. On the other hand, we have

$$\left\| \partial_t \left(A_0 e^{ig(t)} \right) \right\|_{\Sigma} = \left\| A_0 e^{ig(t)} \partial_t g(t) \right\|_{\Sigma}$$

$$= \frac{1}{|t-1|^{\sigma}} \left\| |a_0|^{2\sigma} a_0 \right\|_{\Sigma} \leqslant \frac{C}{|t-1|^{\sigma}},$$

where we have used $\Sigma \hookrightarrow L^{\infty}(\mathbb{R})$. Write

$$\sup_{|t-1| \leqslant \varepsilon^{\beta}} \left\| A^{\varepsilon}(t) - A_0 e^{ig(t)} \right\|_{\Sigma} \leqslant \left\| A^{\varepsilon} \left(1 - \varepsilon^{\beta} \right) - A_0 e^{ig(1-\varepsilon^{\beta})} \right\|_{\Sigma}$$

$$+ \int_{|\tau-1| \leqslant \varepsilon^{\beta}} \|\partial_t A^{\varepsilon}(\tau)\|_{\Sigma} \, d\tau + \int_{|\tau-1| \leqslant \varepsilon^{\beta}} \left\| \partial_t \left(A_0 e^{ig(\tau)} \right) \right\|_{\Sigma} d\tau$$

$$\leqslant o(1) + \int_{|\tau-1| \leqslant \varepsilon^{\beta}} \varepsilon^{-\sigma} d\tau + \int_{|\tau-1| \leqslant \varepsilon^{\beta}} \frac{d\tau}{|\tau-1|^{\sigma}}.$$

Since $t \mapsto |t-1|^{-\sigma}$ is locally integrable, we infer, for all $\beta \in]\sigma, 1[$,

$$\sup_{|t-1| \leqslant \varepsilon^{\beta}} \left\| A^{\varepsilon}(t) - A_0 e^{ig(t)} \right\|_{\Sigma} \xrightarrow[\varepsilon \to 0]{} 0.$$

This allows us to mimic the approach used for $t \leqslant 1 - \varepsilon^{\beta}$, in the region $\{1 + \varepsilon^{\beta} \leqslant t \leqslant T\}$, since in particular,

$$\left\| A^{\varepsilon} \left(1 + \varepsilon^{\beta} \right) - A_0 e^{ig(1+\varepsilon^{\beta})} \right\|_{\Sigma} \xrightarrow[\varepsilon \to 0]{} 0.$$

This concludes the proof of the proposition. $\qquad\square$

7.4   Linear propagation, nonlinear caustic

We now resume the framework of arbitrary space dimension, $n \geqslant 1$, and consider, for $n\sigma > 1$,

$$i\varepsilon \partial_t u^\varepsilon + \frac{\varepsilon^2}{2}\Delta u^\varepsilon = \varepsilon^{n\sigma}\left|u^\varepsilon\right|^{2\sigma} u^\varepsilon \quad ; \quad u^\varepsilon(0,x) = a_0(x)e^{-i|x|^2/(2\varepsilon)}. \quad (7.29)$$

From the discussion in Sec. 6.3, the nonlinear effects are expected to be relevant at leading order in the limit $\varepsilon \to 0$ only near the focal point $(t,x) = (1,0)$. Moreover, in the linear case, the concentration phenomenon occurs at scale ε about the focal point. We blow up the variables at that scale about the focal point:

$$u^\varepsilon(t,x) = \frac{1}{\varepsilon^{n/2}}\psi^\varepsilon\left(\frac{t-1}{\varepsilon}, \frac{x}{\varepsilon}\right).$$

The factor $\varepsilon^{-n/2}$ may be viewed as a normalization in $L^2(\mathbb{R}^n)$: for all t,

$$\|u^\varepsilon(t)\|_{L^2(\mathbb{R}^n)} = \left\|\psi^\varepsilon\left(\frac{t-1}{\varepsilon}\right)\right\|_{L^2(\mathbb{R}^n)}.$$

We first note that ψ^ε satisfies an equation where ε is absent:

$$i\partial_t\psi^\varepsilon + \frac{1}{2}\Delta\psi^\varepsilon = \left|\psi^\varepsilon\right|^{2\sigma}\psi^\varepsilon.$$

However, ψ^ε *does* depend on ε, through its Cauchy data:

$$\psi^\varepsilon\left(\frac{-1}{\varepsilon}, \frac{x}{\varepsilon}\right) = \varepsilon^{n/2}a_0(x)e^{-i|x|^2/(2\varepsilon)},$$

hence

$$\psi^\varepsilon\left(\frac{-1}{\varepsilon}, x\right) = \varepsilon^{n/2}a_0(\varepsilon x)e^{-i\varepsilon|x|^2/2}.$$

Two things must be noticed in the above expression. First, the data are prescribed at a time which is not fixed: it goes to $-\infty$ as $\varepsilon \to 0$. On the other hand, these data become flatter and flatter as $\varepsilon \to 0$, but their L^2 norm is fixed: the wave scatters. These two points of view are reminiscent of scattering theory for dispersive partial differential equations. Before going further into details in the semi-classical analysis of Eq. (7.29), we recall more results on the nonlinear Schrödinger equation.

7.4.1 Elements of scattering theory for the nonlinear Schrödinger equation

The first result we recall concerns the Cauchy problem for nonlinear Schrödinger equations with data prescribed near $t = -\infty$. The proof can be found in [Ginibre (1995, 1997); Ginibre *et al.* (1994)]. We recall that the notion of admissible pair was introduced in Definition 7.4.

Theorem 7.14 (Existence and continuity of wave operators). *Let* $n \geqslant 1$, $t_0 \in [-\infty, 0]$ *and* $\psi_- \in \Sigma$. *Denote* $U_0(t) = e^{i\frac{t}{2}\Delta}$, *and consider the Cauchy problem*

$$i\partial_t \psi + \frac{1}{2}\Delta\psi = |\psi|^{2\sigma}\,\psi \quad ; \quad U_0(-t)\psi(t)\big|_{t=t_0} = \psi_-. \qquad (7.30)$$

If $\frac{2}{n+2} < \sigma < \frac{2}{n-2}$ *($\sigma > 1$ if $n = 1$), then Eq. (7.30) has a unique solution*

$$\psi \in Y := \Big\{\varphi \in C(\mathbb{R};\Sigma) \; ; \; \varphi, \nabla\varphi, (x+it\nabla)\varphi \in L^q(]-\infty, 0]; L^r)$$

$$\text{for all admissible pair } (q, r)\Big\}.$$

The solution ψ *is strongly continuous from* $(t_0, \psi_-) \in [-\infty, 0] \times \Sigma$ *to* Y, *and if we denote* $\widetilde{\psi}(t) = U_0(-t)\psi(t)$, *then* $\widetilde{\psi} \in C([-\infty, 0]; \Sigma)$. *If* $t_0 = -\infty$, *then*

$$\|U_0(-t)\psi(t) - \psi_-\|_\Sigma \xrightarrow[t \to -\infty]{} 0,$$

and the map $W_- : \psi_- \mapsto \psi_{|t=0}$ *is called* wave operator.

Remark 7.15. The above result is *false* as soon as $\sigma \leqslant 1/n$: for instance, if $n = \sigma = 1$ and if $\psi \in C(\mathbb{R}; L^2)$ solves Eq. (7.30) with $t_0 = -\infty$, then necessarily $\psi_- = \psi = 0$. See [Barab (1984); Strauss (1974, 1981)] or [Ginibre (1997)].

The above result shows that it is possible to construct a solution to the nonlinear Schrödinger equation in prescribing an asymptotically free behavior as $t \to -\infty$. This is the first step in the nonlinear scattering theory: proving the existence of wave operators. Now that $\psi \in C(\mathbb{R}; \Sigma)$, the converse question is the following: does ψ behave asymptotically like a solution to the free Schrödinger equation as $t \to +\infty$? One can give a positive answer to this question, up to making an extra assumption on the power σ. See for instance [Cazenave and Weissler (1992); Nakanishi and Ozawa (2002)] or [Cazenave (2003)].

Theorem 7.16 (Asymptotic completeness in Σ). *Let $n \geqslant 1$, $\varphi \in \Sigma$. Consider the Cauchy problem*

$$i\partial_t \psi + \frac{1}{2}\Delta\psi = |\psi|^{2\sigma}\psi \quad ; \quad \psi|_{t=0} = \varphi. \tag{7.31}$$

Assume

$$\sigma \geqslant \sigma_0(n) := \frac{2 - n + \sqrt{n^2 + 12n + 4}}{4n},$$

and in addition, $\sigma < 2/(n-2)$ when $n \geqslant 3$. Then there exists a unique $\psi_+ \in \Sigma$ such that the solution $\psi \in C(\mathbb{R}; \Sigma)$ to Eq. (7.31) satisfies

$$\|U_0(-t)\psi(t) - \psi_+\|_\Sigma \underset{t \to +\infty}{\longrightarrow} 0.$$

Moreover, the map $W_+^{-1} : \varphi \mapsto \psi_+$ is continuous from Σ to itself.

Remark 7.17. We check that $1/n < \sigma_0(n) < 2/n$, and $\sigma_0(n) > 2/(n+2)$ when $n \geqslant 2$.

Recall that the existence of such a solution $\psi \in C(\mathbb{R}; \Sigma)$ to Eq. (7.31) follows from Proposition 1.26.

Definition 7.18. The map $S : \psi_- \mapsto \psi_+$ given by Theorems 7.14 and 7.16 is the (nonlinear) *scattering operator* associated to Eq. (7.30).

The scattering operator can be understood as follows. Since the operator $U_0(t)$ is well-known, one first tries to construct a solution to the nonlinear Schrödinger equation that behaves like $U_0(t)\psi_-$ as $t \to -\infty$ for some prescribed ψ_-. This yields $\varphi = \psi(0)$. Conversely, can we neglect the nonlinearity for $t \to +\infty$ as well? If yes, then $\psi(t)$ behaves like $U_0(t)\psi_+$ for some function ψ_+. See Fig. 7.1. Note that the group $U_0(t)$ is unitary on $H^1(\mathbb{R}^n)$, but not on Σ:

$$x + it\nabla = U_0(t)xU_0(-t) \implies xU_0(t) = U_0(t)(x - it\nabla).$$

This explains why the error estimate is $\|U_0(-t)\psi(t) - \psi_\pm\|_\Sigma$, and not $\|\psi(t) - U_0(t)\psi_\pm\|_\Sigma$.

There is no reason to expect $S\psi_- = \psi_-$. However, besides the existence of the scattering operator S, very few of its properties are known. We can check, however, that at least for small data, it is not trivial; see Sec. 7.4.3.

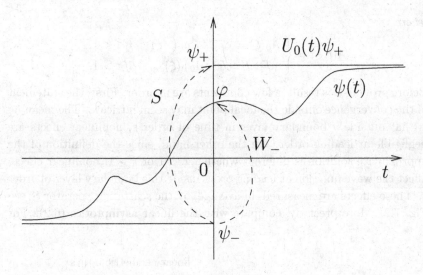

Fig. 7.1   Scattering operator.

7.4.2   *Main result*

Proposition 7.19. *Let $n \geqslant 1$ and $a_0 \in \Sigma$. In addition to the condition $\sigma < 2/(n-2)$ for $n \geqslant 3$, assume that $\sigma \geqslant \sigma_0(n)$, given in Theorem 7.16. Then the following limits hold:*

$$\limsup_{\varepsilon \to 0} \sup_{|t-1| \geqslant \Lambda \varepsilon} \left\| \mathcal{B}^\varepsilon(t) \left(u^\varepsilon(t) - u^\varepsilon_{\mathrm{app}}(t) \right) \right\|_{L^2} \xrightarrow[\Lambda \to +\infty]{} 0, \quad \forall \mathcal{B}^\varepsilon \in \{\mathrm{Id}, \varepsilon\nabla, J^\varepsilon\},$$

where the (discontinuous) function $u^\varepsilon_{\mathrm{app}}$ is given by:

$$u^\varepsilon_{\mathrm{app}}(t,x) = \begin{cases} \dfrac{1}{(1-t)^{n/2}} a_0 \left(\dfrac{x}{1-t} \right) e^{i \frac{|x|^2}{2\varepsilon(t-1)}} & \text{if } t < 1, \\[2mm] \dfrac{e^{-in\pi/4}}{\varepsilon^{n/2}} \left(W_- \circ \mathcal{F} a_0 \right) \left(\dfrac{x}{\varepsilon} \right) & \text{if } t = 1, \\[2mm] \dfrac{e^{-in\pi/2}}{(t-1)^{n/2}} \left(\mathcal{F}^{-1} \circ S \circ \mathcal{F} a_0 \right) \left(\dfrac{x}{1-t} \right) e^{i \frac{|x|^2}{2\varepsilon(t-1)}} & \text{if } t > 1, \end{cases}$$

where W_- and S denote the wave and scattering operators respectively. For $t = 1$, we have

$$\sum_{\mathcal{A}^\varepsilon \in \{\mathrm{Id}, \varepsilon\nabla, x/\varepsilon\}} \left\| \mathcal{A}^\varepsilon \left(u^\varepsilon(1) - u^\varepsilon_{\mathrm{app}}(1) \right) \right\|_{L^2} \xrightarrow[\varepsilon \to 0]{} 0.$$

This implies, in terms of the Lagrangian symbol:

$$\limsup_{\varepsilon \to 0} \sup_{|t-1| \geqslant \Lambda \varepsilon} \left\| A^\varepsilon(t) - \underline{A}(t) \right\|_\Sigma \xrightarrow[\Lambda \to +\infty]{} 0,$$

where

$$\underline{A}(t,\xi) = \begin{cases} A_0(\xi) = e^{-in\pi/4}a_0(-\xi) & \text{if } t < 1, \\ \left(\mathcal{F} \circ S \circ \mathcal{F}^{-1}A_0\right)(\xi) & \text{if } t > 1. \end{cases}$$

Before proving this result, a few comments are in order. First, the statement of the convergence outside the focal point may seem intricate. The meaning is that outside a boundary layer in time of order ε, nonlinear effects are negligible at leading order. On the other hand, since the definition of the approximate solution is different whether $t < 1$ or $t > 1$, nonlinear effects affect the wave function at leading order inside this boundary layer of order ε. These effects are measured, in average, by the scattering operator S; see Fig. 7.2. More precisely, compare with the linear asymptotics (6.2). For

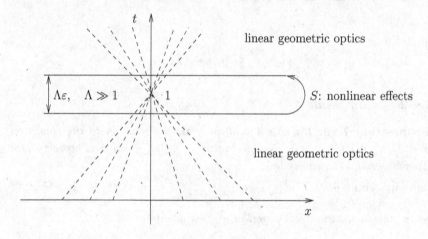

Fig. 7.2   Nonlinear caustic crossing.

$t < 1$, we recover the same asymptotic behavior. For $t = 1$, we still have the phase shift $e^{-in\pi/4}$ (which corresponds to half a Maslov index), but $\mathcal{F}a_0$ is replaced by $W_- \circ \mathcal{F}a_0$: nonlinear effects are not negligible any more. For $t > 1$, we retrieve the linear phenomenon measured by the Maslov index, plus a modification of the amplitude, in terms of S.

Concerning the Lagrangian symbol A^ε, we see that, unless $\mathcal{F}^{-1}A_0$ is a fixed point for S, its limit is discontinuous: this is a typical nonlinear effect, since by construction, the limit of A^ε is continuous in the linear case.

Such a description of caustic crossing in terms of a nonlinear scattering operator first appears in [Bahouri and Gérard (1999)], for the energy-critical wave equation

$$\left(\partial_t^2 - \Delta\right) u + u^5 = 0, \quad (t, x) \in \mathbb{R} \times \mathbb{R}^3. \tag{7.32}$$

For a sequence of initial data $(u^\varepsilon(0), \partial_t u^\varepsilon(0))_\varepsilon$ bounded in the energy space $\dot{H}^1(\mathbb{R}^3) \times L^2(\mathbb{R}^3)$, H. Bahouri and P. Gérard prove that if the weak limit of u^ε is zero, then the only leading order nonlinear effects correspond precisely to the existence of caustics reduced to one point (in general, an infinite number of focal points). In that case, the caustic crossing is described by the scattering operator associated to the nonlinear wave equation. See also [Gallagher and Gérard (2001)] for the case of a wave equation outside a convex obstacle. The proof uses the notion of *profile decomposition*, introduced in [Gérard (1998)]. See Sec. 7.6 for a presentation of this notion in the context of nonlinear Schrödinger equations.

In the case of the semi-linear wave equation (6.5), the case "linear propagation, nonlinear caustic" is studied in [Carles and Rauch (2004a)]. The main result asserts that the caustic crossing is described in terms of a nonlinear scattering operator, and most of the work in [Carles and Rauch (2004a)] is dedicated to constructing this nonlinear scattering operator.

Proof. Resume the change of unknown function introduced in the beginning of this paragraph:

$$u^\varepsilon(t, x) = \frac{1}{\varepsilon^{n/2}} \psi^\varepsilon \left(\frac{t - 1}{\varepsilon}, \frac{x}{\varepsilon}\right).$$

Recall that ψ^ε satisfies

$$i\partial_t \psi^\varepsilon + \frac{1}{2}\Delta\psi^\varepsilon = |\psi^\varepsilon|^{2\sigma} \psi^\varepsilon \; ; \; \psi^\varepsilon\left(\frac{-1}{\varepsilon}, x\right) = \varepsilon^{n/2} a_0(\varepsilon x) e^{-i\varepsilon|x|^2/2}.$$

As suggested by the previous paragraph, we compute

$$U_0\left(\frac{1}{\varepsilon}\right) \psi^\varepsilon\left(\frac{-1}{\varepsilon}, x\right) = \left(\frac{\varepsilon}{2i\pi}\right)^{n/2} \int_{\mathbb{R}^n} e^{i\varepsilon|x-y|^2/2} \varepsilon^{n/2} a_0(\varepsilon y) e^{-i\varepsilon|y|^2/2} dy$$

$$= \frac{e^{-in\pi/4}}{(2\pi)^{n/2}} e^{i\varepsilon|x|^2/2} \int_{\mathbb{R}^n} e^{-ix\cdot y} a_0(y) dy$$

$$= e^{-in\pi/4} e^{i\varepsilon|x|^2/2} \widehat{a_0}(x).$$

By the Dominated Convergence Theorem, we infer:

$$U_0\left(\frac{1}{\varepsilon}\right) \psi^\varepsilon\left(\frac{-1}{\varepsilon}\right) \xrightarrow[\varepsilon \to 0]{} e^{-in\pi/4} \widehat{a_0} \quad \text{in } \Sigma.$$

Define

$$\psi_- := e^{-in\pi/4} \widehat{a_0} \in \Sigma,$$

and introduce the function ψ, solving

$$i\partial_t \psi + \frac{1}{2}\Delta\psi = |\psi|^{2\sigma}\psi \quad ; \quad U_0(-t)\psi(t)\big|_{t=-\infty} = \psi_-.$$

Theorems 7.14 and 7.16 then imply:

$$\sup_{t\in\mathbb{R}} \|U_0(-t)(\psi^\varepsilon(t) - \psi(t))\|_\Sigma \xrightarrow[\varepsilon\to 0]{} 0.$$

Back to the function u^ε, this yields

$$\sum_{\mathcal{B}^\varepsilon\in\{\mathrm{Id},\varepsilon\nabla,J^\varepsilon\}} \sup_{t\in\mathbb{R}} \left\|\mathcal{B}^\varepsilon(t)\left(u^\varepsilon(t,\cdot) - \frac{1}{\varepsilon^{n/2}}\psi\left(\frac{t-1}{\varepsilon},\frac{\cdot}{\varepsilon}\right)\right)\right\|_{L^2} \xrightarrow[\varepsilon\to 0]{} 0.$$

From the scattering for ψ, we infer, for all $\mathcal{B}^\varepsilon \in \{\mathrm{Id}, \varepsilon\nabla, J^\varepsilon\}$:

$$\limsup_{\varepsilon\to 0} \sup_{t\leqslant 1-\Lambda\varepsilon} \left\|\mathcal{B}^\varepsilon(t)\left(u^\varepsilon(t,\cdot) - \frac{1}{\varepsilon^{n/2}}U_0\left(\frac{t-1}{\varepsilon}\right)\psi_-\left(\frac{\cdot}{\varepsilon}\right)\right)\right\|_{L^2} \xrightarrow[\Lambda\to+\infty]{} 0,$$

$$\limsup_{\varepsilon\to 0} \left\|\mathcal{B}^\varepsilon(1)\left(u^\varepsilon(1,\cdot) - \frac{1}{\varepsilon^{n/2}}W_-\psi_-\left(\frac{\cdot}{\varepsilon}\right)\right)\right\|_{L^2} = 0,$$

$$\limsup_{\varepsilon\to 0} \sup_{t\geqslant 1+\Lambda\varepsilon} \left\|\mathcal{B}^\varepsilon(t)\left(u^\varepsilon(t,\cdot) - \frac{1}{\varepsilon^{n/2}}U_0\left(\frac{t-1}{\varepsilon}\right)\psi_+\left(\frac{\cdot}{\varepsilon}\right)\right)\right\|_{L^2} \xrightarrow[\Lambda\to+\infty]{} 0,$$

where $\psi_+ = S\psi_-$. The proposition then stems from the following lemma:

Lemma 7.20. *Let $\varphi \in L^2$. Then the following asymptotics hold in $L^2(\mathbb{R}^n)$, as $t \to \pm\infty$:*

$$U_0(t)\varphi(x) = \frac{e^{-in\pi/4\,\mathrm{sgn}\,t}}{|t|^{n/2}}\widehat{\varphi}\left(\frac{x}{t}\right)e^{i|x|^2/(2t)} + o(1).$$

If in addition $\varphi \in \Sigma$, then for all $\mathcal{A} \in \{\mathrm{Id}, \nabla, x + it\nabla\}$,

$$\left\|\mathcal{A}(t).\left(U_0(t)\varphi - \frac{e^{-in\pi/4\,\mathrm{sgn}\,t}}{|t|^{n/2}}\widehat{\varphi}\left(\frac{\cdot}{t}\right)e^{i|\cdot|^2/(2t)}\right)\right\|_{L^2} \xrightarrow[t\to\pm\infty]{} 0.$$

This lemma can be proved in the same way as Lemma 7.1, by writing

$$U_0(t)\varphi(x) = \frac{1}{(2i\pi t)^{n/2}}\int_{\mathbb{R}^n} e^{i|x-y|^2/(2t)}\varphi(y)dy$$

$$= \frac{1}{(2i\pi t)^{n/2}}e^{i|x|^2/(2t)}\int_{\mathbb{R}^n} e^{-ix\cdot y/t}e^{i|y|^2/(2t)}\varphi(y)dy$$

$$\approx \frac{1}{(2i\pi t)^{n/2}}e^{i|x|^2/(2t)}\int_{\mathbb{R}^n} e^{-ix\cdot y/t}\varphi(y)dy.$$

The proof is omitted.

With this lemma, the proposition follows easily. $\qquad\square$

7.4.3 *On the propagation of Wigner measures*

Since Proposition 7.19 provides a strong convergence of the wave function u^ε, we can infer the expression of the (unique) Wigner measure associated to u^ε. The definition of a Wigner measure was given in Sec. 3.4: it is the limit, up to a subsequence, of the Wigner transform of u^ε,

$$w^\varepsilon(t, x, \xi) = (2\pi)^{-n} \int_{\mathbb{R}^n} u^\varepsilon\left(t, x - \varepsilon\frac{\eta}{2}\right) \overline{u}^\varepsilon\left(t, x + \varepsilon\frac{\eta}{2}\right) e^{i\eta \cdot \xi} d\eta.$$

Proposition 7.19 implies:

$$w^\varepsilon(t, x, \xi) \xrightarrow[\varepsilon \to 0]{} \begin{cases} \mu_-(t, dx, d\xi) & \text{if } t < 1, \\ \mu_+(t, dx, d\xi) & \text{if } t > 1, \end{cases}$$

where

$$\mu_-(t, dx, d\xi) = \frac{1}{(1-t)^n} \left| a_0\left(\frac{x}{1-t}\right) \right|^2 dx \otimes \delta_{\xi = x/(t-1)},$$

$$\mu_+(t, dx, d\xi) = \frac{1}{(t-1)^n} \left| (\mathcal{F}^{-1} \circ S \circ \mathcal{F} a_0)\left(\frac{x}{1-t}\right) \right|^2 dx \otimes \delta_{\xi = x/(t-1)}.$$

In addition,

$$\lim_{t \to 1^-} \mu_-(t, dx, d\xi) = \delta_{x=0} \otimes |a_0(\xi)|^2 d\xi,$$

$$\lim_{t \to 1^+} \mu_+(t, dx, d\xi) = \delta_{x=0} \otimes \left| (\mathcal{F}^{-1} \circ S \circ \mathcal{F} a_0)(\xi) \right|^2 d\xi.$$

We see that, in general, the Wigner measure has a jump at $t = 1$. We can go even further in the analysis, and prove that the propagation of the Wigner measures past the focal point is an ill-posed Cauchy problem: we can find two initial amplitudes a_0 and b_0 such that the corresponding measures μ_- coincide, while the measures μ_+ are different. Consider $b_0(x) = a_0(x)e^{ih(x)}$, where h is a smooth, real-valued, function: the two measures μ_- coincide. We must then compare

$$\left| \mathcal{F}^{-1} \circ S \circ \mathcal{F} a_0 \right|^2 \quad \text{and} \quad \left| \mathcal{F}^{-1} \circ S \circ \mathcal{F}\left(a_0 e^{ih}\right) \right|^2.$$

Note that if h is constant, then the above two functions coincide. The same holds if h is linear in x, since the Fourier transform maps the multiplication by $e^{ix \cdot \xi_0}$ to a translation, and the nonlinear Schrödinger equation is invariant by translation. However, it seems very unlikely that

$$\left| \mathcal{F}^{-1} \circ S \circ \mathcal{F}\left(a_0 e^{ih}\right) \right|^2$$

does not depend on h, for any smooth, real-valued, function h. To prove this vague intuition, we can compute the first two terms of the asymptotic expansion of the scattering operator S near the origin. To simplify the computations, we consider the L^2-critical case $\sigma = 2/n$:

Proposition 7.21. *Let* $n \geqslant 1$, $\sigma = 2/n$, *and* $\psi_- \in L^2(\mathbb{R}^n)$. *Then for* $\delta > 0$ *sufficiently small,* $S(\delta\psi_-)$ *is well defined in* $L^2(\mathbb{R}^n)$, *and, as* $\delta \to 0$:

$$S(\delta\psi_-) = \delta\psi_- - i\delta^{1+4/n} \int_{-\infty}^{+\infty} U_0(-\tau) \left(|U_0(\tau)\psi_-|^{4/n} U_0(\tau)\psi_- \right) d\tau$$
$$+ \mathcal{O}_{L^2}\left(\delta^{1+8/n} \right).$$

Proof. The proof follows the same perturbative analysis as in [Gérard (1996)] (see also [Carles (2001b)] for the nonlinear Schrödinger equation). First, it follows from [Cazenave and Weissler (1989)] that $S(\delta\psi_-)$ is well defined in $L^2(\mathbb{R}^n)$ for $\delta > 0$ sufficiently small. We could also assume that $\psi_- \in \Sigma$, and invoke Theorems 7.14 and 7.16.

Let $\delta \in]0, 1]$, and consider ψ^δ solving:

$$i\partial_t \psi^\delta + \frac{1}{2}\Delta\psi^\delta = |\psi^\delta|^{4/n} \psi^\delta \quad ; \quad U_0(-t)\psi^\delta(t)\big|_{t=-\infty} = \delta\psi_-.$$

Plugging an expansion of the form $\psi^\delta = \delta(\varphi_0 + \delta^{4/n}\varphi_1 + \delta^{4/n}r^\delta)$ into the above equation, and ordering in powers of δ, it is natural to impose the following conditions:

- Leading order: $\mathcal{O}(\delta)$.

$$i\partial_t\varphi_0 + \frac{1}{2}\Delta\varphi_0 = 0 \quad ; \quad U_0(-t)\varphi_0(t)\big|_{t=-\infty} = \psi_-.$$

- First corrector: $\mathcal{O}(\delta^{1+4/n})$.

$$i\partial_t\varphi_1 + \frac{1}{2}\Delta\varphi_1 = |\varphi_0|^{4/n}\varphi_0 \quad ; \quad U_0(-t)\varphi_1(t)\big|_{t=-\infty} = 0.$$

The first equation yields

$$\varphi_0(t) = U_0(t)\psi_-.$$

From the second equation, we have:

$$\varphi_1(t) = -i \int_{-\infty}^{t} U_0(t - \tau) \left(|\varphi_0(\tau)|^{4/n}\varphi_0(\tau) \right) d\tau.$$

Let $\gamma = 2 + 4/n$. Remark that the pair (γ, γ) is admissible (see Definition 7.4), and denote $L_{t,x}^r = L^r(]-\infty, -t] \times \mathbb{R}^n)$. Strichartz estimates ($\varepsilon = 1$ in Lemma 7.6) yield:

$$\|\varphi_0\|_{L^\gamma(\mathbb{R}\times\mathbb{R}^n)} \leqslant C \|\psi_-\|_{L^2},$$

$$\|\varphi_1\|_{L^\gamma(\mathbb{R}\times\mathbb{R}^n)} \leqslant C \left\| |\varphi_0|^{4/n}\varphi_0 \right\|_{L^{\gamma'}(\mathbb{R}\times\mathbb{R}^n)} \leqslant C \|\varphi_0\|_{L^\gamma(\mathbb{R}\times\mathbb{R}^n)}^{1+4/n}$$

$$\leqslant C \|\psi_-\|_{L^2}^{1+4/n},$$

where we have used Hölder's inequality, and the relation $1/\gamma' = (1+4/n)/\gamma$. We also have:

$$i\partial_t r^\delta + \frac{1}{2}\Delta r^\delta = g\left(\varphi_0 + \delta^{4/n}\varphi_1 + \delta^{4/n}r^\delta\right) - g(\varphi_0) \; ; \; U_0(-t)r^\delta(t)\big|_{t=-\infty} = 0,$$

where $g(z) = |z|^{4/n}z$. Strichartz and Hölder inequalities yield

$$\|r^\delta\|_{L_{t,x}^\gamma} \lesssim \left\|\left(|\varphi_0|^{4/n} + \left|\delta^{4/n}\varphi_1\right|^{4/n} + \left|\delta^{4/n}r^\delta\right|^{4/n}\right)\delta^{4/n}\left(|\varphi_1| + |r^\delta|\right)\right\|_{L_{t,x}^{\gamma'}}$$

$$\lesssim \left(\|\varphi_0\|_{L_{t,x}^\gamma}^{4/n} + \left\|\delta^{4/n}\varphi_1\right\|_{L_{t,x}^\gamma}^{4/n} + \left\|\delta^{4/n}r^\delta\right\|_{L_{t,x}^\gamma}^{4/n}\right)\delta^{4/n}\left(\|\varphi_1\|_{L_{t,x}^\gamma} + \|r^\delta\|_{L_{t,x}^\gamma}\right)$$

$$\lesssim \left(1 + \left\|\delta^{4/n}r^\delta\right\|_{L_{t,x}^\gamma}^{4/n}\right)\delta^{4/n}\left(1 + \|r^\delta\|_{L_{t,x}^\gamma}\right)$$

$$\lesssim \delta^{4/n} + \delta^{4/n}\left\|r^\delta\right\|_{L_{t,x}^\gamma} + \left\|\delta^{4/n}r^\delta\right\|_{L_{t,x}^\gamma}^{1+4/n}.$$

The second term on the right hand side is absorbed by the left hand side, provided that δ is sufficiently small (up to doubling the constants). For the last term, we use a standard result:

Lemma 7.22 (Bootstrap argument). *Let $M = M(t)$ be a nonnegative continuous function on $[0,T]$ such that, for every $t \in [0,T]$,*

$$M(t) \leqslant a + bM(t)^\theta,$$

where $a, b > 0$ and $\theta > 1$ are constants such that

$$a < \left(1 - \frac{1}{\theta}\right)\frac{1}{(\theta b)^{1/(\theta-1)}} \; , \quad M(0) \leqslant \frac{1}{(\theta b)^{1/(\theta-1)}} \; .$$

Then, for every $t \in [0,T]$, we have

$$M(t) \leqslant \frac{\theta}{\theta - 1}\, a.$$

This argument shows that for $0 < \delta \ll 1$, $r^\delta \in L^\gamma(\mathbb{R} \times \mathbb{R}^n)$, and

$$\|r^\delta\|_{L^\gamma(\mathbb{R}\times\mathbb{R}^n)} \lesssim \delta^{4/n}.$$

Using Strichartz estimates again, we infer:

$$\|r^\delta\|_{L^\infty(\mathbb{R};L^2(\mathbb{R}^n))} \lesssim \delta^{4/n} + \delta^{4/n}\|r^\delta\|_{L_{t,x}^\gamma} + \left\|\delta^{4/n}r^\delta\right\|_{L_{t,x}^\gamma}^{1+4/n} \lesssim \delta^{4/n}.$$

Therefore,

$$U_0(-t)\psi^\delta(t) = \delta U_0(-t)\varphi_0(t) + \delta^{1+4/n}U_0(-t)\varphi_1(t) + \delta^{1+4/n}U_0(-t)r^\delta(t)$$

$$= \delta\psi_- - i\delta^{1+4/n}\int_{-\infty}^t U_0(-\tau)\left(|\varphi_0(\tau)|^{4/n}\varphi_0(\tau)\right)d\tau$$

$$+ \mathcal{O}_{L^2}\left(\delta^{1+8/n}\right),$$

where the last term is uniform with respect to $t \in \mathbb{R}$. Letting $t \to +\infty$, the result follows. $\square$

We proceed as in [Carles (2001b)]. Denote

$$P(\psi_-) = -i \int_{-\infty}^{+\infty} U_0(-\tau) \left(|U_0(\tau)\psi_-|^{4/n} U_0(\tau)\psi_- \right) d\tau.$$

Obviously,

$$\left| \mathcal{F} \circ S\left(\delta\psi_-\right) \right|^2 = \delta^2 \left| \widehat{\psi}_- \right|^2 + 2\delta^{2+4/n} \operatorname{Re}\left(\overline{\widehat{\psi}_-} \, \widehat{P\psi_-} \right) + \mathcal{O}\left(\delta^{2+8/n} \right),$$

and we have to prove that we can find $\psi_- \in \Sigma$, and h smooth and real-valued, such that

$$\operatorname{Re}\left(\overline{\mathcal{F}\psi_-} \mathcal{F}\left(P\psi_-\right) \right) \neq \operatorname{Re}\left(\overline{\mathcal{F}\left(\psi_h\right)} \mathcal{F}\left(P\left(\psi_h\right)\right) \right) =: R(\psi_-, h),$$

where ψ_h is defined by

$$\widehat{\psi}_h(\xi) = e^{ih(\xi)} \widehat{\psi}_-(\xi).$$

If this was not true, then for every $\psi_- \in \Sigma$, the differential of the map $h \mapsto R(\psi_-, h)$ would be zero at every smooth, real-valued function h. An elementary but tedious computation shows that

$$D_h R(\psi_-, 0)(h) \not\equiv 0,$$

with $h(x) = |x|^2/2$ and $\psi_-(x) = e^{-|x|^2/2}$. Indeed with this choice, computations are explicit:

$$\psi_-(x) = \widehat{\psi}_-(x) = e^{-|x|^2/2}.$$

With $h_a(x) = a|x|^2/2$, $a \in \mathbb{R}$, we introduce ψ_a such that $\widehat{\psi}_a = e^{ih_a}\widehat{\psi}_-$. It is given by

$$\psi_a(x) = (1 + ia)^{-n/2} e^{-|x|^2/(2(1+ia))}.$$

The evolution of Gaussian functions under the action of the free Schrödinger group can be computed explicitly, and we find:

$$U_0(t)\psi_a(x) = (1 + i(a+t))^{-n/2} e^{-|x|^2/(2(1+i(a+t)))}.$$

Therefore,

$$R(\psi_-, h_a) = \operatorname{Im}\left(e^{-(1+ia)\frac{|x|^2}{2}} \times \right.$$

$$\left. \times \int_{-\infty}^{+\infty} \frac{(1 + i(a+t))^{-n/2}}{1 + (a+t)^2} \left(1 - it\zeta(a)\right)^{-n/2} e^{-\frac{\zeta(a)}{1-it\zeta(a)} \frac{|x|^2}{2}} dt \right),$$

$$\text{where } \zeta(a) = \frac{4/n + 1 - i(a+t)}{1 + (a+t)^2}.$$

To prove the above claim, we differentiate this quantity with respect to a, and assess the result at $a = 0$. Considering for simplicity $x = 0$, we check that $D_h R(\psi_-, 0)(h) \not\equiv 0$. This shows that the caustic crossing is an ill-posed Cauchy problem as far as Wigner measures are concerned:

Proposition 7.23. *Let $n \geqslant 1$ and $\sigma = 2/n$.*
(1) There exists $a_0 \in \Sigma$ such that the Wigner measure associated to u^ε, the solution of Eq. (7.29), is discontinuous at $t = 1$:

$$\lim_{t \to 1^-} \mu_-(t, dx, d\xi) \neq \lim_{t \to 1^+} \mu_+(t, dx, d\xi).$$

(2) There exist a_0 and $\widetilde{a}_0$ in Σ, such that if u^ε and $\widetilde{u}^\varepsilon$ denote the solutions to Eq. (7.29) with these data, we have:

$$\mu_-(t, dx, d\xi) = \widetilde{\mu}_-(t, dx, d\xi), \quad t < 1,$$
$$\mu_+(t, dx, d\xi) \neq \widetilde{\mu}_+(t, dx, d\xi), \quad t > 1,$$

where $\mu_\pm$ and $\widetilde{\mu}_\pm$ denote the Wigner measures associated to u^ε and $\widetilde{u}^\varepsilon$, respectively.

Remark 7.24. This result is not specific to the value $\sigma = 2/n$. It is stated in this case because we have studied the asymptotic expansion of the scattering operator near the origin for $\sigma = 2/n$: Proposition 7.21 could be extended to other values of σ, with a slightly longer proof, allowing the extension of Proposition 7.23 to other values of σ.

This reveals a difference between WKB régime and caustic crossing for the propagation of Wigner measures. We have seen in Chap. 2 that in the critical case of the WKB régime, the Wigner measures are propagated like in the linear case; the leading order nonlinear effect does not affect the Wigner measure. On the other hand, the above discussion shows that as soon as nonlinear effects affect the wave function u^ε at leading order ($\alpha = n\sigma > 1$), the propagation of the Wigner measure undergoes nonlinear phenomena.

7.5 Nonlinear propagation, nonlinear caustic

In the previous paragraph, we have seen that when $\alpha = n\sigma > 1$, the caustic crossing is described in terms of a nonlinear scattering operator. Suppose that this aspect remains when $\alpha = n\sigma = 1$. Then because $\alpha = 1$, we know that also outside the focal point, the nonlinearity cannot be neglected at leading order (see Chap. 2). This suggests that it is not possible to compare the dynamics of the nonlinear Schrödinger equation to such a simple dynamics as that of the free Schrödinger equation; see also Remark 7.15. To compare the nonlinear dynamics with a simpler one (but necessarily not "too" simple), we need the notion of *long range scattering*.

We suppose $n = 1$. Note that removing this assumption is everything but easy [Ginibre and Ozawa (1993)]. The existence of modified wave operators (wave operators adapted to the long range scattering framework) was first established in [Ozawa (1991)]. A notion of asymptotic completeness appears in [Hayashi and Naumkin (1998)]. The most advanced results (so far) on the long range scattering for the one-dimensional, cubic, nonlinear Schrödinger equation, can be found in [Hayashi and Naumkin (2006)].

We present the main result of [Carles (2001a)] without giving all the details. The main technical idea is to work with oscillatory integrals, like we did in Sec. 7.3. When $n = \sigma = 1$, we modify the initial data, and consider:

$$\begin{cases} i\varepsilon\partial_t u^\varepsilon + \dfrac{\varepsilon^2}{2}\partial_x^2 u^\varepsilon = \varepsilon\left|u^\varepsilon\right|^2 u^\varepsilon, \\ u^\varepsilon(0,x) = e^{-ix^2/(2\varepsilon)-i|a_0(x)|^2\log\varepsilon}a_0(x). \end{cases} \tag{7.33}$$

The new term in the phase is closely related to the fact that we need long range scattering to describe the caustic crossing. It is also suggested by a formal computation. Recall that the expected limiting equation for the Lagrangian amplitude A^ε is given by Eq. (7.25). In the current case $n = \sigma = 1$, this equation is:

$$i\partial_t A(t,\xi) = \frac{1}{|t-1|}\left|A(t,\xi)\right|^2 A(t,\xi).$$

For some general initial data $\widetilde{A}_0$ (not necessarily equal to A_0 derived in Lemma 7.1), we find, for $t < 1$:

$$A(t,\xi) = \widetilde{A}_0(\xi)e^{i|\widetilde{A}_0(\xi)|^2\log(1-t)}.$$

Applying stationary phase formula like in §7.3, we get, for $t < 1$, on a formal level:

$$\begin{aligned} u^\varepsilon(t,x) &\approx \frac{e^{i\pi/4}}{\sqrt{1-t}}A\left(t,\frac{x}{t-1}\right)e^{i\frac{|x|^2}{2\varepsilon(t-1)}} \\ &\approx \frac{e^{i\pi/4}}{\sqrt{1-t}}\widetilde{A}_0\left(\frac{x}{t-1}\right)e^{i|\widetilde{A}_0(x/(t-1))|^2\log(1-t)}e^{i\frac{|x|^2}{2\varepsilon(t-1)}}. \end{aligned}$$

Motivated by the approach of Sec. 7.4, change the unknown function u^ε to ψ^ε, where

$$u^\varepsilon(t,x) = \frac{1}{\sqrt{\varepsilon}}\psi^\varepsilon\left(\frac{t-1}{\varepsilon}, \frac{x}{\varepsilon}\right).$$

In the present case, ψ^ε is given by:

$$\psi^\varepsilon(t,x) = \frac{e^{i\pi/4}}{\sqrt{|t|}}\widetilde{A}_0\left(\frac{x}{t}\right)e^{i|\widetilde{A}_0(x/t)|^2\log(\varepsilon|t|)}e^{i\varepsilon|x|^2/(2t)}, \quad t < 0.$$

Passing to the strong limit in L^2, the last exponential becomes negligible as $\varepsilon \to 0$. On the other hand, the term $\log(\varepsilon|t|)$ causes a weak convergence to zero, if $\widetilde{A}_0$ is independent of ε. Since the L^2-norm of the wave functions u^ε and ψ^ε is independent of time, this convergence cannot be strong. If, instead of considering $\widetilde{A}_0$ independent of ε, we choose

$$\widetilde{A}_0(\xi) = A_0(\xi)e^{-i|A_0(\xi)|^2\log\varepsilon},$$

where A_0 is given by Lemma 7.1, then we have:

$$\psi^\varepsilon(t,x) = \frac{e^{i\pi/4}}{\sqrt{|t|}}A_0\left(\frac{x}{t}\right)e^{i|A_0(x/t)|^2\log|t|}e^{i\varepsilon|x|^2/(2t)}.$$

This function converges strongly in L^2, to

$$\psi(t,x) = \frac{e^{i\pi/4}}{\sqrt{|t|}}A_0\left(\frac{x}{t}\right)e^{i|A_0(x/t)|^2\log|t|}.$$

This formal approach explains why it is convenient to modify the initial data, like we did in Eq. (7.33). Note that the above phase term in $\log|t|$ is exactly the modification which is needed in long range scattering; see also $S^\pm$ below. With this preliminary explanation, we can state our main result:

Proposition 7.25 ([Carles (2001a)]). *Assume $n = \alpha = \sigma = 1$.*
1. We can define a modified scattering operator for

$$i\partial_t\psi + \frac{1}{2}\partial_x^2\psi = |\psi|^2\psi, \tag{7.34}$$

and data in $\mathcal{F}(\mathcal{H})$, where:

$$\mathcal{H} = \{f \in H^3(\mathbb{R}); \ xf \in H^2(\mathbb{R})\}.$$

More precisely, there exists $\delta > 0$ such that if $\psi_- \in \mathcal{F}(\mathcal{H})$ with $\|\psi_-\|_\Sigma < \delta$, we can find unique $\psi \in C(\mathbb{R}_t, \Sigma)$ solving (7.34), and $\psi_+ \in L^2$, such that

$$\left\|\psi(t) - e^{iS^\pm(t)}U_0(t)\psi_\pm\right\|_{L^2} \xrightarrow[t\to\pm\infty]{} 0,$$

where $S^{\pm}$ are defined by:

$$S^{\pm}(t,x) := \left|\widehat{\psi_{\pm}}\left(\frac{x}{t}\right)\right|^2 \log|t|.$$

Denote $S : \psi_- \mapsto \psi_+$.

2. *Let $a_0 \in \mathcal{H}$, with $\|a_0\|_{\Sigma}$ sufficiently small. Let u^{ε} be the solution of (7.33) (which is in $C(\mathbb{R}; L^2)$). Define $\psi_- = e^{-i\pi/4}\widehat{a_0}$. The following asymptotics hold in L^2:*

- *If $t < 1$, then:*

$$u^{\varepsilon}(t,x) \underset{\varepsilon \to 0}{\sim} \frac{e^{i\pi/4}}{\sqrt{1-t}} e^{i\frac{x^2}{2\varepsilon(t-1)} + i\left|\widehat{\psi_-}\left(\frac{x}{t-1}\right)\right|^2 \log\frac{1-t}{\varepsilon}} \widehat{\psi_-}\left(\frac{x}{t-1}\right).$$

- *If $t > 1$, then:*

$$u^{\varepsilon}(t,x) \underset{\varepsilon \to 0}{\sim} \frac{e^{-i\pi/4}}{\sqrt{t-1}} e^{i\frac{x^2}{2\varepsilon(t-1)} + i\left|\widehat{\psi_+}\left(\frac{x}{t-1}\right)\right|^2 \log\frac{t-1}{\varepsilon}} \widehat{\psi_+}\left(\frac{x}{t-1}\right),$$

where $\psi_+ = S\psi_-$.

The $-\pi/2$ phase shift between the two asymptotics (before and after focusing) is the Maslov index. The change in the amplitude, measured by a scattering operator, is like in Proposition 7.19. The new phenomenon is the phase shift

$$\left|\widehat{\psi_+}\left(\frac{x}{t-1}\right)\right|^2 \log\frac{t-1}{\varepsilon} - \left|\widehat{\psi_-}\left(\frac{x}{t-1}\right)\right|^2 \log\frac{t-1}{\varepsilon},$$

which appears when comparing the asymptotics for the wave function u^{ε} before and after the focal point. It is "highly nonlinear", and depends on ε. Following an idea due to Guy Métivier, we called it a "random" phase shift: it depends on the subsequence ε going to zero which is considered.

In [Carles (2001a)], this result is proved by revisiting the approach of [Ozawa (1991)] for the asymptotics on $t \in [0,1[$, so that we can use the result of [Hayashi and Naumkin (1998)] to describe the asymptotic behavior of u^{ε} on $]1,T]$ for any $T > 1$. In view of the improvement of [Hayashi and Naumkin (2006)] (the domain and range of the above operator S are improved), it should be possible to relax the assumption $a_0 \in \mathcal{H}$, and require less regularity for a_0. We shall not pursue this issue.

We shall merely outline the proof of Proposition 7.25. We leave out the proof of the first point, which is now a consequence of [Hayashi and Naumkin (2006)], and we focus our attention on the second point. We have already seen that a natural candidate as an approximate solution for $t < 1$ is given by:

$$u_{\text{app}}^{\varepsilon}(t, x) = \frac{1}{\sqrt{2\pi\varepsilon}} \int_{\mathbb{R}} e^{-i\frac{t-1}{2\varepsilon}|\xi|^2 + i\frac{x\cdot\xi}{\varepsilon}} e^{i|A_0(\xi)|^2 \log\frac{1-t}{\varepsilon}} A_0(\xi) d\xi.$$

The last point of Lemma 7.12 shows that since $a_0 \in \mathcal{H}$,

$$i\varepsilon\partial_t u_{\text{app}}^{\varepsilon} + \frac{\varepsilon^2}{2}\partial_x^2 u_{\text{app}}^{\varepsilon} = \varepsilon \left|u_{\text{app}}^{\varepsilon}\right|^2 u_{\text{app}}^{\varepsilon} - \varepsilon r^{\varepsilon},$$

with

$$\|r^{\varepsilon}(t)\|_{L^2} \lesssim \frac{\varepsilon}{(1-t)^2} \left(\log\frac{1-t}{\varepsilon}\right)^2,$$

$$\sum_{\mathcal{B}^{\varepsilon} \in \{\varepsilon\partial_x, J^{\varepsilon}\}} \|\mathcal{B}^{\varepsilon}(t)r^{\varepsilon}(t)\|_{L^2} \lesssim \frac{\varepsilon}{(1-t)^2} \left(\log\frac{1-t}{\varepsilon}\right)^3.$$

Introduce $w^{\varepsilon} = u^{\varepsilon} - u_{\text{app}}^{\varepsilon}$. Lemma 7.12 also yields

$$\|w^{\varepsilon}(0)\|_{L^2} \lesssim \varepsilon \left(\log\frac{1}{\varepsilon}\right)^2,$$

$$\sum_{\mathcal{B}^{\varepsilon} \in \{\varepsilon\partial_x, J^{\varepsilon}\}} \|\mathcal{B}^{\varepsilon}(0)w^{\varepsilon}(0)\|_{L^2} \lesssim \varepsilon \left(\log\frac{1}{\varepsilon}\right)^3.$$

Let $\delta > 0$. Lemma 7.12 and weighted Gagliardo–Nirenberg inequality (7.6) show that there exists C_* depending on $\|a_0\|_{\mathcal{H}}$ and δ such that for $1 - t \geqslant C_*\varepsilon$,

$$\left\|u_{\text{app}}^{\varepsilon}(t)\right\|_{L^{\infty}} \leqslant \frac{\|a_0\|_{L^{\infty}} + \delta}{\sqrt{1-t}},$$

$$\left\|\varepsilon\partial_x u_{\text{app}}^{\varepsilon}(t)\right\|_{L^{\infty}} \leqslant \frac{C}{\sqrt{1-t}},$$

$$\left\|J^{\varepsilon}(t)u_{\text{app}}^{\varepsilon}(t)\right\|_{L^{\infty}} \leqslant \frac{C}{\sqrt{1-t}} \log\frac{1-t}{\varepsilon},$$

where C depends on $\|a_0\|_{\mathcal{H}}$. The error term w^{ε} solves:

$$i\varepsilon\partial_t w^{\varepsilon} + \frac{\varepsilon^2}{2}\partial_x^2 w^{\varepsilon} = \varepsilon |u^{\varepsilon}|^2 u^{\varepsilon} - \varepsilon \left|u_{\text{app}}^{\varepsilon}\right|^2 u_{\text{app}}^{\varepsilon} + \varepsilon r^{\varepsilon}$$

$$= \varepsilon \left(|u^{\varepsilon}|^2 w^{\varepsilon} + \left(\left|w^{\varepsilon} + u_{\text{app}}^{\varepsilon}\right|^2 - \left|u_{\text{app}}^{\varepsilon}\right|^2\right) u_{\text{app}}^{\varepsilon}\right) + \varepsilon r^{\varepsilon}.$$

From the above estimates and (7.6), there exists $\varepsilon_0 > 0$ such that for $\varepsilon \in]0, \varepsilon_0]$,

$$\|w^{\varepsilon}(0)\|_{L^{\infty}} \leqslant \frac{\delta}{2}.$$

By continuity, there exists $t^\varepsilon > 0$ such that

$$\|w^\varepsilon(\tau)\|_{L^\infty} \leqslant \frac{\delta}{\sqrt{1-\tau}} \tag{7.35}$$

for $t \in [0, t^\varepsilon]$. So long as (7.35) holds, Lemma 1.2 yields:

$$\|w^\varepsilon(t)\|_{L^2} \leqslant \|w^\varepsilon(0)\|_{L^2} + \int_0^t \|r^\varepsilon(\tau)\|_{L^2} \, d\tau$$

$$+ \int_0^t \left(2 \left\|u_{\mathrm{app}}^\varepsilon(\tau)\right\|_{L^\infty}^2 + \left\|u_{\mathrm{app}}^\varepsilon(\tau)\right\|_{L^\infty} \|w^\varepsilon(\tau)\|_{L^\infty}\right) \|w^\varepsilon(\tau)\|_{L^2} \, d\tau$$

$$\leqslant \|w^\varepsilon(0)\|_{L^2} + C \int_0^t \frac{\varepsilon}{(1-\tau)^2} \left(\log \frac{1-\tau}{\varepsilon}\right)^2 d\tau$$

$$+ C_0 \int_0^t \|w^\varepsilon(\tau)\|_{L^2} \frac{d\tau}{1-\tau},$$

where

$$C_0 = 2 \left(\|a_0\|_{L^\infty} + \delta\right)^2 + \delta \left(\|a_0\|_{L^\infty} + \delta\right).$$

Gronwall lemma yields, so long as (7.35) holds:

$$\|w^\varepsilon(t)\|_{L^2} \leqslant \frac{\|w^\varepsilon(0)\|_{L^2}}{(1-t)^{C_0}} + C \int_0^t \frac{\varepsilon}{(1-\tau)^2} \left(\log \frac{1-\tau}{\varepsilon}\right)^2 \left(\frac{1-\tau}{1-t}\right)^{C_0} d\tau.$$

Rewrite the last term as

$$\int_0^t \frac{\varepsilon}{(1-\tau)^2} \left(\log \frac{1-\tau}{\varepsilon}\right)^2 \left(\frac{1-\tau}{1-t}\right)^{C_0} d\tau = \left(\frac{\varepsilon}{1-t}\right)^{C_0} \int_{(1-t)/\varepsilon}^{1/\varepsilon} \frac{(\log \tau)^2}{\tau^{2-C_0}} d\tau.$$

For $C_0 < 1$, an integration by parts shows that

$$\int_a^b \frac{\log \tau}{\tau^{2-C_0}} d\tau = \mathcal{O}\left(\frac{(\log a)^2}{a^{1-C_0}}\right) \quad \text{as } b \geqslant a \to +\infty.$$

Therefore, if $C_0 < 1$ and $1 - t \geqslant C_* \varepsilon$ with C_* sufficiently large,

$$\|w^\varepsilon(t)\|_{L^2} \leqslant \frac{C}{(1-t)^{C_0}} \varepsilon \left(\log \frac{1}{\varepsilon}\right)^2 + C \frac{\varepsilon}{1-t} \left(\log \left(\frac{1-t}{\varepsilon}\right)\right)^2$$

$$\lesssim \frac{\varepsilon}{1-t} \left(\log \left(\frac{1-t}{\varepsilon}\right)\right)^2.$$

With this first estimate, we can infer

$$\sum_{\mathcal{B}^\varepsilon \in \{\varepsilon \partial_x, J^\varepsilon\}} \|\mathcal{B}^\varepsilon(t) w^\varepsilon(t)\|_{L^2} \leqslant C \frac{\varepsilon}{1-t} \left(\log \left(\frac{1-t}{\varepsilon}\right)\right)^3.$$

Using the weighted Gagliardo–Nirenberg inequality (7.6), we deduce

$$\|w^\varepsilon(t)\|_{L^2} \leqslant \frac{C}{\sqrt{1-t}} \frac{\varepsilon}{1-t} \left(\log \left(\frac{1-t}{\varepsilon} \right) \right)^{5/2}.$$

Therefore, if we choose C_* sufficiently large, there exists $\varepsilon_* > 0$ such that for all $\varepsilon \in]0, \varepsilon_*]$, (7.35) holds on $[0, 1 - C_*\varepsilon]$. Applying Lemma 7.12 to $u^\varepsilon_{\mathrm{app}}$ yields the asymptotic behavior of u^ε for $t < 1$ in Proposition 7.25. Note that the smallness condition that we have used so far is $C_0 < 1$. Since $\delta > 0$ is arbitrarily small, this assumption boils down to

$$\|a_0\|_{L^\infty} < \frac{1}{\sqrt{2}}.$$

For the asymptotics when $t > 1$, the smallness condition ceases to be explicit, since we have to use the results of [Hayashi and Naumkin (1998)] or [Hayashi and Naumkin (2006)]. Introduce ψ^ε given by

$$u^\varepsilon(t, x) = \frac{1}{\varepsilon^{n/2}} \psi^\varepsilon \left(\frac{t-1}{\varepsilon}, \frac{x}{\varepsilon} \right).$$

It is easy to deduce from the above analysis that

$$\sup_{t \leqslant -C_*} \|U_0(-t)\left(\psi^\varepsilon(t) - \psi(t)\right)\|_\Sigma \xrightarrow[\varepsilon \to 0]{} 0,$$

where ψ is the unique solution to Eq. (7.34) with

$$\left\| \psi(t) - e^{iS^-(t)} U_0(t) \psi_- \right\|_{L^2} \xrightarrow[t \to -\infty]{} 0.$$

The local well-posedness for Eq. (7.34) shows that for all $T > 0$,

$$\sup_{t \leqslant T} \|U_0(-t)\left(\psi^\varepsilon(t) - \psi(t)\right)\|_\Sigma \xrightarrow[\varepsilon \to 0]{} 0.$$

Theorem 7.26 ([Hayashi and Naumkin (1998)]). *Let* $\varphi \in \Sigma$, *with* $\|\varphi\|_\Sigma = \delta' \leqslant \delta$, *where* δ *is sufficiently small. Let* $\psi \in C(\mathbb{R}_t, \Sigma)$ *be the solution of the initial value problem*

$$i\partial_t \psi + \frac{1}{2} \partial_x^2 \psi = |\psi|^2 \psi \quad ; \quad \psi_{|t=0} = \varphi.$$

There exists a unique pair $(W, \phi) \in \left(L^2 \cap L^\infty\right) \times L^\infty$ *such that for* $t \geqslant 1$,

$$\left\| \mathcal{F}\left(U_0(-t)\psi\right)(t) \exp\left(-i \int_1^t |\widehat{\psi}(\tau)|^2 \frac{d\tau}{\tau} \right) - W \right\|_{L^2 \cap L^\infty} \leqslant C\delta' t^{-\alpha + C(\delta')^2},$$

$$\left\| \int_1^t |\widehat{\psi}(\tau)|^2 \frac{d\tau}{\tau} - |W|^2 \log t - \phi \right\|_{L^\infty} \leqslant C\delta' t^{-\alpha + C(\delta')^2},$$

where $C\delta' < \alpha < 1/4$, and ϕ is a real valued function. Furthermore we have the asymptotic formula for large time,

$$\psi(t,x) = \frac{1}{(it)^{1/2}} W\left(\frac{x}{t}\right) \exp\left(i\frac{x^2}{2t} + i\left|W\left(\frac{x}{t}\right)\right|^2 \log t + i\phi\left(\frac{x}{t}\right)\right)$$
$$+ \mathcal{O}_{L^2}\left(\delta' t^{-1/2-\alpha+C(\delta')^2}\right),$$

and the estimate

$$\left\|\mathcal{F}\left(U_0(-t)\psi(t)\right) - W\exp(i|W|^2 \log t + i\phi)\right\|_{L^2 \cap L^\infty} \leqslant C\delta' t^{-\alpha+C(\delta')^2}.$$

Finally, the map $\varphi \mapsto (W,\phi)$ is continuous on the above spaces.

First, note that in view of Lemma 7.20, the above result yields the first part of Proposition 7.25 with

$$\psi_+ = \mathcal{F}^{-1}\left(We^{i\phi}\right).$$

We now briefly explain how to conclude the proof of Proposition 7.25. Inspired by the linear long range scattering theory (see e.g. [Dereziński and Gérard (1997)]), introduce

$$\phi^\varepsilon(t,\xi) = \int_{-1/\varepsilon}^{(t-1)/\varepsilon} |\psi(\tau,\tau\xi)|^2 d\tau + |a_0(-\xi)|^2 \log\frac{1}{\varepsilon}.$$

Write

$$A^\varepsilon(t,\xi) = e^{i\phi^\varepsilon(t,\xi)} B^\varepsilon(t,\xi),$$

so that

$$u^\varepsilon(t,x) = \frac{1}{\sqrt{2\pi\varepsilon}} \int_{\mathbb{R}} e^{-i\frac{t-1}{2\varepsilon}|\xi|^2 + i\frac{x\cdot\xi}{\varepsilon}} e^{i\phi^\varepsilon(t,\xi)} B^\varepsilon(t,\xi) d\xi.$$

The asymptotics for $t < 1$ and Lemma 7.20 imply, in $L^\infty_{\text{loc}}(]0,1[; L^\infty(\mathbb{R}))$,

$$\phi^\varepsilon(t,\xi) = |a_0(-\xi)|^2 \log\frac{1-t}{\varepsilon} + o(1),$$

hence

$$\|B^\varepsilon(t,\cdot) - A_0\|_{L^2} \xrightarrow[\varepsilon\to 0]{} 0,$$

with $A_0(\xi) = e^{-i\pi/4}a_0(-\xi)$. For $t > 1$, Theorem 7.26 yields $W \in L^2 \cap L^\infty$ and $H \in L^\infty$ such that, in $L^\infty_{\text{loc}}(]1,+\infty[; L^\infty(\mathbb{R}))$,

$$\phi^\varepsilon(t,\xi) = |W(\xi)|^2 \log\frac{t-1}{\varepsilon} + H(\xi) + o(1).$$

By construction,

$$B^\varepsilon(t,\xi) = e^{-i\phi^\varepsilon(t,\xi)} \mathcal{F}\left(U_0\left(\frac{1-t}{\varepsilon}\right)\psi^\varepsilon\left(\frac{t-1}{\varepsilon}\right)\right).$$

Since the map $\varphi \mapsto (W,\phi)$ of Theorem 7.26 is continuous, we conclude:

$$B^\varepsilon(t,\xi) \xrightarrow[\varepsilon\to 0]{} e^{-iH(\xi)+i\phi(\xi)} W(\xi) = e^{-iH(\xi)} \widehat{\psi}_+(\xi) \quad \text{in } L^\infty_{\text{loc}}(]1,+\infty[; L^2(\mathbb{R})).$$

Using Lemma 7.12, we infer Proposition 7.25. Note that the function H is not present in the limit, due to some cancellations.

7.6 Why initial quadratic oscillations?

In this chapter, except in the case "nonlinear propagation, nonlinear caustic", we have considered initial data with quadratic oscillations, exactly of the form

$$u^\varepsilon(0, x) = a_0(x)e^{-i|x|^2/(2\varepsilon)}. \qquad (7.36)$$

We have seen that in order to observe nonlinear effects at leading order near the focal point, we have to impose $\alpha \leqslant n\sigma$. In this section, we consider the critical case

$$i\varepsilon\partial_t u^\varepsilon + \frac{\varepsilon^2}{2}\Delta u^\varepsilon = \varepsilon^{n\sigma}|u^\varepsilon|^{2\sigma}u^\varepsilon, \qquad (7.37)$$

with $n\sigma > 1$ (linear propagation). The formal computations of Chap. 6 suggest that since a focal point is expected to concentrate the maximum of energy when a caustic is formed, any other caustic should be "linear". All in all, this means that we expect nonlinear effects to be visible at leading order only when a focal point is present. In this paragraph, we show that this intuition can be made rigorous at least when $\sigma \geqslant 2/n$. The complete proofs appear in [Carles *et al.* (2003)] for the case $\sigma > 2/n$, and in [Carles and Keraani (2007)] for the case $\sigma = 2/n$. Since the papers are rather technical, we only give a flavor of their content, and invite the reader to consult the articles for details. Throughout this section, we assume that if $n \geqslant 3$, $\sigma < 2/(n-2)$ (the nonlinearity is H^1-subcritical).

7.6.1 *Notion of linearizability*

The notion of linearizability that we shall use is the analogue for nonlinear Schrödinger equations of the concept introduced by P. Gérard [Gérard (1996)] in the case of the semi-linear wave equation (7.32). Consider the linear evolution of the initial data for u^ε:

$$i\varepsilon\partial_t v^\varepsilon + \frac{\varepsilon^2}{2}\Delta v^\varepsilon = 0 \quad ; \quad v^\varepsilon_{|t=0} = u^\varepsilon_{|t=0}. \qquad (7.38)$$

Roughly speaking, the nonlinearity in Eq. (7.37) is relevant at leading order if and only if the relation $u^\varepsilon(t) - v^\varepsilon(t) = o(v^\varepsilon(t))$ ceases to hold for some $t > 0$ (we consider only forward in time propagation here, since backward propagation is similar). To make this vague statement precise, we clarify our assumptions on the initial data.

Assumption 7.27. The initial data $u^\varepsilon_{|t=0} = u^\varepsilon_0$ belong to $H^1(\mathbb{R}^n)$, uniformly in the following sense: if we denote

$$\|f^\varepsilon\|_{H^1_\varepsilon} = \|f^\varepsilon\|_{L^2} + \|\varepsilon\nabla f^\varepsilon\|_{L^2},$$

then

$$\sup_{0<\varepsilon\leqslant 1} \|u_0^\varepsilon\|_{H_\varepsilon^1} < \infty.$$

This assumption is satisfied for data of the form (7.36), provided that we assume $a_0 \in \Sigma$. More generally, if u_0^ε is of the form considered in WKB régime,

$$u_0^\varepsilon(x) = a_0(x)e^{i\phi_0(x)/\varepsilon},$$

where ϕ_0 is smooth and subquadratic, and $a_0 \in \Sigma$, then Assumption 7.27 is satisfied.

Definition 7.28 (Linearizability). *Let I^ε be an interval of $\mathbb{R}$, possibly depending on ε, containing the origin; u^ε is linearizable on I^ε if*

$$\limsup_{\varepsilon\to 0} \sup_{t\in I^\varepsilon} \|u^\varepsilon(t) - v^\varepsilon(t)\|_{H_\varepsilon^1} = 0.$$

Recall the conservations of mass and energy in this case:

$$\frac{d}{dt}\|u^\varepsilon(t)\|_{L^2}^2 = \frac{d}{dt}\|v^\varepsilon(t)\|_{L^2}^2 = 0,$$
$$\frac{d}{dt}\|\varepsilon\nabla v^\varepsilon(t)\|_{L^2}^2 = 0, \tag{7.39}$$
$$\frac{d}{dt}\left(\frac{1}{2}\|\varepsilon\nabla u^\varepsilon(t)\|_{L^2}^2 + \frac{\varepsilon^{n\sigma}}{\sigma+1}\|u^\varepsilon(t)\|_{L^{2\sigma+2}}^{2\sigma+2}\right) = 0.$$

We have the first result:

Lemma 7.29. *Let u_0^ε satisfying Assumption 7.27. Assume in addition that initially, the potential energy goes to zero:*

$$\varepsilon^{n\sigma}\|u_0^\varepsilon\|_{L^{2\sigma+2}}^{2\sigma+2} \xrightarrow[\varepsilon\to 0]{} 0.$$

Let $T > 0$. If u^ε is linearizable on $[0,T]$, then

$$\limsup_{\varepsilon\to 0} \sup_{0\leqslant t\leqslant T} \varepsilon^{n\sigma}\|v^\varepsilon(t)\|_{L^{2\sigma+2}}^{2\sigma+2} = 0.$$

Proof. The proof follows the same lines as for the energy-critical wave equation [Gérard (1996)]. Let

$$R := \limsup_{\varepsilon\to 0} \sup_{0\leqslant t\leqslant T} \left| \frac{1}{2}\|\varepsilon\nabla u^\varepsilon(t)\|_{L^2}^2 + \frac{\varepsilon^{n\sigma}}{\sigma+1}\|u^\varepsilon(t)\|_{L^{2\sigma+2}}^{2\sigma+2} \right.$$
$$\left. - \frac{1}{2}\|\varepsilon\nabla v^\varepsilon(t)\|_{L^2}^2 - \frac{\varepsilon^{n\sigma}}{\sigma+1}\|v^\varepsilon(t)\|_{L^{2\sigma+2}}^{2\sigma+2} \right|.$$

On the one hand, linearizability implies that

$$R \leqslant \limsup_{\varepsilon \to 0} \sup_{0 \leqslant t \leqslant T} \frac{\varepsilon^{n\sigma}}{\sigma+1} \int_{\mathbb{R}^n} \left| |u^\varepsilon(t,x)|^{2\sigma+2} - |v^\varepsilon(t,x)|^{2\sigma+2} \right| dx.$$

Writing

$$\left| |u^\varepsilon(t,x)|^{2\sigma+2} - |v^\varepsilon(t,x)|^{2\sigma+2} \right| \lesssim$$
$$\lesssim \left(|u^\varepsilon(t,x)|^{2\sigma+1} + |v^\varepsilon(t,x)|^{2\sigma+1} \right) |u^\varepsilon(t,x) - v^\varepsilon(t,x)|,$$

Hölder's inequality yields

$$R \lesssim \limsup_{\varepsilon \to 0} \sup_{0 \leqslant t \leqslant T} \varepsilon^{n\sigma} \|u^\varepsilon(t) - v^\varepsilon(t)\|_{L^{2\sigma+2}} \left(\|u^\varepsilon(t)\|_{L^{2\sigma+2}} + \|v^\varepsilon(t)\|_{L^{2\sigma+2}} \right)^{2\sigma+1}.$$

Assumption 7.27, and the conservation of linear and nonlinear energy for v^ε and u^ε respectively, yield, along with Gagliardo–Nirenberg inequality (for v^ε):

$$\|u^\varepsilon(t)\|_{L^{2\sigma+2}}^{2\sigma+2} + \|v^\varepsilon(t)\|_{L^{2\sigma+2}}^{2\sigma+2} \lesssim \varepsilon^{-n\sigma}.$$

Using Gagliardo–Nirenberg inequality, this implies

$$R \lesssim \limsup_{\varepsilon \to 0} \sup_{0 \leqslant t \leqslant T} \varepsilon^{n\sigma} \|u^\varepsilon(t) - v^\varepsilon(t)\|_{L^{2\sigma+2}} \varepsilon^{-n\sigma \frac{2\sigma+1}{2\sigma+2}}$$
$$\lesssim \limsup_{\varepsilon \to 0} \sup_{0 \leqslant t \leqslant T} \varepsilon^{\delta(2\sigma+2)} \|u^\varepsilon(t) - v^\varepsilon(t)\|_{L^{2\sigma+2}}$$
$$\lesssim \limsup_{\varepsilon \to 0} \sup_{0 \leqslant t \leqslant T} \|u^\varepsilon(t) - v^\varepsilon(t)\|_{L^2}^{1-\delta(2\sigma+2)} \|\varepsilon \nabla u^\varepsilon(t) - \varepsilon \nabla v^\varepsilon(t)\|_{L^2}^{\delta(2\sigma+2)}.$$

We conclude from the linearizability assumption that $R = 0$. On the other hand, the conservation of energy (7.39) yields

$$R = \limsup_{\varepsilon \to 0} \sup_{0 \leqslant t \leqslant T} \frac{\varepsilon^{n\sigma}}{\sigma+1} \left| \|u_0^\varepsilon\|_{L^{2\sigma+2}}^{2\sigma+2} - \|v^\varepsilon(t)\|_{L^{2\sigma+2}}^{2\sigma+2} \right|.$$

By assumption, the first term of the right hand side goes to zero, and the lemma follows.   □

The above lemma announces an important idea in the linearizability criterion, first introduced in [Gérard (1996)]: to assess the nonlinear effects affecting u^ε, we consider the evolution of a quantity involving the solution v^ε to the companion *linear* equation only.

It turns out that this necessary condition for linearizability is sufficient when $\sigma > 2/n$. It is not so when $\sigma = 2/n$. We present here a general sufficient condition for linearizability, which yields the result when $\sigma > 2/n$, and which is also a necessary condition when $\sigma = 2/n$. This condition, and its proof, may be viewed as a simplification of the approach presented in [Carles *et al.* (2003)]. Instead of the usual inhomogeneous Strichartz estimates (7.14), we use similar estimates for pairs (q_j, r_j) not necessarily admissible. Such estimates were derived in [Cazenave and Weissler (1992)], and generalized in [Foschi (2005)]:

Lemma 7.30. *Let (q, r) be an admissible pair with $r > 2$. Let $k > q/2$, and define $\widetilde{k}$ by*

$$\frac{1}{\widetilde{k}} + \frac{1}{k} = \frac{2}{q}.$$

There exists C depending only on n, r and k such that for all interval I,

$$\varepsilon^{2/q} \left\| \int_{I \cap \{\tau \leqslant t\}} U_0^\varepsilon(t - \tau) F(\tau) d\tau \right\|_{L^k(I;L^r)} \leqslant C \|F\|_{L^{\widetilde{k}'}(I;L^{r'})},$$

for all $F \in L^{\widetilde{k}'}(I; L^{r'})$.

Proof. In view of the dispersive estimate

$$\|U_0^\varepsilon(t)\|_{L^{r'} \to L^r} \lesssim |\varepsilon t|^{-\delta(r)},$$

we have

$$\left\| \int_{I \cap \{\tau \leqslant t\}} U_0^\varepsilon(t - \tau) F(\tau) d\tau \right\|_{L^r} \lesssim \int_0^t \varepsilon^{-\delta(r)} (t - \tau)^{-\delta(r)} \|F(\tau)\|_{L^{r'}} d\tau.$$

By assumption, $\delta(r) = 2/q$. The lemma then follows from Riesz potential inequalities (see e.g. [Stein (1993)]). $\qquad \square$

Proposition 7.31. *Let Assumption 7.27 be satisfied. Assume that*

$$\sigma > \sigma_0(n) = \frac{2 - n + \sqrt{n^2 + 12n + 4}}{4n},$$

where $\sigma_0(n)$ appeared in Theorem 7.16. Let

$$r = 2\sigma + 2 \quad ; \quad q = \frac{4\sigma + 4}{n\sigma} \quad ; \quad k = \frac{4\sigma(\sigma + 1)}{2 - \sigma(n - 2)}.$$

If

$$\varepsilon^{2/q - 1/k} \|v^\varepsilon\|_{L^k(I^\varepsilon; L^r)} \xrightarrow[\varepsilon \to 0]{} 0, \tag{7.40}$$

then u^ε is linearizable on I^ε.

Proof. We first note that (q, r) is admissible, and that $k > q/2$ if and only if $\sigma > \sigma_0(n)$. Define $w^\varepsilon = u^\varepsilon - v^\varepsilon$. It solves

$$i\varepsilon\partial_t w^\varepsilon + \frac{\varepsilon^2}{2}\Delta w^\varepsilon = \varepsilon^{n\sigma}|u^\varepsilon|^{2\sigma}u^\varepsilon \quad ; \quad w^\varepsilon_{|t=0} = 0.$$

Writing $|u^\varepsilon|^{2\sigma}u^\varepsilon = |u^\varepsilon|^{2\sigma}u^\varepsilon - |v^\varepsilon|^{2\sigma}v^\varepsilon + |v^\varepsilon|^{2\sigma}v^\varepsilon$, Lemma 7.30 yields:

$$\|w^\varepsilon\|_{L^k L^r} \lesssim \varepsilon^{n\sigma-1-2/q}\left\||u^\varepsilon|^{2\sigma}u^\varepsilon - |v^\varepsilon|^{2\sigma}v^\varepsilon\right\|_{L^{\tilde{k}'}L^{r'}}$$
$$+ \varepsilon^{n\sigma-1-2/q}\left\||v^\varepsilon|^{2\sigma}v^\varepsilon\right\|_{L^{\tilde{k}'}L^{r'}},$$

where $L^j L^s$ stands for $L^j(I^\varepsilon; L^s)$. Note that

$$\frac{1}{r'} = \frac{2\sigma+1}{r} \quad ; \quad \frac{1}{\tilde{k}'} = \frac{2\sigma+1}{k}.$$

From Taylor formula, Hölder's inequality and the relation $u^\varepsilon = w^\varepsilon + v^\varepsilon$, we infer:

$$\|w^\varepsilon\|_{L^k L^r} \lesssim \varepsilon^{n\sigma-1-2/q}\left(\|w^\varepsilon\|^{2\sigma}_{L^k L^r} + \|v^\varepsilon\|^{2\sigma}_{L^k L^r}\right)\|w^\varepsilon\|_{L^k L^r}$$
$$+ \varepsilon^{n\sigma-1-2/q}\|v^\varepsilon\|^{2\sigma+1}_{L^k L^r}.$$

Again, note that

$$n\sigma - 1 - \frac{1}{k} = (2\sigma+1)\left(\frac{2}{q} - \frac{1}{k}\right),$$

to deduce:

$$\varepsilon^{2/q-1/k}\|w^\varepsilon\|_{L^k L^r} \lesssim \left(\varepsilon^{2/q-1/k}\|w^\varepsilon\|_{L^k L^r}\right)^{2\sigma}\varepsilon^{2/q-1/k}\|w^\varepsilon\|_{L^k L^r}$$
$$+ \left(\varepsilon^{2/q-1/k}\|v^\varepsilon\|_{L^k L^r}\right)^{2\sigma}\varepsilon^{2/q-1/k}\|w^\varepsilon\|_{L^k L^r}$$
$$+ \left(\varepsilon^{2/q-1/k}\|v^\varepsilon\|_{L^k L^r}\right)^{2\sigma+1}.$$

By assumption, the second term on the right hand side can be absorbed by the left hand side, provided that ε is sufficiently small:

$$\varepsilon^{2/q-1/k}\|w^\varepsilon\|_{L^k L^r} \lesssim \left(\varepsilon^{2/q-1/k}\|w^\varepsilon\|_{L^k L^r}\right)^{2\sigma}\varepsilon^{2/q-1/k}\|w^\varepsilon\|_{L^k L^r}$$
$$+ \left(\varepsilon^{2/q-1/k}\|v^\varepsilon\|_{L^k L^r}\right)^{2\sigma+1}.$$

Using the bootstrap argument of Lemma 7.22, we infer, for ε sufficiently small:

$$\varepsilon^{2/q-1/k}\|w^\varepsilon\|_{L^k L^r} \lesssim \left(\varepsilon^{2/q-1/k}\|v^\varepsilon\|_{L^k L^r}\right)^{2\sigma+1} \ll 1.$$

Note that this implies, along with (7.40),

$$\varepsilon^{2/q-1/k} \, \|u^\varepsilon\|_{L^k L^r} \ll 1.$$

This estimate allows us to conclude, by applying Strichartz estimates. Indeed, we first find

$$\|w^\varepsilon\|_{L^q L^r} \lesssim \varepsilon^{n\sigma-1-2/q} \left\| |u^\varepsilon|^{2\sigma} u^\varepsilon - |v^\varepsilon|^{2\sigma} v^\varepsilon \right\|_{L^{q'} L^{r'}}$$
$$+ \varepsilon^{n\sigma-1-2/q} \left\| |v^\varepsilon|^{2\sigma} v^\varepsilon \right\|_{L^{q'} L^{r'}}.$$

Noticing that

$$\frac{1}{r'} = \frac{2\sigma+1}{r} \quad ; \quad \frac{1}{q'} = \frac{1}{q} + \frac{2\sigma}{k},$$

Taylor formula and Hölder's inequality yield

$$\|w^\varepsilon\|_{L^q L^r} \lesssim \varepsilon^{n\sigma-1-2/q} \left(\|w^\varepsilon\|_{L^k L^r}^{2\sigma} + \|v^\varepsilon\|_{L^k L^r}^{2\sigma} \right) \|w^\varepsilon\|_{L^q L^r}$$
$$+ \varepsilon^{n\sigma-1-2/q} \|v^\varepsilon\|_{L^k L^r}^{2\sigma} \|v^\varepsilon\|_{L^q L^r}.$$

Again, since

$$n\sigma - 1 - \frac{1}{q} = 2\sigma \left(\frac{2}{q} - \frac{1}{k} \right) + \frac{1}{q},$$

we infer

$$\varepsilon^{1/q} \|w^\varepsilon\|_{L^q L^r} \lesssim \left(\varepsilon^{2/q-1/k} \|w^\varepsilon\|_{L^k L^r} \right)^{2\sigma} \varepsilon^{1/q} \|w^\varepsilon\|_{L^q L^r}$$
$$+ \left(\varepsilon^{2/q-1/k} \|v^\varepsilon\|_{L^k L^r} \right)^{2\sigma} \varepsilon^{1/q} \|w^\varepsilon\|_{L^q L^r}$$
$$+ \left(\varepsilon^{2/q-1/k} \|v^\varepsilon\|_{L^k L^r} \right)^{2\sigma} \varepsilon^{1/q} \|v^\varepsilon\|_{L^q L^r}$$
$$\lesssim \left(\varepsilon^{2/q-1/k} \|w^\varepsilon\|_{L^k L^r} \right)^{2\sigma} \varepsilon^{1/q} \|w^\varepsilon\|_{L^q L^r}$$
$$+ \left(\varepsilon^{2/q-1/k} \|v^\varepsilon\|_{L^k L^r} \right)^{2\sigma} \varepsilon^{1/q} \|w^\varepsilon\|_{L^q L^r}$$
$$+ \left(\varepsilon^{2/q-1/k} \|v^\varepsilon\|_{L^k L^r} \right)^{2\sigma} \|u_0^\varepsilon\|_{L^2},$$

where we have used the homogeneous Strichartz inequality for v^ε. Therefore, the first two terms of the right hand side can be absorbed by the left hand side for ε sufficiently small, and

$$\varepsilon^{1/q} \|w^\varepsilon\|_{L^q L^r} \ll 1.$$

Applying Strichartz inequality again, we have

$$
\begin{aligned}
\|w^\varepsilon\|_{L^\infty L^2} &\lesssim \varepsilon^{n\sigma-1-1/q} \left\| |u^\varepsilon|^{2\sigma} u^\varepsilon - |v^\varepsilon|^{2\sigma} v^\varepsilon \right\|_{L^{q'} L^{r'}} \\
&\quad + \varepsilon^{n\sigma-1-2/q} \left\| |v^\varepsilon|^{2\sigma} v^\varepsilon \right\|_{L^{q'} L^{r'}} \\
&\lesssim \left(\varepsilon^{2/q-1/k} \|w^\varepsilon\|_{L^k L^r} \right)^{2\sigma} \varepsilon^{1/q} \|w^\varepsilon\|_{L^q L^r} \\
&\quad + \left(\varepsilon^{2/q-1/k} \|v^\varepsilon\|_{L^k L^r} \right)^{2\sigma} \varepsilon^{1/q} \|w^\varepsilon\|_{L^q L^r} \\
&\quad + \left(\varepsilon^{2/q-1/k} \|v^\varepsilon\|_{L^k L^r} \right)^{2\sigma} \|u_0^\varepsilon\|_{L^2},
\end{aligned}
$$

so

$$
\|w^\varepsilon\|_{L^\infty L^2} \ll 1,
$$

which is exactly the first part of the definition of linearizability. Differentiating the equation for w^ε and applying Strichartz inequalities, we find

$$
\begin{aligned}
\varepsilon^{1/q} \|\varepsilon \nabla w^\varepsilon\|_{L^q L^r} &\lesssim \varepsilon^{n\sigma-1-2/q} \varepsilon^{1/q} \left\| |u^\varepsilon|^{2\sigma} \varepsilon \nabla u^\varepsilon \right\|_{L^{q'} L^{r'}} \\
&\lesssim \varepsilon^{n\sigma-1-2/q} \|u^\varepsilon\|_{L^k L^r}^{2\sigma} \varepsilon^{1/q} \|\varepsilon \nabla u^\varepsilon\|_{L^q L^r} \\
&\lesssim \varepsilon^{n\sigma-1-2/q} \|u^\varepsilon\|_{L^k L^r}^{2\sigma} \varepsilon^{1/q} \|\varepsilon \nabla w^\varepsilon\|_{L^q L^r} \\
&\quad + \varepsilon^{n\sigma-1-2/q} \|u^\varepsilon\|_{L^k L^r}^{2\sigma} \varepsilon^{1/q} \|\varepsilon \nabla v^\varepsilon\|_{L^q L^r} \\
&\lesssim \left(\varepsilon^{2/q-1/k} \|u^\varepsilon\|_{L^k L^r} \right)^{2\sigma} \varepsilon^{1/q} \|\varepsilon \nabla w^\varepsilon\|_{L^q L^r} \\
&\quad + \left(\varepsilon^{2/q-1/k} \|u^\varepsilon\|_{L^k L^r} \right)^{2\sigma} \|\varepsilon \nabla u_0^\varepsilon\|_{L^2},
\end{aligned}
$$

where we have used the homogeneous Strichartz estimate for $\varepsilon \nabla v^\varepsilon$. Thus,

$$
\varepsilon^{1/q} \|\varepsilon \nabla w^\varepsilon\|_{L^q L^r} \ll 1.
$$

Using Strichartz inequality again, we conclude

$$
\|\varepsilon \nabla w^\varepsilon\|_{L^\infty L^2} \ll 1,
$$

which completes the proof of the proposition. $\qquad\square$

We now distinguish the cases $\sigma > 2/n$ and $\sigma = 2/n$.

7.6.2 *The L^2-supercritical case: $\sigma > 2/n$*

From Lemma 7.29 and Proposition 7.31, we infer:

Corollary 7.32. *Let $\sigma > 2/n$ and u_0^ε satisfying Assumption 7.27. Assume in addition that initially, the potential energy goes to zero:*

$$
\varepsilon^{n\sigma} \|u_0^\varepsilon\|_{L^{2\sigma+2}}^{2\sigma+2} \xrightarrow[\varepsilon\to 0]{} 0. \tag{7.41}
$$

Let $T > 0$. Then u^ε is linearizable on $[0, T]$, if and only if

$$\limsup_{\varepsilon \to 0} \sup_{0 \leqslant t \leqslant T} \varepsilon^{n\sigma} \|v^\varepsilon(t)\|_{L^{2\sigma+2}}^{2\sigma+2} = 0. \tag{7.42}$$

Proof.   In view of Lemma 7.29, we need only prove that the above condition is sufficient for linearizability. Since $\sigma > 2/n$, we have $k > q$. Using Strichartz inequality, we obtain

$$\varepsilon^{2/q-1/k} \|v^\varepsilon\|_{L^k L^r} \leqslant \left(\varepsilon^{2/q} \|v^\varepsilon\|_{L^\infty L^r} \right)^{1-q/k} \left(\varepsilon^{1/q} \|v^\varepsilon\|_{L^q L^r} \right)^{q/k}$$

$$\lesssim \left(\varepsilon^{2/q} \|v^\varepsilon\|_{L^\infty L^r} \right)^{1-q/k} \|u_0^\varepsilon\|_{L^2}^{q/k}$$

$$\lesssim \left(\varepsilon^{2/q} \|v^\varepsilon\|_{L^\infty L^r} \right)^{1-q/k}.$$

Recalling that $r = 2\sigma + 2$ and $2/q = n\sigma/(2\sigma + 2)$, (7.42) implies

$$\varepsilon^{2/q-1/k} \|v^\varepsilon\|_{L^k L^r} \ll 1,$$

which in turn yields linearizability, from Proposition 7.31. $\square$

The next step in the analysis consists in answering the following question: when is (7.42) violated? As pointed out before, it must be noticed that Corollary 7.32 turns the analysis of a nonlinear problem into the analysis of the behavior of v^ε, solving a *linear* Schrödinger equation.

To study the negation of (7.42), our approach relies on the notion of *profile decomposition*, introduced by P. Gérard [Gérard (1998)] to measure the lack of compactness of critical Sobolev embeddings, inspired by the approach of [Métivier and Schochet (1998)]. For the linear Schrödinger equation, we use more precisely the decomposition in the homogeneous Sobolev space $\dot{H}^1$, due to S. Keraani in the case $n = 3$ [Keraani (2001)]. It can easily be generalized to any spatial dimension. For simplicity, we assume $n = 3$.

Theorem 7.33 ([Keraani (2001)], Theorem 1.6). *Assume $n = 3$. Let $V^\varepsilon = V^\varepsilon(s, y)$ solve*

$$i\partial_s V^\varepsilon + \frac{1}{2} \Delta V^\varepsilon = 0, \tag{7.43}$$

with $V^\varepsilon_{|t=0} = V_0^\varepsilon$, where the family $(V_0^\varepsilon)_{0 < \varepsilon \leqslant 1}$ is bounded in $\dot{H}^1(\mathbb{R}^3)$. Up to extracting a subsequence, we have:

$$V^\varepsilon(s, y) = \sum_{j=0}^{\ell} \frac{1}{\sqrt{\eta_j^\varepsilon}} V_j \left(\frac{s - s_j^\varepsilon}{(\eta_j^\varepsilon)^2}, \frac{y - y_j^\varepsilon}{\eta_j^\varepsilon} \right) + W_\ell^\varepsilon(s, y), \tag{7.44}$$

where $\eta_j^\varepsilon \in \mathbb{R}_+ \setminus \{0\}$ *are the scales of concentration, satisfying the following orthogonality condition:*

$$\forall j \neq k, \quad either \quad \limsup_{\varepsilon \to 0} \frac{\eta_j^\varepsilon}{\eta_k^\varepsilon} + \frac{\eta_k^\varepsilon}{\eta_j^\varepsilon} = +\infty,$$

$$or \quad \eta_j^\varepsilon = \eta_k^\varepsilon \quad and \quad \limsup_{\varepsilon \to 0} \frac{|s_j^\varepsilon - s_k^\varepsilon| + |y_j^\varepsilon - y_k^\varepsilon|}{\eta_j^\varepsilon} = +\infty.$$

The remainder W_ℓ^ε satisfies

$$\limsup_{\varepsilon \to 0} \|W_\ell^\varepsilon\|_{L^q(\mathbb{R}, L^r)} \xrightarrow[\ell \to \infty]{} 0, \tag{7.45}$$

for $\dfrac{2}{q} + \dfrac{3}{r} = \dfrac{1}{2}$, *with $r < +\infty$. Such (q, r) are said to be $\dot{H}^1$-admissible (as opposed to the L^2-admissible pairs of Definition 7.4).*

Finally the V_j's and W_ℓ^ε are solutions to Eq. (7.43) in $L^\infty(\mathbb{R}, \dot{H}^1)$.

We define the rescaled function (recall that $n = 3$)

$$V^\varepsilon(s, y) := \varepsilon^{3/2} v^\varepsilon(\varepsilon s, \varepsilon y),$$

which satisfies the linear equation (7.43) with data

$$V_0^\varepsilon(y) := \varepsilon^{3/2} u_0^\varepsilon(\varepsilon y).$$

Clearly, $(V_0^\varepsilon)_{0 < \varepsilon \leqslant 1}$ is bounded in $\dot{H}^1(\mathbb{R}^3)$. Applying Theorem 7.33, we prove that up to a subsequence, the scales η_j^ε all have a non-zero, finite limit. For that, we use the notion of ε_n-oscillatory sequences. For more details on the subject, we refer to [Bahouri and Gérard (1999); Gérard *et al.* (1997)].

Definition 7.34. Let $(\varepsilon_n)_{n \in \mathbb{N}}$ be a given sequence in $\mathbb{R}_+ \setminus \{0\}$, and let (V^n) be a bounded sequence in $\dot{H}^1$. The sequence (V^n) is ε_n-oscillatory if the following property holds:

$$\limsup_{n \to \infty} \int_{\varepsilon_n |\xi| \leqslant R^{-1}} |\xi|^2 |\mathcal{F}(V^n)(\xi)|^2 \, d\xi + \int_{\varepsilon_n |\xi| \geqslant R} |\xi|^2 |\mathcal{F}(V^n)(\xi)|^2 \, d\xi \xrightarrow[R \to +\infty]{} 0.$$

Remark 7.35. For a time–dependent sequence (V^n), uniformly bounded in $L^\infty(\mathbb{R}_+, \dot{H}^1)$, the definition holds taking the limit uniformly in time.

It is easy to see (see [Gérard (1998)] or [Keraani (2001)]) that V^ε is η_j^ε-oscillatory for every sequence η_j^ε appearing in the decomposition (7.44).

Lemma 7.36. *Suppose the sequence (V^ε) is η^ε-oscillatory for some sequence η^ε. Then up to a subsequence, $\eta^\varepsilon = \lambda$ for some $\lambda > 0$.*

Proof.   We can write, uniformly in time,

$$\|\nabla V^\varepsilon(s)\|_{L^2}^2 \lesssim \int_{R^{-1}\leqslant \eta^\varepsilon|\xi|\leqslant R} |\xi|^2 |\mathcal{F}(V^\varepsilon)|^2\, d\xi + \delta(\varepsilon, R)$$

$$\lesssim \left(\frac{R}{\eta^\varepsilon}\right)^2 \|V^\varepsilon(s)\|_{L^2}^2 + \delta(\varepsilon, R),$$

where $\limsup_{\varepsilon\to 0} \delta(\varepsilon, R) \to 0$ as $R \to \infty$. The conservation of the energy yields

$$\|\nabla V^\varepsilon(s)\|_{L^2}^2 = \|\nabla V^\varepsilon(0)\|_{L^2}^2 = \|\varepsilon\nabla u_0^\varepsilon\|_{L^2}^2.$$

Up to a subsequence, we can suppose that this quantity is bounded from below by some $c > 0$ independent of $\varepsilon \in\,]0, 1]$ (otherwise, Condition (7.42) would not be violated, from Gagliardo–Nirenberg inequality). Fixing R such that

$$\limsup_{\varepsilon\to 0} \delta(\varepsilon, R) \leqslant \frac{c}{2},$$

yields, up to an extraction,

$$\infty > \limsup_{\varepsilon\to 0} \eta^\varepsilon = \lambda \geqslant 0.$$

Now suppose that $\lambda = 0$. Write, for all time,

$$V^\varepsilon = V_R^\varepsilon + W_R^\varepsilon, \quad \text{with} \quad \mathcal{F}V_R^\varepsilon(t, \xi) := \mathbf{1}_{R^{-1}\leqslant \eta^\varepsilon|\xi|\leqslant R}\mathcal{F}V^\varepsilon(t, \xi),$$

and for all $\delta > 0$, if R is large enough uniformly in ε and η^ε, we have

$$\|W_R^\varepsilon\|_{L^\infty(\mathbb{R}, H^1)} \leqslant \delta.$$

Gagliardo–Nirenberg inequality implies that $\|W_R^\varepsilon\|_{L^\infty(\mathbb{R}, L^{2\sigma+2})}$ can be chosen arbitrarily small if R is large enough, uniformly in ε and η^ε. Thus,

$$\|V^\varepsilon\|_{L^\infty([0,T], L^{2\sigma+2})}^{2\sigma+2} \lesssim \|V_R^\varepsilon\|_{L^\infty([0,T], L^{2\sigma+2})}^{2\sigma+2} + o(1) \qquad (7.46)$$

$$\lesssim \|V_R^\varepsilon\|_{L^\infty([0,T], L^2)}^{(2\sigma+2)(1-\delta(2\sigma+2))} + o(1),$$

where the second inequality is due again to Gagliardo–Nirenberg inequality and the boundedness of V^ε in H^1.

Frequency localization implies that for all $s \in [0, T]$,

$$\|V_R^\varepsilon(s)\|_{L^2}^2 \lesssim (\eta^\varepsilon)^2 \int_{\frac{1}{R}\leqslant \eta^\varepsilon\xi\leqslant R} |\xi|^2\, |\mathcal{F}V_R^\varepsilon(s, \xi)|^2\, d\xi.$$

So the result follows, since by assumption the left hand side in (7.46) does not go to zero and $(2\sigma + 2)(1 - \delta(2\sigma + 2)) = 2 - \sigma > 0$ when $n = 3$.   □

Lemma 7.36 has several important consequences. First, the V_j's and W_ℓ^ε are bounded in $L^\infty(\mathbb{R}, H^1)$. Indeed, by the orthogonality properties, one has

$$V^\varepsilon(s + s_j^\varepsilon, y + y_j^\varepsilon) \rightharpoonup V_j(s, y) \quad \text{in} \quad \mathcal{D}'(\mathbb{R} \times \mathbb{R}^3).$$

But V^ε is bounded in $L^\infty(\mathbb{R}, L^2)$, so it follows that for all $j \in \mathbb{N}$, the profiles V_j are bounded in $L^\infty(\mathbb{R}, L^2)$. This implies also that

$$(W_\ell^\varepsilon)_{0<\varepsilon\leqslant 1} \quad \text{is bounded in} \quad L^\infty(\mathbb{R}, L^2), \quad \text{uniformly in} \quad \ell \in \mathbb{N}. \quad (7.47)$$

Second,

$$\limsup_{\varepsilon \to 0} \|W_\ell^\varepsilon\|_{L^\infty(\mathbb{R}; L^{2\sigma+2})} \xrightarrow[\ell \to \infty]{} 0.$$

This follows from (7.45) with $q = +\infty$, (7.47) and Hölder's inequality.

Note also that the family $(V_j)_{j \in \mathbb{N}}$ is not trivial. Indeed,

$$\|V^\varepsilon\|_{L^\infty([0,T],L^{2\sigma+2})} \leqslant \sum_{j=1}^{\ell} \|V_j\|_{L^\infty([0,T],L^{2\sigma+2})} + \|W_\ell^\varepsilon\|_{L^\infty([0,T],L^{2\sigma+2})},$$

so since (7.42) is not satisfied, the above limit implies that all of the V_j's cannot be zero.

Back to the definition of v^ε, it follows that one can write

$$u_0^\varepsilon(x) = \sum_{j=0}^{\ell} \frac{1}{\varepsilon^{3/2}} V_j \left(-\frac{t_j^\varepsilon}{\varepsilon}, \frac{x - x_j^\varepsilon}{\varepsilon} \right) + w_\ell^\varepsilon(x),$$

where we have set

$$t_j^\varepsilon := \varepsilon s_j^\varepsilon, \quad x_j^\varepsilon := \varepsilon y_j^\varepsilon \quad \text{and} \quad w_\ell^\varepsilon := \frac{1}{\varepsilon^{3/2}} W_\ell^\varepsilon \left(0, \frac{\cdot}{\varepsilon} \right).$$

Note that w_ℓ^ε is such that

$$\limsup_{\varepsilon \to 0} \varepsilon^{n\sigma} \|w_\ell^\varepsilon\|_{L^{2\sigma+2}}^{2\sigma+2} \xrightarrow[\ell \to \infty]{} 0.$$

Using the assumption (7.41), one can prove

$$\limsup_{\varepsilon \to 0} \frac{t_j^\varepsilon}{\varepsilon} = +\infty.$$

Finally, quadratic oscillations appear like in Lemma 7.20, up to a rescaling, and we obtain (see [Carles *et al.* (2003)] for details):

Theorem 7.37. *Let $\sigma > 2/n$. Let Assumption 7.27 be satisfied, and assume that (7.41) holds. Assume that (7.42) is not satisfied for some $T > 0$.*

Then up to the extraction of a subsequence, there exist an orthogonal family $(t_j^\varepsilon, x_j^\varepsilon)_{j \in \mathbb{N}}$ in $\mathbb{R}_+ \times \mathbb{R}^n$, that is

$$\limsup_{\varepsilon \to 0} \left(\frac{|t_j^\varepsilon - t_k^\varepsilon|}{\varepsilon} + \frac{|x_j^\varepsilon - x_k^\varepsilon|}{\varepsilon} \right) = \infty \quad \forall j \neq k,$$

a family $(\Psi_\ell^\varepsilon)_{\ell \in \mathbb{N}}$, bounded in $H_\varepsilon^1(\mathbb{R}^n)$, and a (non-trivial) family $(\varphi_j)_{j \in \mathbb{N}}$, bounded in $\mathcal{F}(H^1)$, such that:

$$u_0^\varepsilon(x) = \Psi_\ell^\varepsilon(x) + r_\ell^\varepsilon(x), \quad with \limsup_{\varepsilon \to 0} \varepsilon^{n\sigma} \|U_0^\varepsilon(t) r_\ell^\varepsilon\|_{L^\infty(\mathbb{R}_+, L^{2\sigma+2})}^{2\sigma+2} \xrightarrow[\ell \to \infty]{} 0,$$

and for every $\ell \in \mathbb{N}$, the following asymptotics holds in $L^2(\mathbb{R}^n)$, as $\varepsilon \to 0$,

$$\Psi_\ell^\varepsilon(x) = \sum_{j=0}^{\ell} \frac{1}{(t_j^\varepsilon)^{n/2}} \varphi_j \left(\frac{x - x_j^\varepsilon}{t_j^\varepsilon} \right) e^{-i|x - x_j^\varepsilon|^2/(2\varepsilon t_j^\varepsilon)} + o(1).$$

Moreover, we have

$$\limsup_{\varepsilon \to 0} \frac{t_j^\varepsilon}{\varepsilon} = +\infty \quad and \quad \limsup_{\varepsilon \to 0} t_j^\varepsilon \in [0, T], \quad \forall j \in \mathbb{N}.$$

In view of Corollary 7.32, the condition on r_ℓ^ε means that in the limit $\ell \to \infty$, the evolution of r_ℓ^ε is essentially linear. Therefore, the only obstruction to linearizability stems from Ψ_ℓ^ε. The asymptotic behavior of Ψ_ℓ^ε then shows that linearizability fails to hold because of the presence of quadratic oscillations in the initial data.

7.6.3 *The L^2-critical case: $\sigma = 2/n$*

When $\sigma = 2/n$, it is easy to see that not only initial data of Theorem 7.37 have a truly nonlinear evolution. Let U solve

$$i\partial_t U + \frac{1}{2}\Delta U = |U|^{4/n} U, \tag{7.48}$$

with $U_{|t=0} = \phi$. If $\phi \in \Sigma$, then U is defined globally in time, $U \in C(\mathbb{R}_t; \Sigma)$. Let $(t_0, x_0) \in \mathbb{R} \times \mathbb{R}^n$. Then the function

$$u^\varepsilon(t, x) = \frac{1}{\varepsilon^{n/4}} U \left(t - t_0, \frac{x - x_0}{\sqrt{\varepsilon}} \right) \tag{7.49}$$

solves

$$i\varepsilon \partial_t u^\varepsilon + \frac{\varepsilon^2}{2}\Delta u^\varepsilon = \varepsilon^2 |u^\varepsilon|^{4/n} u^\varepsilon. \tag{7.50}$$

Moreover, $u^\varepsilon(0, \cdot)$ and $\varepsilon \nabla u^\varepsilon(0, \cdot)$ are bounded in $L^2(\mathbb{R}^n)$. This peculiar solution is such that the nonlinearity in (7.50) has a leading order influence for any (finite) time, near $x = x_0$.

We check that the solutions (7.49) are deduced from those of Theorem 7.37 by the scaling

$$\widetilde{U}(t,x) = \lambda^{n/2} U \left(\lambda^2 t, \lambda x \right),$$

with $\lambda = \sqrt{\varepsilon}$. This scaling leaves Eq. (7.48) invariant (for any positive λ). We saw above that one of the key steps in the proof of Theorem 7.37 consists in singling out the scale $\eta_j^\varepsilon = 1$ by showing that scales going to zero or to infinity yield a linearizable evolution. This step must be modified when $\sigma = 2/n$.

As a matter of fact, we have to resume the study from the very start. It is easy to check that in the above example, (7.42) is verified, even though the solution u^ε is obviously not linearizable. The conclusion of Corollary 7.32 is no longer true in the case $\sigma = 2/n$, and the limitation $\sigma > 2/n$ in the proof of Corollary 7.32 was not a technical artifice. We have the following new linearizability criterion:

Theorem 7.38 ([Carles and Keraani (2007)]). *Suppose* $\sigma = 2/n$, *with* $n = 1$ *or* 2, *and that Assumption 7.27 is satisfied. Let* I^ε *be a time interval containing the origin. Then* u^ε *is linearizable on* I^ε *if and only if*

$$\limsup_{\varepsilon \to 0} \varepsilon \|v^\varepsilon\|_{L^{2+4/n}(I^\varepsilon \times \mathbb{R}^n)}^{2+4/n} = 0. \qquad (7.51)$$

The proof that (7.51) implies linearizability is a direct consequence of Proposition 7.31, since when $\sigma = 2/n$, we have

$$r = q = k = \frac{4}{n} + 2.$$

It is in the proof of the converse that the assumption $n = 1$ or 2 appears in [Carles and Keraani (2007)]. This assumption could be removed, in view of the results in [Bégout and Vargas (2007)]. This part of the proof relies on a profile decomposition, in $L^2(\mathbb{R}^n)$, as opposed to the profile decomposition in $\dot{H}^1(\mathbb{R}^n)$ which was used in the case $\sigma > 2/n$. Moreover, we apply this technique not only to solutions to the linear Schrödinger equation, but also to solutions of the nonlinear equation, like in [Bahouri and Gérard (1999); Keraani (2001)]. This part therefore relies on the existence of these two decompositions, which are established after improved Strichartz estimates. In the case $n = 2$, such estimates were proved after the work of J. Bourgain [Bourgain (1995)], in [Bourgain (1998); Moyua et al. (1999)], and profile decomposition were given by F. Merle and L. Vega [Merle and Vega (1998)]. The case $n = 1$ was established by S. Keraani in his PhD thesis, and appears in [Carles and Keraani (2007)]. For $n \geqslant 3$, improved Strichartz estimates are proved in [Bégout and Vargas (2007)]; as noted in [Carles and Keraani (2007)], once such estimates are available, a general technique to prove profile decompositions yields the result.

Since the proof of the two profile decompositions and the proof of Theorem 7.38 are rather technical, we leave them out here, and invite the interested reader to consult directly [Carles and Keraani (2007)] and [Bégout and Vargas (2007)]. To make the comparison with the L^2 supercritical case complete, we conclude this paragraph by stating the analogue of Theorem 7.37 in the L^2-critical case.

Definition 7.39. If $(h_j^\varepsilon, t_j^\varepsilon, x_j^\varepsilon, \xi_j^\varepsilon)_{j \in \mathbb{N}}$ is a family of sequences in $\mathbb{R}_+ \setminus \{0\} \times \mathbb{R} \times \mathbb{R}^n \times \mathbb{R}^n$, then we say that $(h_j^\varepsilon, t_j^\varepsilon, x_j^\varepsilon, \xi_j^\varepsilon)_{j \in \mathbb{N}}$ is an orthogonal family if

$$\limsup_{\varepsilon \to 0} \left(\frac{h_j^\varepsilon}{h_k^\varepsilon} + \frac{h_k^\varepsilon}{h_j^\varepsilon} + \frac{|t_j^\varepsilon - t_k^\varepsilon|}{(h_j^\varepsilon)^2} + \left| \frac{x_j^\varepsilon - x_k^\varepsilon}{h_j^\varepsilon} + \frac{t_j^\varepsilon \xi_j^\varepsilon - t_k^\varepsilon \xi_k^\varepsilon}{h_j^\varepsilon} \right| \right) = \infty, \quad \forall j \neq k.$$

Theorem 7.40. *Assume $n = 1$ or 2, and let Assumption 7.27 be satisfied. Let $T > 0$ and assume that (7.51) is not satisfied with $I^\varepsilon = [0, T]$. Then up to the extraction of a subsequence, there exist an orthogonal family $(h_j^\varepsilon, t_j^\varepsilon, x_j^\varepsilon, \xi_j^\varepsilon)_{j \in \mathbb{N}}$, a family $(\phi_j)_{j \in \mathbb{N}}$, bounded in $L^2(\mathbb{R}^n)$, such that:*

$$u_0^\varepsilon(x) = \sum_{j=1}^{\ell} \widetilde{H}_j^\varepsilon(\phi_j)(x) + w_\ell^\varepsilon(x),$$

where $\widetilde{H}_j^\varepsilon(\phi_j)(x) = e^{ix \cdot \xi_j^\varepsilon / \sqrt{\varepsilon}} e^{-i\varepsilon \frac{t_j^\varepsilon}{2} \Delta} \left(\frac{1}{(h_j^\varepsilon \sqrt{\varepsilon})^{n/2}} \phi_j \left(\frac{x - x_j^\varepsilon}{h_j^\varepsilon \sqrt{\varepsilon}} \right) \right),$

and $\limsup_{\varepsilon \to 0} \varepsilon \, \|U_0^\varepsilon(t) w_\ell^\varepsilon\|_{L^{2+4/n}(\mathbb{R} \times \mathbb{R}^n)}^{2+4/n} \xrightarrow[\ell \to +\infty]{} 0.$

We have $\liminf t_j^\varepsilon / (h_j^\varepsilon)^2 \neq -\infty$, $\liminf (T - t_j^\varepsilon)/(h_j^\varepsilon)^2 \neq -\infty$ *(as $\varepsilon \to 0$), and $\sqrt{\varepsilon} \leqslant h_j^\varepsilon \leqslant 1$ for every $j \in \mathbb{N}$.*
If $t_j^\varepsilon / (h_j^\varepsilon)^2 \to +\infty$ as $\varepsilon \to 0$, then we also have, in $L^2(\mathbb{R}^n)$:

$$\widetilde{H}_j^\varepsilon(\phi_j)(x) = e^{ix \cdot \xi_j^\varepsilon / \sqrt{\varepsilon} + in\pi/4 - i|x - x_j^\varepsilon|^2 / (2\varepsilon t_j^\varepsilon)}$$

$$\times \left(\frac{h_j^\varepsilon}{t_j^\varepsilon \sqrt{\varepsilon}} \right)^{n/2} \widehat{\phi}_j \left(-\frac{h_j^\varepsilon}{t_j^\varepsilon \sqrt{\varepsilon}} (x - x_j^\varepsilon) \right) + o(1) \text{ as } \varepsilon \to 0.$$

A few comments are in order. First, w_ℓ^ε plays the same role as r_ℓ^ε in Theorem 7.37: for large ℓ, it does not see nonlinear effects at leading order, since it satisfies the linearizability condition (now Eq. (7.51)) with $I^\varepsilon = \mathbb{R}$. In the last case of the theorem, we recover the quadratic oscillations. However, there are other obstructions to linearizability, since we may have $t_j^\varepsilon = T/2$ and $h_j^\varepsilon = 1$: this is the case in the example given in the beginning of this paragraph, Eq. (7.49). Roughly speaking, the above result shows that this is the other borderline case: the scales of concentration, $h_j^\varepsilon \sqrt{\varepsilon}$, lie between ε and $\sqrt{\varepsilon}$. Therefore, (7.49) and (7.36) are essentially the two extreme cases when $\sigma = 2/n$. Finally, almost all the conclusions of the above theorem remain true, if instead of Assumption 7.27, we simply suppose that $(u_0^\varepsilon)_{0 < \varepsilon \leqslant 1}$ is bounded in $L^2(\mathbb{R}^n)$. The only difference in the conclusions is that we lose the lower bound for the scales h_j^ε: we cannot say more than $0 < h_j^\varepsilon \leqslant 1$.

7.6.4 *Nonlinear superposition*

When Theorems 7.37 and 7.40 are available, one can ask: how does an initial sum of data with quadratic oscillations evolve under the nonlinear dynamics of Eq. (7.37)? The answer, both when $\sigma > 2/n$ and when $\sigma = 2/n$, is that each part of the initial data evolves independently of the others at leading order, when $\varepsilon \to 0$. In other words, there exists a superposition principle, even though we consider nonlinear equations.

Heuristically, the explanation is rather simple. We know that in the linear case, each part of the data with the form

$$a_j\left(x - x_j\right) e^{-i|x-x_j|^2/(2\varepsilon t_j)}$$

focuses at the point $(t, x) = (t_j, x_j)$. It is of order $\mathcal{O}(1)$ away from the focus, and of order $\varepsilon^{-n/2}$ at the focal point, in a neighborhood of order ε. In the nonlinear case, since $n\sigma > 1$, the nonlinearity is negligible in WKB régime. Therefore, outside the focal points, the nonlinear superposition principle is simply the usual linear superposition principle. Near the focal points, nonlinear effects become relevant at leading order. The decoupling of nonlinear interactions is due to the orthogonality of the cores, denoted $(t_j^\varepsilon, x_j^\varepsilon)$. Roughly speaking, concentration at focal points occurs in balls which do not intersect. From this point of view, the nonlinear superposition is a consequence of precise geometric properties.

In the L^2-supercritical case, the rigorous justification of this statement relies on a precise use of the linearizability criterion (7.42). In the L^2-critical case, the nonlinear superposition principle is a direct consequence of the nonlinear profile decomposition, which is at the heart of the proof of Theorem 7.38.

7.7   Focusing on a line

To conclude this long chapter, we mention a result where focusing at one point is replaced by focusing on a line. Consider the Cauchy problem, in space dimension $n = 2$,

$$i\varepsilon\partial_t u^\varepsilon + \frac{\varepsilon^2}{2}\Delta u^\varepsilon = \varepsilon^\alpha |u^\varepsilon|^{2\sigma} u^\varepsilon \quad ; \quad u^\varepsilon(0,x) = a_0(x)e^{-x_1^2/(2\varepsilon)}, \qquad (7.52)$$

with $x = (x_1, x_2) \in \mathbb{R}^2$. The rays of geometric optics are given by

$$\dot{x} = \xi \quad ; \quad \dot{\xi} = 0 \quad ; \quad x_{|t=0} = y \quad ; \quad \xi_{1|t=0} = -y_1 \quad ; \quad \xi_{2|t=0} = 0.$$

Therefore, we have

$$x_1(t) = y_1(1-t) \quad ; \quad x_2(t) = y_2.$$

Rays meet on the line $\{x_1 = 0\}$ at time $t = 1$, see Fig. 7.3. Heuristically, x_2

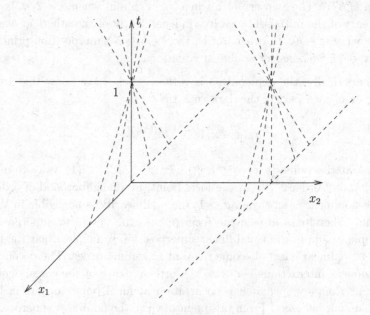

Fig. 7.3   Focusing on a line in $\mathbb{R}^2$.

plays the role of a parameter, and the geometric information is carried in the x_1 variable. Roughly speaking, there is focusing at a point ($x_1 = 0$) along a continuous range of a parameter ($x_2 \in \mathbb{R}$). This suggests the following distinctions for Eq. (7.52):

	$\alpha > 1$	$\alpha = 1$
$\alpha > \sigma$	linear propagation linear caustic	nonlinear propagation linear caustic
$\alpha = \sigma$	linear propagation, nonlinear caustic	nonlinear propagation nonlinear caustic

This is exactly the general table considered in the beginning of this chapter, in the case $n = 1$. This table is justified in [Carles (2000a)], where the case "nonlinear propagation, nonlinear caustic", is not studied. In particular, in the case "linear propagation, nonlinear caustic", the caustic crossing is described by a nonlinear scattering operator, associated to the equation:

$$i\partial_t\psi + \frac{1}{2}\partial_{x_1}^2\psi = |\psi|^{2\sigma}\psi \quad ; \quad e^{-i\frac{t}{2}\partial_{x_1}^2}\psi(t, x_1, x_2)\big|_{t=\infty} = \psi_-(x_1, x_2).$$

By working in suitable spaces, the usual scattering theory for the nonlinear Schrödinger equation, recalled in Theorems 7.14 and 7.16, is adapted for $\sigma > \sigma_0(1)$, the critical power for scattering in space dimension one. Like in the case of a single focal point, the crossing of the line caustic is described in terms of this operator, in the limit $\varepsilon \to 0$. We invite the interested reader to consult [Carles (2000a)] for technical issues.

Chapter 8

Focal Point in the Presence of an External Potential

In this chapter, we continue the analysis of the previous one, in the presence of a non-trivial external potential:

$$i\varepsilon\partial_t u^\varepsilon + \frac{\varepsilon^2}{2}\Delta u^\varepsilon = V u^\varepsilon + \varepsilon^\alpha \left|u^\varepsilon\right|^{2\sigma} u^\varepsilon.$$

We consider two cases:

- When the initial data are independent of ε, and the external potential is harmonic.
- When the initial data are concentrated at scale ε, and the external potential is more general.

We will see that in the first case, it is easy to include rapid plane oscillations in the initial data (Remark 8.3). In the second case, we present a rather complete picture when the external potential is exactly a polynomial of degree at most two. On the other hand, for general subquadratic potentials, we give partial results only.

8.1 Isotropic harmonic potential

First we consider the case of an isotropic harmonic potential, with ε-independent initial data:

$$i\varepsilon\partial_t u^\varepsilon + \frac{\varepsilon^2}{2}\Delta u^\varepsilon = \frac{|x|^2}{2}u^\varepsilon + \varepsilon^\alpha \left|u^\varepsilon\right|^{2\sigma} u^\varepsilon \quad ; \quad u^\varepsilon(0,x) = a_0(x). \qquad (8.1)$$

Rays of geometric optics are given by the Hamiltonian system

$$\dot{x} = \xi \quad ; \quad \dot{\xi} = -x \quad ; \quad x(0,y) = y \quad ; \quad \xi(0,y) = 0. \qquad (8.2)$$

We find: $\ddot{x} + x = 0$, along with the initial conditions $x(0,y) = y$ and $\dot{x}(0,y) = 0$. Therefore, $x(t,y) = y\cos t$: rays are sinusoids, which meet at

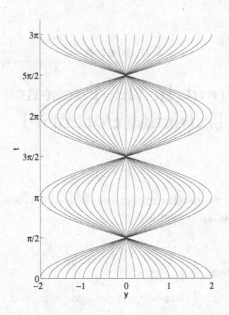

Fig. 8.1   Rays of geometric optics: sinusoids.

the origin for $t \in \pi/2 + \pi\mathbb{Z}$, see Fig. 8.1. Geometrically, this is quite the same thing as in the previous paragraph (no external potential, but initial quadratic oscillations), repeated indefinitely many times. The parallel is even analytical. Consider the linear analogue of Eq. (8.1):

$$i\varepsilon\partial_t v^\varepsilon + \frac{\varepsilon^2}{2}\Delta v^\varepsilon = \frac{|x|^2}{2}v^\varepsilon \quad ; \quad v^\varepsilon(0,x) = a_0(x). \tag{8.3}$$

Because the potential is exactly quadratic, this solution is known explicitly in terms of an oscillatory integral, given by the *Mehler's formula* (see e.g. [Feynman and Hibbs (1965); Hörmander (1995)]): for $|t| < \pi$,

$$v^\varepsilon(t,x) = \frac{1}{(2i\pi\varepsilon\sin t)^{n/2}}\int_{\mathbb{R}^n} e^{i\left(\frac{|x|^2+|y|^2}{2}\cos t - x\cdot y\right)/(\varepsilon\sin t)}a_0(y)dy. \tag{8.4}$$

For $t \in \pi + 2\pi\mathbb{Z}$, the fundamental solution is singular, and some extra phase shifts must be included in the above formula when $|t| > \pi$; see e.g. [Yajima (1996)] and references therein. For $0 < t < \pi/2$, we can apply stationary phase formula in (8.4). We find:

$$v^\varepsilon(t,x) \underset{\varepsilon\to 0}{\sim} \frac{1}{(\cos t)^{n/2}}a_0\left(\frac{x}{\cos t}\right)e^{-i|x|^2\tan t/(2\varepsilon)}. \tag{8.5}$$

Note that we retrieve the solutions to the eikonal equation and to the transport equation of WKB analysis, derived in Examples 1.12 and 1.18 respectively. For $t = \pi/2$, we find directly:

$$v^\varepsilon\left(\frac{\pi}{2}, x\right) = \frac{1}{(2i\pi\varepsilon)^{n/2}} \int_{\mathbb{R}^n} e^{-ix\cdot y/\varepsilon} a_0(y)dy = \frac{e^{-in\pi/4}}{\varepsilon^{n/2}} \widehat{a}_0\left(\frac{x}{\varepsilon}\right). \quad (8.6)$$

The general picture we obtain for the solution of the linear equation is therefore closely akin to what we got in the case of a single focal point, with no external potential, see (6.2). Thus, the parallel is not only geometrical, it is also analytical. As a consequence, the same table as in Chap. 7 is expected. We shall consider only the critical case "linear propagation, nonlinear caustic", that is

$$i\varepsilon\partial_t u^\varepsilon + \frac{\varepsilon^2}{2}\Delta u^\varepsilon = \frac{|x|^2}{2}u^\varepsilon + \varepsilon^{n\sigma}|u^\varepsilon|^{2\sigma}u^\varepsilon \quad ; \quad u^\varepsilon(0, x) = a_0(x), \quad (8.7)$$

with $n\sigma > 1$. This case was studied in [Carles (2003b)]. As mentioned there, the proof in the case $\alpha = n\sigma > 1$ would make it possible to show that when $\alpha > \max(n\sigma, 1)$, the nonlinear term is indeed negligible on any (finite) time interval, in the limit $\varepsilon \to 0$. Since we have given a rather detailed proof of this fact in the case of a focal point without external potential, we do not pursue this issue here.

To introduce two useful operators, resume one of the points of view that led us to the introduction of the operator $x/\varepsilon + i(t - 1)\nabla$ in the previous chapter. Using stationary phase argument on Mehler's formula (8.4), or solving directly the eikonal equation, we see that outside the focal point, the rapid oscillations of the linear solution are given by

$$\phi_{\text{eik}}(t, x) = -\frac{|x|^2}{2}\tan t.$$

To recover the fact that the L^p norm of the approximation of v^ε is independent of ε by using Gagliardo–Nirenberg inequalities, we replace the usual operator ∇ by

$$e^{i\phi_{\text{eik}}(t,x)/\varepsilon}\nabla\left(e^{-i\phi_{\text{eik}}(t,x)/\varepsilon}\cdot\right) = e^{-i|x|^2\tan t/(2\varepsilon)}\nabla\left(e^{i|x|^2\tan t/(2\varepsilon)}\cdot\right).$$

Also, to compensate the scaling factor $\cos t$ in the approximation of v^ε, multiply the above operator by $\cos t$. Up to an irrelevant factor i, we therefore consider

$$J^\varepsilon(t) = \frac{1}{i}\cos t\, e^{-i|x|^2\tan t/(2\varepsilon)}\nabla\left(e^{i|x|^2\tan t/(2\varepsilon)}\cdot\right) = \frac{x}{\varepsilon}\sin t - i\cos t\nabla. \quad (8.8)$$

Like in the case of Chap. 7 with $x + it\nabla$, we note that this operator has been known for quite a long time (*Heisenberg derivative*, see e.g. [Robert (1987); Thirring (1981)]). Denote

$$U^\varepsilon(t) = \exp\left(-i\frac{t}{2\varepsilon}\left(-\varepsilon^2\Delta + |x|^2\right)\right)$$

the propagator associated to (8.3). Then we also have

$$J^\varepsilon(t) = U^\varepsilon(t)\left(\frac{1}{i}\nabla\right)U^\varepsilon(-t). \tag{8.9}$$

Therefore, it commutes with the linear part of the equation:

$$\left[i\varepsilon\partial_t + \frac{\varepsilon^2}{2}\Delta - \frac{|x|^2}{2}, J^\varepsilon(t)\right] = 0.$$

Similarly, another important Heisenberg derivative can be computed:

$$\begin{aligned}H^\varepsilon(t) &= U^\varepsilon(t)xU^\varepsilon(-t) = x\cos t + i\varepsilon\sin t\nabla \\ &= i\varepsilon\sin t\, e^{i|x|^2/(2\varepsilon\tan t)}\nabla\left(e^{-i|x|^2/(2\varepsilon\tan t)}\cdot\right).\end{aligned} \tag{8.10}$$

By the first relation, H^ε also commutes with the linear part of the equation. The phase $\phi(t,x) = |x|^2/(2\tan t)$ solves the eikonal equation

$$\partial_t\phi + \frac{1}{2}|\nabla\phi|^2 + \frac{|x|^2}{2} = 0 \quad ; \quad \phi\left(\frac{\pi}{2}, x\right) = 0.$$

Note that, as Heisenberg derivatives or as linear combinations of x and ∇ with time dependent coefficients, J^ε and H^ε are well-defined for all time. On the other hand, the factorization with a phase solving the eikonal equation is valid for almost all time only. Finally, note that the operators J^ε and H^ε make it possible to rewrite the (conserved) energy associated to Eq. (8.3) (which is the kinetic part of the energy associated to Eq. (8.1)):

$$\begin{aligned}E_{\text{lin}}^\varepsilon(0) = E_{\text{lin}}^\varepsilon(t) &= \frac{1}{2}\left\|\varepsilon\nabla v^\varepsilon(t)\right\|_{L^2}^2 + \frac{1}{2}\int_{\mathbb{R}^n}|x|^2\left|v^\varepsilon(t,x)\right|^2 dx \\ &= \frac{1}{2}\left\|\varepsilon J^\varepsilon(t)v^\varepsilon(t)\right\|_{L^2}^2 + \frac{1}{2}\left\|H^\varepsilon(t)v^\varepsilon(t)\right\|_{L^2}^2.\end{aligned}$$

In view of the nonlinear analysis, we list the most interesting properties of these operators below.

Lemma 8.1. *The operators J^ε and H^ε defined by (8.8) and (8.10) respectively satisfy the following properties.*
(1) *They commute with the linear part of the equation, since*

$$J^\varepsilon(t) = U^\varepsilon(t)\left(\frac{1}{i}\nabla\right)U^\varepsilon(-t) \; ; \; H^\varepsilon(t) = U^\varepsilon(t)xU^\varepsilon(-t).$$

(2) *There are two real-valued functions* $\phi_1(t,x)$ *and* $\phi_2(t,x)$ *such that we can write*

$$J^\varepsilon(t) = -i\cos t e^{i\phi_1(t,x)/\varepsilon} \nabla \left(e^{-i\phi_1(t,x)/\varepsilon} \cdot \right), \qquad t \notin \frac{\pi}{2} + \pi\mathbb{Z},$$

$$H^\varepsilon(t) = i\varepsilon \sin t e^{i\phi_2(t,x)/\varepsilon} \nabla \left(e^{-i\phi_2(t,x)/\varepsilon} \cdot \right), \qquad t \notin \pi\mathbb{Z}.$$

(3) *Weighted Gagliardo–Nirenberg estimates are available: for* $0 \leqslant \delta(p) < 1$, *there exists* C_p *such that for all* $u \in \Sigma$,

$$\|u\|_{L^p} \leqslant \frac{C_p}{|\cos t|^{\delta(p)}} \|u\|_{L^2}^{1-\delta(p)} \|J^\varepsilon(t)u\|_{L^2}^{\delta(p)}, \qquad t \notin \frac{\pi}{2} + \pi\mathbb{Z}, \qquad (8.11)$$

$$\|u\|_{L^p} \leqslant \frac{C_p}{|\varepsilon \sin t|^{\delta(p)}} \|u\|_{L^2}^{1-\delta(p)} \|H^\varepsilon(t)u\|_{L^2}^{\delta(p)}, \qquad t \notin \pi\mathbb{Z}. \qquad (8.12)$$

(4) *They act on gauge invariant nonlinearities like derivatives. If* $G(z) = F\left(|z|^2\right) z$ *is* C^1, *then*

$$J^\varepsilon(t)G(u) = \partial_z G(u) J^\varepsilon(t)u - \partial_{\bar{z}} G(u) \overline{J^\varepsilon(t)u},$$

$$H^\varepsilon(t)F(u) = \partial_z G(u) H^\varepsilon(t)u - \partial_{\bar{z}} G(u) \overline{H^\varepsilon(t)u}.$$

The last two points are direct consequences of the second. Remark that the third point implies that J^ε yields good L^p estimates away from focuses, while H^ε is better suited near focal points. This lemma shows that these operators enjoy similar properties to that of Killing vector-fields, whose use has proven efficient in the study of the nonlinear wave equation; see e.g. [Klainerman (1985)]. Before stating our main result, we point out the following formal approximation, which turns out to be not so formal during the proof:

$$J^\varepsilon(t) \underset{t \to \pi/2}{\sim} \frac{x}{\varepsilon} + i\left(t - \frac{\pi}{2}\right)\nabla \quad ; \quad H^\varepsilon(t) \underset{t \to \pi/2}{\sim} i\varepsilon\nabla.$$

Up to replacing the focusing time $t = 1$ by $t = \pi/2$, we retrieve the two operators used in Chap. 7. The geometrical interpretation is that near the focuses, sinusoids can be approximated by straight lines. The analytical interpretation is that near the focuses, the harmonic potential becomes negligible.

Theorem 8.2. *Let* $1 \leqslant n \leqslant 5$, $a_0 \in \Sigma$ *and* $\sigma > 1/2$ *with* $\sigma_0(n) \leqslant \sigma < \frac{2}{n-2}$, *where* $\sigma_0(n)$ *is defined in Theorem 7.16. Let* $k \in \mathbb{N}$. *Then the following asymptotics holds for* u^ε *when* $\pi/2 + (k-1)\pi < a \leqslant b < \pi/2 + k\pi$: *for all* $\mathcal{B}^\varepsilon \in \{\mathrm{Id}, J^\varepsilon, H^\varepsilon\}$,

$$\sup_{a \leqslant t \leqslant b} \left\| \mathcal{B}^\varepsilon(t)\left(u^\varepsilon(t) - \right.\right.$$

$$\left.\left. -\frac{e^{-ink\pi/2}}{|\cos t|^{n/2}} \left(\mathcal{F} \circ S^k \circ \mathcal{F}^{-1}\right) a_0 \left(\frac{\cdot}{\cos t}\right) e^{-i|\cdot|^2 \tan t/(2\varepsilon)} \right) \right\|_{L^2} \underset{\varepsilon \to 0}{\longrightarrow} 0,$$

where S^k stands for the kth iterate of the scattering operator S associated to the nonlinear Schrödinger equation

$$i\partial_t \psi + \frac{1}{2}\Delta\psi = |\psi|^{2\sigma}\psi. \tag{8.13}$$

At focal points, for all $\mathcal{A}^\varepsilon \in \{\mathrm{Id}, \frac{x}{\varepsilon}, \varepsilon\nabla_x\}$,

$$\left\| \mathcal{A}^\varepsilon \left(u^\varepsilon\left(\frac{\pi}{2} + k\pi\right) - \frac{e^{-in\pi/4 - ink\pi/2}}{\varepsilon^{n/2}} \left(W_- \circ S^k \circ \mathcal{F}^{-1}\right) a_0\left(\frac{\cdot}{\varepsilon}\right)\right) \right\|_{L^2} \xrightarrow[\varepsilon \to 0]{} 0,$$

where W_- is the wave operator associated to (8.13) (see Theorem 7.14).

Remark 8.3. We can infer a similar result for $\widetilde{u}^\varepsilon$ solving

$$i\varepsilon\partial_t \widetilde{u}^\varepsilon + \frac{\varepsilon^2}{2}\Delta\widetilde{u}^\varepsilon = \frac{|x|^2}{2}\widetilde{u}^\varepsilon + \varepsilon^{n\sigma}|\widetilde{u}^\varepsilon|^{2\sigma}\widetilde{u}^\varepsilon \quad ; \quad \widetilde{u}^\varepsilon(0,x) = a_0(x)e^{ix\cdot\xi_0/\varepsilon},$$

for $\xi_0 \in \mathbb{R}^n$. Indeed, we check that

$$u^\varepsilon(t,x) = \widetilde{u}^\varepsilon\left(t, x + \xi_0\sin t\right)e^{-i\left(x + \frac{\xi_0}{2}\sin t\right)\cdot\xi_0\cos t/\varepsilon}$$

solves Eq. (8.7).

We now comment on Theorem 8.2. First, we assume $\sigma > 1/2$ so that the nonlinearity $z \mapsto |z|^{2\sigma}z$ is twice differentiable. Since on the other hand, we have to assume $\sigma < 2/(n-2)$, we suppose $n \leqslant 5$. If we consider the case $k = 0$ in Theorem 8.2, we see that the asymptotic behavior of u^ε is described by the right hand side of (8.5), which is exactly the approximate solution provided by WKB analysis for the linear equation (8.3). This shows that for $|t| < \pi/2$, the nonlinearity is negligible at leading order in Eq. (8.7), and the geometry of the propagation is dictated by the harmonic oscillator. The same is true for any $k \in \mathbb{N}$: outside the focal points, nonlinear effects are negligible. On the other hand, nonlinear effects are relevant at leading order at every focus. Each caustic crossing is described by the Maslov index, plus a change in the amplitude, measured by the scattering operator associated to Eq. (8.13). From this respect, the result is quite similar to Proposition 7.19. It should be noticed though, that the harmonic oscillator is absent from the description of the caustic crossing. The explanation is the following: because of the influence of the harmonic potential, u^ε focuses at the origin as $t \to \pi/2$. For $t \approx \pi/2$, u^ε is concentrated at scale ε, in the same fashion as in Eq. (8.6) (but with a different profile). Therefore, the "right" space variable is x/ε, and not x. Writing

$$|x|^2 u^\varepsilon = \varepsilon^2 \left|\frac{x}{\varepsilon}\right|^2 u^\varepsilon,$$

this explains why the harmonic potential becomes negligible near the focus. Note finally that under our assumptions on a_0 and σ, $\left(\mathcal{F} \circ S^k \circ \mathcal{F}^{-1}\right) a_0$ is well defined as a function of Σ, for all $k \in \mathbb{N}$. Therefore, two dynamics dominate alternatingly in the behavior of u^ε: the linear dynamics of the harmonic oscillator outside the focal points, and the nonlinear dynamics of Eq. (8.13) near the focal points. It is remarkable that these dynamics act in a rather decoupled way.

Remark 8.4. A similar result was established by S. Ibrahim in the case of a nonlinear wave equation [Ibrahim (2004)]. In this case, the geometry of the propagation is not dictated by an external potential but by the fact that the space variable lies on a sphere. This causes several focusing phenomena, like in the present case, and for suitable scalings, each caustic crossing is described by a scattering operator.

In view of the next section, we restate Theorem 8.2 by considering $t = \pi/2$ as the initial time: after the change of variable $t \mapsto t - \pi/2$, we have

Corollary 8.5. *Suppose that* $1 \leqslant n \leqslant 5$ *and* $\sigma > 1/2$. *Let* $\varphi \in \Sigma$, *and* $\sigma_0(n) \leqslant \sigma < \frac{2}{n-2}$. *Assume that* u^ε *solves*

$$\begin{cases} i\varepsilon \partial_t u^\varepsilon + \dfrac{\varepsilon^2}{2} \Delta u^\varepsilon = \dfrac{|x|^2}{2} u^\varepsilon + \varepsilon^{n\sigma} |u^\varepsilon|^{2\sigma} u^\varepsilon, \\ u^\varepsilon(0,x) = \dfrac{1}{\varepsilon^{n/2}} \varphi \left(\dfrac{x}{\varepsilon} \right) + \dfrac{1}{\varepsilon^{n/2}} r^\varepsilon \left(\dfrac{x}{\varepsilon} \right), \end{cases} \tag{8.14}$$

with $\|r^\varepsilon\|_\Sigma \to 0$ *as* $\varepsilon \to 0$. *Denote* $\psi_\pm = W_\pm^{-1} \varphi$, *where* $W_\pm$ *are the wave operators associated to Eq. (8.13). Then if* $0 \leqslant \delta(r) < 1$, *the following asymptotics hold in* $L^2 \cap L^r$:

- *For* $-\pi < t < 0$, $\quad u^\varepsilon(t,x) \underset{\varepsilon \to 0}{\sim} \dfrac{e^{in\pi/4}}{|\sin t|^{n/2}} \widehat{\psi_-} \left(\dfrac{x}{\sin t} \right) e^{i|x|^2/(2\varepsilon \tan t)}$.

- *For* $0 < t < \pi$, $\quad u^\varepsilon(t,x) \underset{\varepsilon \to 0}{\sim} \dfrac{e^{-in\pi/4}}{(\sin t)^{n/2}} \widehat{\psi_+} \left(\dfrac{x}{\sin t} \right) e^{i|x|^2/(2\varepsilon \tan t)}$.

Remark 8.6. In [Nier (1996)], the author considers equations which can be compared to Eq. (8.7), that is

$$i\varepsilon \partial_t v^\varepsilon + \dfrac{\varepsilon^2}{2} \Delta v^\varepsilon = V(x) v^\varepsilon + U \left(\dfrac{x}{\varepsilon} \right) v^\varepsilon \ ; \ v^\varepsilon(0,x) = \dfrac{1}{\varepsilon^{n/2}} \varphi \left(\dfrac{x}{\varepsilon} \right), \tag{8.15}$$

where U is a short range potential. The potential V in that case cannot be the harmonic potential, for it has to be bounded as well as all its derivatives. In that paper, the author proves that under suitable assumptions, the influence of U occurs near $t = 0$ and is localized near the origin, while only the value $V(0)$ of V at the origin is relevant in this régime. For times $\varepsilon \ll |t| \leqslant T_*$, the situation is different: the potential U becomes negligible, while V dictates the propagation. As in our paper, the transition between these two régimes is measured by the scattering operator associated to U.

Our assumption $\sigma \geqslant \sigma_0(n) > 1/n$ makes the nonlinear term short range. With our scaling for the nonlinearity, this perturbation is relevant only near the focus, where the harmonic potential is negligible, while the opposite occurs for $\varepsilon \ll |t| < \pi$.

Remark 8.7. Since near the focal point, the description of the wave function u^ε does not involve the external potential at leading order, the phenomenon is the same as in Sec. 7.4. In particular, the discussion of Sec. 7.4.3 can be repeated: the conclusions of Proposition 7.23 remain valid in the present case, up to replacing the focusing time. So, the Cauchy problem for the propagation of Wigner measures is ill-posed.

To simplify the presentation, we sketch the argument of the proof of Theorem 8.2 at a formal level only. To justify these computations, the key point in [Carles (2003b)] consists in introducing the analogues of Proposition 7.8 and Corollary 7.9 (in space dimension $n \geqslant 1$), in two cases:

- With the same operator $U_0^\varepsilon(t) = e^{i\varepsilon \frac{t}{2}\Delta}$. This is because near the focal points, the harmonic potential can be neglected in Eq. (8.7).

- When U_0^ε is replaced by U^ε. Indeed, Eq. (8.4) shows that U^ε enjoys the same dispersive properties as U_0^ε, *locally in time* (and only locally in time, because $-\varepsilon^2\Delta + |x|^2$ has eigenvalues). Therefore, Strichartz inequalities are available for U^ε. The only difference is that in the homogeneous Strichartz estimate, the time interval $\mathbb{R}$ must be replaced by a *finite* time interval, and in this case as well as in the inhomogeneous estimate (7.14), the constants C depend on the *finite* time interval I.

The organization of the proof of Theorem 8.2 is the following:

- Justify WKB analysis until t is as close as possible to $\pi/2$: this allows $t \leqslant \pi/2 - \Lambda\varepsilon$, in the limit $\Lambda \to +\infty$.
- Show that there exists a transition régime for $t = \pi/2 - \Lambda\varepsilon$, when $\Lambda \to +\infty$: the harmonic potential becomes negligible. We then approximate u^ε by essentially the same approximate solution as the one studied in the proof of Proposition 7.19.

- For $|t - \pi/2| \leqslant \Lambda\varepsilon$, we proceed as in the proof of Proposition 7.19. However, some extra source terms appear because of the harmonic potential. This forces us to consider smoother initial data and use a density argument relying on the global well-posedness for Eq. (8.13). This is where we need to assume that $z \mapsto |z|^{2\sigma}z$ is twice differentiable.
- For $t \geqslant \pi/2 + \Lambda\varepsilon$, we can repeat the analysis of the second, then of the first point.

Before the first focus

For $t \in [0, \pi/2[$, our natural candidate as an approximate solution is v^ε, solution of Eq. (8.3). We can also approximate v^ε, by

$$v_{\text{app}}^\varepsilon(t,x) = \frac{1}{(\cos t)^{n/2}} a_0 \left(\frac{x}{\cos t}\right) e^{-i|x|^2 \tan t/(2\varepsilon)}.$$

The error term $w_{\text{lin}}^\varepsilon = v^\varepsilon - v_{\text{app}}^\varepsilon$ satisfies

$$i\varepsilon\partial_t w_{\text{lin}}^\varepsilon + \frac{\varepsilon^2}{2}\Delta w_{\text{lin}}^\varepsilon = \frac{|x|^2}{2}w_{\text{lin}}^\varepsilon + \left(\frac{\varepsilon}{\cos t}\right)^2 \frac{e^{-i|x|^2 \tan t/(2\varepsilon)}}{2(\cos t)^{n/2}} (\Delta a_0)\left(\frac{x}{\cos t}\right),$$

along with the initial condition $w_{\text{lin}}^\varepsilon(0, \cdot) = 0$. Assume that a_0 is relatively smooth:

$$a_0 \in \mathcal{H} = \{f \in H^3(\mathbb{R}^n); \quad xf \in H^2(\mathbb{R}^n)\}.$$

The basic energy estimate of Lemma 1.2 then yields:

$$\|w_{\text{lin}}^\varepsilon(t)\|_{L^2} \leqslant \frac{1}{2}\int_0^t \frac{\varepsilon}{(\cos\tau)^2}\|\Delta a_0\|_{L^2}\, d\tau.$$

Apply J^ε to the equation satisfied by $w_{\text{lin}}^\varepsilon$:

$$i\varepsilon\partial_t J^\varepsilon w_{\text{lin}}^\varepsilon + \frac{\varepsilon^2}{2}\Delta J^\varepsilon w_{\text{lin}}^\varepsilon = \frac{|x|^2}{2}J^\varepsilon w_{\text{lin}}^\varepsilon$$
$$-i\cos t \left(\frac{\varepsilon}{\cos t}\right)^2 \frac{e^{-i|x|^2 \tan t/(2\varepsilon)}}{2(\cos t)^{n/2}}\nabla(\Delta a_0)\left(\frac{x}{\cos t}\right),$$

by the definition of J^ε given in Eq. (8.8). We infer

$$\|J^\varepsilon(t)w_{\text{lin}}^\varepsilon(t)\|_{L^2} \leqslant \frac{1}{2}\int_0^t \frac{\varepsilon}{(\cos\tau)^2}\|a_0\|_{H^3}\, d\tau.$$

Finally, apply H^ε to the equation satisfied by $w^\varepsilon_{\text{lin}}$:

$$i\varepsilon \partial_t H^\varepsilon w^\varepsilon_{\text{lin}} + \frac{\varepsilon^2}{2} \Delta H^\varepsilon w^\varepsilon_{\text{lin}} = \frac{|x|^2}{2} H^\varepsilon w^\varepsilon_{\text{lin}}$$

$$+ x \cos t \left(\frac{\varepsilon}{\cos t}\right)^2 \frac{e^{-i|x|^2 \tan t/(2\varepsilon)}}{2(\cos t)^{n/2}} (\Delta a_0) \left(\frac{x}{\cos t}\right)$$

$$+ i\varepsilon \sin t \left(\frac{\varepsilon}{\cos t}\right)^2 \nabla \left(\frac{e^{-i|x|^2 \tan t/(2\varepsilon)}}{2(\cos t)^{n/2}} (\Delta a_0) \left(\frac{x}{\cos t}\right)\right).$$

We deduce:

$$\|H^\varepsilon(t) w^\varepsilon_{\text{lin}}(t)\|_{L^2} \lesssim \int_0^t \frac{\varepsilon}{(\cos \tau)^2} \|x a_0\|_{H^2} \, d\tau + \int_0^t \frac{\varepsilon^2 \sin \tau}{(\cos \tau)^3} \|a_0\|_{H^3} \, d\tau.$$

Therefore, if $a_0 \in \mathcal{H}$, we have:

$$\sum_{\mathcal{B}^\varepsilon \in \{\text{Id}, J^\varepsilon, H^\varepsilon\}} \|\mathcal{B}^\varepsilon(t) w^\varepsilon_{\text{lin}}(t)\|_{L^2} \lesssim \int_0^t \frac{\varepsilon}{(\cos \tau)^2} d\tau + \left(\frac{\varepsilon}{\cos t}\right)^2$$

$$\lesssim \frac{\varepsilon}{\pi/2 - t} + \left(\frac{\varepsilon}{\pi/2 - t}\right)^2.$$

A density argument yields:

Lemma 8.8. *Let $a_0 \in \Sigma$. We have*

$$\sum_{\mathcal{B}^\varepsilon \in \{\text{Id}, J^\varepsilon, H^\varepsilon\}} \limsup_{\varepsilon \to 0} \sup_{0 \leqslant t \leqslant \pi/2 - \Lambda \varepsilon} \|\mathcal{B}^\varepsilon(t) \left(v^\varepsilon(t) - v^\varepsilon_{\text{app}}(t)\right)\|_{L^2} \xrightarrow[\Lambda \to +\infty]{} 0.$$

We now have to compare u^ε and v^ε. Let $w^\varepsilon = u^\varepsilon - v^\varepsilon$. It solves

$$i\varepsilon \partial_t w^\varepsilon + \frac{\varepsilon^2}{2} \Delta w^\varepsilon = \frac{|x|^2}{2} w^\varepsilon + \varepsilon^{n\sigma} |u^\varepsilon|^{2\sigma} u^\varepsilon.$$

The initial datum for w^ε is zero. In view of the analysis for $t > \pi/2$, notice that it suffices to assume

$$\sum_{\mathcal{B}^\varepsilon \in \{\text{Id}, J^\varepsilon, H^\varepsilon\}} \|\mathcal{B}^\varepsilon(0) w^\varepsilon_{\text{lin}}(0)\|_{L^2} \xrightarrow[\varepsilon \to 0]{} 0.$$

It is this assumption which is needed when we iterate the argument, from $t \in [0, \pi]$ to $t \in [\pi, 2\pi]$, and so on (finitely many times). Proceeding the same way as in Sec. 7.4, thanks to the operator J^ε which provides sharp L^p estimates before the focus (sharp in term of the dependence upon t and ε), we can prove:

Lemma 8.9. *Let $a_0 \in \Sigma$, and $\sigma > \max(1/n, 2/(n+2))$, with $\sigma < 2/(n-2)$ if $n \geqslant 3$. We have*

$$\sum_{\mathcal{B}^\varepsilon \in \{\text{Id}, J^\varepsilon, H^\varepsilon\}} \limsup_{\varepsilon \to 0} \sup_{0 \leqslant t \leqslant \pi/2 - \Lambda \varepsilon} \|\mathcal{B}^\varepsilon(t) \left(u^\varepsilon(t) - v^\varepsilon(t)\right)\|_{L^2} \xrightarrow[\Lambda \to +\infty]{} 0.$$

Matching linear and nonlinear régimes

Lemmas 8.8 and 8.9 show that up to $t = \pi/2 - \Lambda\varepsilon$, u^ε can be approximated by v_{app}^ε as $\varepsilon \to 0$, in the limit $\Lambda \to \infty$. In this régime, we have:

$$v_{app}^\varepsilon(t,x) = \frac{1}{(\cos t)^{n/2}} a_0 \left(\frac{x}{\cos t}\right) e^{-i|x|^2 \tan t/(2\varepsilon)}$$

$$\approx \frac{1}{(\pi/2-t)^{n/2}} a_0 \left(\frac{x}{\pi/2-t}\right) e^{-i|x|^2/(2\varepsilon(\pi/2-t))}.$$

Up to replacing $\pi/2$ by 1, we retrieve the same approximation as in Sec. 7.4, before the focus: this is a hint that the harmonic potential is becoming negligible at the approach of the focal point. Resume the notation

$$\psi_- = e^{-in\pi/4}\widehat{a_0}.$$

We can then prove:

Proposition 8.10. *Let $a_0 \in \Sigma$, and σ as in Lemma 8.9. We have*

$$\limsup_{\varepsilon\to 0} \left\| \mathcal{A}_\Lambda^\varepsilon \left(u^\varepsilon\left(\frac{\pi}{2} - \Lambda\varepsilon\right) - \frac{1}{\varepsilon^{n/2}} (U_0(-\Lambda)\psi_-)\left(\frac{\cdot}{\varepsilon}\right) \right) \right\|_{L^2} \xrightarrow[\Lambda\to+\infty]{} 0,$$

for all $\mathcal{A}_\Lambda^\varepsilon \in \{\mathrm{Id}, \varepsilon\nabla, \frac{x}{\varepsilon} - i\Lambda\varepsilon\nabla\}$.

Two things must be said about this proposition. First, we match the evolution of u^ε with a large time asymptotics for the free Schrödinger equation (without potential), after some rescaling, as in Sec. 7.4. This indicates that the harmonic potential loses its influence at leading order. In addition, we do not measure the error in terms of the operators J^ε and H^ε, but in terms of their counterparts used in Sec. 7.4. The two measurements are actually equivalent at leading order, and computations show that we can approximate the rays of geometric optics by lines, near $t = \pi/2$. See Fig. 8.2.

Inside the boundary layer

Inside the boundary layer in time about $t = \pi/2$, of width $\Lambda\varepsilon$, we neglect the harmonic potential. As in Sec. 7.4, introduce ψ solution to

$$i\partial_t\psi + \frac{1}{2}\Delta\psi = |\psi|^{2\sigma}\psi \quad ; \quad U_0(-t)\psi(t)\big|_{t=-\infty} = \psi_- := e^{-in\pi/4}\widehat{a_0}.$$

It scatters like $U_0(t)\psi_+$ as $t \to +\infty$, for some $\psi_+ \in \Sigma$, provided that $\sigma \geqslant \sigma_0(n)$ (Theorems 7.14 and 7.16). Rescale this function as follows:

$$u_{app}^\varepsilon(t,x) = \frac{1}{\varepsilon^{n/2}} \psi\left(\frac{t-\pi/2}{\varepsilon}, \frac{x}{\varepsilon}\right).$$

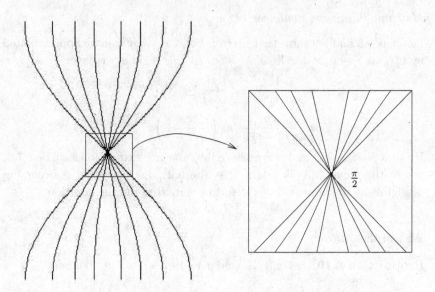

Fig. 8.2   Rays of geometric optics are straightened near $t = \frac{\pi}{2}$.

In [Carles (2003b)], the proof proceeds in two steps: first, it is shown that the exact solution u^ε can be truncated in a neighborhood of the origin of size ε^γ, for any $0 < \gamma < 1$. Then, this truncated solution is shown to be close to $u^\varepsilon_{\text{app}}$. More precisely, let $\chi \in C_0^\infty(\mathbb{R}^n; \mathbb{R}_+)$, with $\chi = 1$ on $B(0,1)$ and $\operatorname{supp}\chi \subset B(0,2)$. For $\lambda \in \,]0,1[$, let

$$u^\varepsilon_\lambda(t,x) = \chi\left(\frac{x}{\varepsilon^\lambda}\right) u^\varepsilon(t,x).$$

Introduce the error $w^\varepsilon_\lambda = u^\varepsilon - u^\varepsilon_\lambda$. It has small initial data at time $\pi/2 - \Lambda\varepsilon$, in the sense of Proposition 8.10, and solves

$$i\varepsilon\partial_t w^\varepsilon_\lambda + \frac{\varepsilon^2}{2}\Delta w^\varepsilon_\lambda = \frac{|x|^2}{2}w^\varepsilon_\lambda + \varepsilon^{n\sigma}|u^\varepsilon|^{2\sigma}w^\varepsilon_\lambda - \frac{\varepsilon^{2-\lambda}}{2}\nabla\chi\left(\frac{x}{\varepsilon^\lambda}\right)\cdot\nabla u^\varepsilon$$
$$-\,\varepsilon^{2-2\lambda}\Delta\chi\left(\frac{x}{\varepsilon^\lambda}\right)u^\varepsilon.$$

If $a_0 \in \mathcal{H}$, we can establish extra bounds on u^ε that make it possible to show that w^ε_λ, $J^\varepsilon w^\varepsilon_\lambda$ and $H^\varepsilon w^\varepsilon_\lambda$ are small in L^2. It is in proving that such a regularity is well propagated that we need to assume that the nonlinearity $z \mapsto |z|^{2\sigma}z$ is C^2, hence $\sigma > 1/2$. The last two terms of the right hand side are then treated like small source terms. Here, we use Strichartz estimates associated to the group U^ε.

The second step consists in considering $\widetilde{w}_\lambda^\varepsilon = u_\lambda^\varepsilon - u_{\mathrm{app}}^\varepsilon$. In view of the identity

$$\chi(x) = \chi\left(\frac{x}{2}\right)\chi(x),$$

it solves:

$$i\varepsilon\partial_t\widetilde{w}_\lambda^\varepsilon + \frac{\varepsilon^2}{2}\Delta\widetilde{w}_\lambda^\varepsilon = \chi\left(\frac{x}{2\varepsilon^\lambda}\right)\frac{|x|^2}{2}u_\lambda^\varepsilon + \varepsilon^{n\sigma}\left(|u^\varepsilon|^{2\sigma}u_\lambda^\varepsilon - \left|u_{\mathrm{app}}^\varepsilon\right|^{2\sigma}u_{\mathrm{app}}^\varepsilon\right)$$

$$+ \frac{\varepsilon^{2-\lambda}}{2}\nabla\chi\left(\frac{x}{\varepsilon^\lambda}\right)\cdot\nabla u^\varepsilon + \varepsilon^{2-2\lambda}\Delta\chi\left(\frac{x}{\varepsilon^\lambda}\right)u^\varepsilon.$$

Here, we use Strichartz estimates associated to U_0^ε, and view the first term of the right hand side as a small source term:

$$\chi\left(\frac{x}{2\varepsilon^\lambda}\right)\frac{|x|^2}{2} = \frac{\varepsilon^{2\lambda}}{2}\chi\left(\frac{x}{2\varepsilon^\lambda}\right)\left|\frac{x}{\varepsilon^\lambda}\right|^2 = \varepsilon^{2\lambda}\widetilde{\chi}\left(\frac{x}{2\varepsilon^\lambda}\right),\ \text{ where } \widetilde{\chi}\in C_0^\infty(\mathbb{R}^n;\mathbb{R}).$$

Then it is possible to factor out $\widetilde{w}_\lambda^\varepsilon$ in the difference of the two nonlinear terms, up to an extra, small, source term, which appears when replacing u^ε with u_λ^ε. The approximation $a_0 \in \mathcal{H}$ is removed by a density argument for $\sigma \geqslant \sigma_0(n)$, since (8.13) is globally well-posed in Σ. We can prove finally:

Proposition 8.11. *Under the assumptions of Theorem 8.2,*

$$\sum_{\mathcal{B}^\varepsilon\in\{\mathrm{Id},J^\varepsilon,H^\varepsilon\}}\limsup_{\varepsilon\to 0}\ \sup_{|\pi/2-t|\leqslant\Lambda\varepsilon}\left\|\mathcal{B}^\varepsilon(t)\left(u^\varepsilon(t) - u_{\mathrm{app}}^\varepsilon(t)\right)\right\|_{L^2}\xrightarrow[\Lambda\to+\infty]{} 0.$$

The proof of this result, barely sketched above, is fairly technical, and is not reproduced here for the sake of readability.

Past the first boundary layer

For $t = \pi/2 + \Lambda\varepsilon$, we can then prove the analogue of Proposition 8.10, with $-\Lambda$ replaced by $+\Lambda$, and ψ_- replaced by ψ_+. Then, we can mimic the proofs and results of Lemmas 8.8 and 8.9, by using the remark that u^ε and the solution of the linear equation don't have to match exactly at the initial time, but only up to a small error in L^2, when any of the operators Id, J^ε or H^ε acts on the difference. Using this remark again, we see that on $[\pi, 2\pi]$, we can repeat what was done on $[0, \pi]$. Hence Theorem 8.2.

8.2   General quadratic potentials

When the external potential $V = V(x)$ is time-independent and is exactly a polynomial of degree at most two, we can extend Corollary 8.5. We consider

$$\begin{cases} i\varepsilon\partial_t u^\varepsilon + \dfrac{\varepsilon^2}{2}\Delta u^\varepsilon = V u^\varepsilon + \varepsilon^{n\sigma}\,|u^\varepsilon|^{2\sigma}\,u^\varepsilon, \\[2mm] u^\varepsilon(0,x) = \dfrac{1}{\varepsilon^{n/2}}\varphi\left(\dfrac{x-x_0}{\varepsilon}\right)e^{ix\cdot\xi_0/\varepsilon}, \end{cases} \tag{8.16}$$

for some $x_0, \xi_0 \in \mathbb{R}^n$. In this paragraph, we make the following assumptions:

Assumption 8.12. We suppose that $\sigma_0(n) \leqslant \sigma < 2/(n-2)$, $\sigma > 1/2$, and therefore $1 \leqslant n \leqslant 5$.
The initial profile φ is in Σ.
The potential V is time-independent, and is exactly a polynomial of degree at most two:

$$\partial^3_{jk\ell} V \equiv 0, \quad \forall j,k,\ell \in \{1,\ldots,n\}.$$

Introduce the Hamiltonian system with initial data (x_0, ξ_0):

$$\dot{x}(t) = \xi(t) \quad ; \quad \dot{\xi}(t) = -\nabla V(x(t)) \quad ; \quad x(0) = x_0 \quad ; \quad \xi(0) = \xi_0. \tag{8.17}$$

Note that under our assumption on V, $x(t)$ and $\xi(t)$ can be computed exactly (see Eq. (8.20) below). Introduce the solution ψ to

$$i\partial_t \psi + \frac{1}{2}\Delta\psi = |\psi|^{2\sigma}\psi \quad ; \quad \psi_{|t=0} = \varphi. \tag{8.18}$$

Assumption 8.12 and Theorem 7.16 show that there exist $\psi_\pm \in \Sigma$ such that

$$\|U_0(-t)\psi(t) - \psi_\pm\|_\Sigma \underset{t\to\pm\infty}{\longrightarrow} 0.$$

To state our main result, introduce the quantity

$$\Phi(t,x) = x\cdot\xi(t) - \frac{1}{2}\left(x(t)\cdot\xi(t) - x_0\cdot\xi_0\right).$$

Proposition 8.13. *Under Assumption 8.12, suppose in addition that V is of the form*

$$V(x) = \frac{1}{2}\sum_{j=1}^n \delta_j\omega_j^2 x_j^2 + \sum_{j=1}^n b_j x_j,$$

for some real constants b_j, where $\omega_j > 0$, $\delta_j \in \{-1,0,+1\}$, and $\delta_j b_j = 0$ for all j. There exists $T > 0$ independent of $\varepsilon \in]0,1]$ and $\varepsilon(T) > 0$ such that for $0 < \varepsilon \leqslant \varepsilon(T)$, Eq. (8.16) has a unique solution $u^\varepsilon \in C([-T,T];\Sigma)$.

In addition, its behavior on $[-T, T]$ is given by the following régimes:

(1) *For any $\Lambda > 0$,*

$$\limsup_{\varepsilon \to 0} \sup_{|t| \leqslant \Lambda\varepsilon} \left(\left\| u^\varepsilon(t) - u^\varepsilon_{\mathrm{app}}(t) \right\|_{L^2} + \left\| \varepsilon\nabla u^\varepsilon(t) - \varepsilon\nabla u^\varepsilon_{\mathrm{app}}(t) \right\|_{L^2} \right.$$

$$\left. + \left\| \frac{x - x(t)}{\varepsilon} \left(u^\varepsilon(t) - u^\varepsilon_{\mathrm{app}}(t) \right) \right\|_{L^2} \right) = 0,$$

where $u^\varepsilon_{\mathrm{app}}(t, x) = \dfrac{1}{\varepsilon^{n/2}} \psi \left(\dfrac{t}{\varepsilon}, \dfrac{x}{\varepsilon} \right) e^{i\Phi(t,x)/\varepsilon}$, and ψ solves Eq. (8.18).

(2) *Beyond this boundary layer, we have:*

$$\limsup_{\varepsilon \to 0} \sup_{\Lambda\varepsilon \leqslant \pm t \leqslant T} \left(\left\| u^\varepsilon(t) - v^\varepsilon_\pm(t) \right\|_{L^2} + \left\| \varepsilon\nabla u^\varepsilon(t) - \varepsilon\nabla v^\varepsilon_\pm(t) \right\|_{L^2} \right.$$

$$\left. + \left\| (x - x(t)) \left(u^\varepsilon(t) - v^\varepsilon_\pm(t) \right) \right\|_{L^2} \right) \xrightarrow[\Lambda \to +\infty]{} 0,$$

where $v^\varepsilon_\pm$ solve the linear equations

$$i\varepsilon\partial_t v^\varepsilon_\pm + \frac{\varepsilon^2}{2}\Delta v^\varepsilon_\pm = V v^\varepsilon_\pm \quad ; \quad v^\varepsilon_\pm(0, x) = \frac{1}{\varepsilon^{n/2}}\psi_\pm \left(\frac{x - x_0}{\varepsilon} \right) e^{ix \cdot \xi_0/\varepsilon}.$$

Like in the case of the isotropic harmonic potential, we obtain a nonlinear extension of the result of [Nier (1996)] (see Remark 8.6). We compare this result with yet another problem at the end of this section, when the nonlinearity is focusing instead of defocusing. Proposition 8.13 is proved in [Carles and Miller (2004)]. We present the main steps of the proof only. Before doing so, we point out that in some cases, more can be said on the time T in Proposition 8.13. Essentially, if there is no refocusing at one point, then T can be taken arbitrarily large: the nonlinearity remains negligible off $\{t = 0\}$. On the other hand, if the potential V causes at least one of the solutions $v^\varepsilon_\pm$ to refocus at one point for $\pm t > 0$, then so does u^ε. Nonlinear effects then affect u^ε at leading order again, like in Sec. 8.1. Note that the refocusing phenomenon has to occur at one point: for any other caustic, the nonlinearity is subcritical. Only a focal point can ignite the nonlinearity $\varepsilon^{n\sigma}|u^\varepsilon|^{2\sigma}u^\varepsilon$ at leading order. More precisely, such refocusing can occur only if $\delta_j = +1$ for all j and the ω_j's are pairwise rationally dependent; see below.

We now explain why the assumption on the form of the potential made in Proposition 8.13 is not really one. Up to an orthonormal change of basis in $\mathbb{R}^n$ (which leaves the Laplacian invariant), we can assume that V is of the form

$$V(x) = \frac{1}{2} \sum_{j=1}^n \delta_j \omega_j^2 x_j^2 + \sum_{j=1}^n b_j x_j + c, \tag{8.19}$$

for some real constants b_j and c, where $\omega_j > 0$, and $\delta_j \in \{-1, 0, +1\}$. This form is obtained by diagonalizing the quadratic part of V. Up to completing the square and changing the origin, we may even assume $\delta_j b_j = 0$ for all j. By changing the origin, we may modify the form of the initial data, by a multiplicative factor $e^{ia/\varepsilon}$ for some constant $a \in \mathbb{R}$. Finally, we may assume that $a = c = 0$ by considering $u^\varepsilon(t, x)e^{i(ct+a)/\varepsilon}$ instead of $u^\varepsilon(t, x)$.

Note that the real numbers $\delta_j \omega_j^2/2$ are the eigenvalues of the quadratic part of V. Then refocusing at one point can happen only if $\delta_j = +1$ for all j and the ω_j's are pairwise rationally dependent:

$$\frac{\omega_j}{\omega_k} \in \mathbb{Q} \quad \forall j, k \in \{1, \ldots, n\}.$$

This case is very similar to that described in Sec. 8.1. Note that before refocusing at one point, there may be focusing on an affine space of dimension at least one.

Example 8.14. Assume $n = 2$ and $x_0 = \xi_0 = 0$. If $\omega_1 = 2\omega_2$ for instance, then at time $t = \pi/\omega_1$, u^ε focuses on the line $\{x_1 = 0\}$, and at time $t = 2\pi/\omega_1 = \pi/\omega_2$, u^ε refocuses at the origin. We have seen in Sec. 7.7 that the critical indexes for focusing on a line correspond to focusing at one point in space dimension one, $\alpha = \sigma$. Since we consider the case $\alpha = 2\sigma$, the nonlinearity is negligible when the wave focuses on the line.

When u^ε focuses on a set which is not reduced to a point, the only effect at leading order is linear, and is measured by the Maslov index. When u^ε focuses at one point, nonlinear effects are described in terms of scattering operator.

The first step in the proof of Proposition 8.13 consists in reducing the analysis to the case $x_0 = \xi_0 = 0$. This can be achieved because V is a polynomial of degree at most two, and time-independent. Indeed, solve the above Hamiltonian system when V has the form (8.19). Introduce

$$g_j(t) = \begin{cases} \dfrac{\sin(\omega_j t)}{\omega_j}, & \text{if } \delta_j = 1, \\ t, & \text{if } \delta_j = 0, \\ \dfrac{\sinh(\omega_j t)}{\omega_j}, & \text{if } \delta_j = -1. \end{cases} \quad ; \quad h_j(t) = \begin{cases} \cos(\omega_j t), & \text{if } \delta_j = 1, \\ 1, & \text{if } \delta_j = 0, \\ \cosh(\omega_j t), & \text{if } \delta_j = -1. \end{cases}$$

Then the solution of Eq. (8.17) is given by

$$\begin{aligned} x_j(t) &= h_j(t)x_{0j} + g_j(t)\xi_{0j} - \frac{1}{2}b_j t^2, \\ \xi_j(t) &= h_j(t)\xi_{0j} - \delta_j \omega_j^2 g_j(t)x_{0j} - b_j t. \end{aligned} \tag{8.20}$$

Assume that u^ε solves Eq. (8.16). Introduce

$$\widetilde{u}^\varepsilon(t, x) = u^\varepsilon\left(t, x + x(t)\right) e^{-i\Phi(t, x + x(t))/\varepsilon}.$$

We check that $\widetilde{u}^\varepsilon$ solves Eq. (8.16) with $x_0 = \xi_0 = 0$. Without loss of generality, we therefore assume that u^ε solves Eq. (8.16) with $x_0 = \xi_0 = 0$.

A second consequence of the fact that V is polynomial is the existence of "nice" operators. In the case of a focal point at $(t, x) = (1, 0)$, we have used the operators $x/\varepsilon + i(t-1)\nabla$ and $\varepsilon\nabla$. In the case of the isotropic harmonic potential, we have used the operators J^ε and H^ε, defined in Eq. (8.8) and Eq. (8.10) respectively. The definition of "nice" operators is somehow summarized in Lemma 8.1. Essentially, they commute with the linear part of the equation, they act on gauge invariant nonlinearities like derivatives, and they provide sharp (for the limit $\varepsilon \to 0$) weighted Gagliardo–Nirenberg inequalities. We recall that the second point of Lemma 8.1 is just stated in order to infer the last two points without computations. In the present case, we introduce:

$$J^\varepsilon(t) := U^\varepsilon(t)\frac{x}{\varepsilon}U^\varepsilon(-t) \quad ; \quad H^\varepsilon(t) := U^\varepsilon(t)i\varepsilon\nabla_x U^\varepsilon(-t),$$

where we denote

$$U^\varepsilon(t) = \exp\left(-i\frac{t}{\varepsilon}\left(-\frac{\varepsilon^2}{2}\Delta + V\right)\right).$$

By computing commutators, we check:

$$\partial_t J^\varepsilon(t) = U^\varepsilon(t)i\nabla U^\varepsilon(-t) = \frac{1}{\varepsilon}H^\varepsilon(t) \quad ; \quad \partial_t H^\varepsilon(t) = -U^\varepsilon(t)\nabla V U^\varepsilon(-t).$$

Therefore,

$$\partial_t^2 J_j^\varepsilon(t) = -\frac{1}{\varepsilon}U^\varepsilon(t)\partial_j V U^\varepsilon(-t)$$
$$= -\delta_j\omega_j^2 U^\varepsilon(t)\frac{x_j}{\varepsilon}U^\varepsilon(-t) - \frac{b_j}{\varepsilon} = -\delta_j\omega_j^2 J_j^\varepsilon(t) - \frac{b_j}{\varepsilon}.$$

We thus have explicitly,

$$J_j^\varepsilon(t) = \frac{x_j}{\varepsilon}h_j(t) + ig_j(t)\partial_j - \frac{b_j}{2\varepsilon}t^2,$$
$$H_j^\varepsilon(t) = -\delta_j\omega_j^2 x_j g_j(t) + ih_j(t)\varepsilon\partial_j - b_j t. \tag{8.21}$$

Remark 8.15. In view of the definition *via* the group U^ε, the notations J^ε and H^ε may not seem consistent with Eqs. (8.9) and (8.10) (first expression) in the case where V is an isotropic harmonic potential. However, it is easily checked that Eq. (8.21) agrees with the expressions (8.8) and (8.10) (second expression). The explanation lies in the fact that we have changed the origin of time, from 0 to $\pi/2$, if we compare with Sec. 8.1. More precisely, we check we following identities:

$$U_{\text{i.h.}}^{\varepsilon}(t)\left(\frac{1}{i}\nabla\right)U_{\text{i.h.}}^{\varepsilon}(-t) = U_{\text{i.h.}}^{\varepsilon}\left(t - \frac{\pi}{2}\right)\frac{x}{\varepsilon}U_{\text{i.h.}}^{\varepsilon}\left(\frac{\pi}{2} - t\right),$$

$$U_{\text{i.h.}}^{\varepsilon}(t)xU_{\text{i.h.}}^{\varepsilon}(-t) = U_{\text{i.h.}}^{\varepsilon}\left(t - \frac{\pi}{2}\right)(i\varepsilon\nabla)U_{\text{i.h.}}^{\varepsilon}\left(\frac{\pi}{2} - t\right),$$

where $U_{\text{i.h.}}^{\varepsilon}$ is the group associated to the semi-classical isotropic harmonic potential

$$U_{\text{i.h.}}^{\varepsilon}(t) = \exp\left(-i\frac{t}{2\varepsilon}\left(-\varepsilon^2\Delta + |x|^2\right)\right).$$

These identities are consequences of the fact that the harmonic oscillator rotates the phase space at angular velocity one (see Eq. (8.2)), so after $\pi/2$ time units, x has become $-\xi$, and ξ has become x.

These operators inherit interesting properties which we list below.

Lemma 8.16. *The operators J^ε and H^ε satisfy the following properties.*
(1) They commute with the linear part of (8.16).
(2) Denote

$$\phi_1(t,x) := \frac{1}{2}\sum_{k=1}^{n}\left(\frac{h_k(t)}{g_k(t)}x_k^2 - b_k tx_k - \frac{t^3}{12}b_k^2\right),$$

$$\phi_2(t,x) := -\frac{1}{2}\sum_{k=1}^{n}\left(\delta_k\omega_k^2\frac{g_k(t)}{h_k(t)}x_k^2 + 2b_k tx_k + \frac{t^3}{3}b_k^2\right).$$

Then ϕ_1 and ϕ_2 are well-defined and smooth for almost all t, and

$$J_j^\varepsilon(t) = ig_j(t)e^{i\phi_1(t,x)/\varepsilon}\partial_j\left(e^{-i\phi_1(t,x)/\varepsilon}\;\cdot\right),$$
$$H_j^\varepsilon(t) = i\varepsilon h_j(t)e^{i\phi_2(t,x)/\varepsilon}\partial_j\left(e^{-i\phi_2(t,x)/\varepsilon}\;\cdot\right). \tag{8.22}$$

(3) Let $r \geqslant 2$ such that $\delta(r) < 1$. Define $P^\varepsilon(t)$ by

$$P^\varepsilon(t) := \prod_{j=1}^{n}\left(|g_j(t)| + \varepsilon|h_j(t)|\right)^{1/n}.$$

There exists C_r such that, for any $u \in \Sigma$,

$$\|u\|_{L^r} \leqslant \frac{C_r}{P^\varepsilon(t)^{\delta(r)}}\|u\|_{L^2}^{1-\delta(r)}\left(\|J^\varepsilon(t)u\|_{L^2} + \|H^\varepsilon(t)u\|_{L^2}\right)^{\delta(r)}. \tag{8.23}$$

(4) The operators J^ε and H^ε act on gauge invariant nonlinearities like derivatives (see the last point of Lemma 8.1).

The first point stems from the definition of the operators J^ε and H^ε. The second point can easily be checked, and the last two points are direct consequences of the second.

Remark 8.17 (Avron–Herbst formula). *In the form* (8.19), *we could have assumed* $b_j = 0$ *for all* j. *As noticed in [Carles and Nakamura (2004)], since the nonlinearity is gauge invariant, Avron–Herbst formula reduces the case* $\delta_j b_j = 0$ *to the case* $b_j = 0$. *For* $b = (b_1, \ldots, b_n)$, *we check that* $\widetilde{u}^\varepsilon$ *solves Eq.* (8.16) *with* $b_j = 0$ *for all* j *in Eq.* (8.19), *where* $\widetilde{u}^\varepsilon$ *is given by:*

$$\widetilde{u}^\varepsilon(t,x) = u^\varepsilon\left(t, x - \frac{t^2}{2}b\right)\exp\left(i\left(tb\cdot x - \frac{t^3}{3}|b|^2\right)/\varepsilon\right).$$

We can now explain why, unless all the ω_j's are pairwise rationally dependent, there is no refocusing at one point off $\{t = 0\}$. Since $n\sigma > 1$, the geometry of the propagation is the same as in the linear case: nonlinear effects become relevant only near focal points. We can then use the estimate provided by Eq. (8.23). Let v^ε solve

$$i\varepsilon\partial_t v^\varepsilon + \frac{\varepsilon^2}{2}\Delta v^\varepsilon = Vv^\varepsilon \quad ; \quad v^\varepsilon(0,x) = \frac{1}{\varepsilon^{n/2}}\psi_0\left(\frac{x}{\varepsilon}\right).$$

From the first point of Lemma 8.16, we infer that not only the L^2 norm of v^ε is time independent, but also the L^2 norms of $J^\varepsilon v^\varepsilon$ and $H^\varepsilon v^\varepsilon$. Therefore, if $0 \leqslant \delta(r) < 1$, the following estimate holds uniformly in time:

$$\|v^\varepsilon(t)\|_{L^r} \lesssim P^\varepsilon(t)^{-\delta(r)}.$$

The case of a focal point corresponds to the situation where

$$\|v^\varepsilon(t)\|_{L^r} \approx \varepsilon^{-\delta(r)}.$$

In view of the definition of the weight P^ε, this shows that there is focusing at one point $(t,x) = (t_0, x_0)$ if and only if $g_j(t_0) = 0$ for all j. Notice that whenever $g_j(t) = 0$, $h_j(t) = 1$. Of course, Eq. (8.20) shows that $g_j(t_0) = 0$ for all j and $t_0 \neq 0$ is possible only if $\delta_j = +1$ for all j: the quadratic part of V is positive definite. Finally, $g_j(t_0) = 0$ for all j and $t_0 \neq 0$ if and only if $\omega_j t_0 \in \pi\mathbb{Z}$ for all j, which in turn is equivalent to the property announced above.

After these preliminary reductions and properties, the proof of Proposition 8.13 becomes very similar to the proof of Theorem 8.2. We first show that for $|t| \leqslant \Lambda\varepsilon$, we can neglect the external potential (recall that up to a phase shift of the form $e^{ict/\varepsilon}$ in u^ε, we have assumed $V(0) = 0$). To do this, we can assume that φ is in the Schwartz class, rather than just $\varphi \in \Sigma$ (this is where we assume that $z \mapsto |z|^{2\sigma} z$ is C^2). Then, the global well-posedness for Eq. (8.18) makes it possible to use a density argument. This yields

$$\limsup_{\varepsilon \to 0} \sup_{|t| \leqslant \Lambda \varepsilon} \left(\left\| u^\varepsilon(t) - u^\varepsilon_{\mathrm{app}}(t) \right\|_{L^2} + \left\| J^\varepsilon(t) \left(u^\varepsilon(t) - u^\varepsilon_{\mathrm{app}}(t) \right) \right\|_{L^2} \right.$$
$$\left. + \left\| H^\varepsilon(t) \left(u^\varepsilon(t) - u^\varepsilon_{\mathrm{app}}(t) \right) \right\|_{L^2} \right) = 0.$$

Note also that it is equivalent to use the operators J^ε and H^ε or the operators $x/\varepsilon + it\nabla$ and $\varepsilon\nabla$, as we have seen in the case of the isotropic harmonic potential; see Fig. 8.2. To see that this implies the first point of Proposition 8.13, recall that we have assumed $x(t) \equiv 0$, and that if we also suppose $b = 0$, then

$$\begin{pmatrix} J^\varepsilon_j \\ H^\varepsilon_j \end{pmatrix} = \begin{pmatrix} h_j & g_j/\varepsilon \\ -\varepsilon\delta_j\omega_j^2 g_j & h_j \end{pmatrix} \begin{pmatrix} x_j/\varepsilon \\ i\varepsilon\partial_j \end{pmatrix}.$$

The determinant of the above matrix is $h_j^2 + \delta_j\omega_j^2 g_j^2 \equiv 1$, and we have

$$\frac{x_j}{\varepsilon} = h_j(t) J^\varepsilon_j(t) - \frac{g_j(t)}{\varepsilon} H^\varepsilon_j(t) \quad ; \quad i\varepsilon\partial_j = \varepsilon\delta_j\omega_j^2 g_j(t) J^\varepsilon_j(t) + h_j(t) H^\varepsilon_j(t).$$

Since $g_j(t) = \mathcal{O}(t)$ as t goes to zero, the first point of Proposition 8.13 follows.

Suppose $t > 0$, since the case $t < 0$ is similar. For the matching region $\{t = \Lambda\varepsilon\}$, we can show that v^ε_+ can be approximated by $\tilde{v}^\varepsilon_+$, solutions to the free equation (recall that we now assume $x_0 = \xi_0 = 0$):

$$i\varepsilon\partial_t\tilde{v}^\varepsilon_+ + \frac{\varepsilon^2}{2}\Delta\tilde{v}^\varepsilon_+ = 0 \quad ; \quad \tilde{v}^\varepsilon_+(0, x) = \frac{1}{\varepsilon^{n/2}}\psi_+\left(\frac{x}{\varepsilon}\right).$$

We can prove that for all $\Lambda > 0$,

$$\sum_{\mathcal{B}^\varepsilon \in \{\mathrm{Id}, J^\varepsilon, H^\varepsilon\}} \limsup_{\varepsilon \to 0} \sup_{0 \leqslant t \leqslant \Lambda\varepsilon} \left\| \mathcal{B}^\varepsilon(t) \left(v^\varepsilon_+(t) - \tilde{v}^\varepsilon_+(t) \right) \right\|_{L^2} = 0.$$

This means that on the linear level, V is negligible for $0 \leqslant t \leqslant \Lambda\varepsilon$. Lemma 7.20 shows that $u^\varepsilon_{\mathrm{app}}$ and $\tilde{v}^\varepsilon_+$ match at time $t = \Lambda\varepsilon$ in the limit $\varepsilon \to 0$ for large Λ:

$$\sum_{\mathcal{B}^\varepsilon \in \{\mathrm{Id}, J^\varepsilon, H^\varepsilon\}} \limsup_{\varepsilon \to 0} \left\| \mathcal{B}^\varepsilon(\Lambda\varepsilon) \left(u^\varepsilon_{\mathrm{app}}(\Lambda\varepsilon) - \tilde{v}^\varepsilon_+(\Lambda\varepsilon) \right) \right\|_{L^2} \xrightarrow[\Lambda \to +\infty]{} 0.$$

Gathering all the informations together, we infer

$$\sum_{\mathcal{B}^\varepsilon \in \{\mathrm{Id}, J^\varepsilon, H^\varepsilon\}} \limsup_{\varepsilon \to 0} \left\| \mathcal{B}^\varepsilon(\Lambda\varepsilon) \left(u^\varepsilon(\Lambda\varepsilon) - v^\varepsilon_+(\Lambda\varepsilon) \right) \right\|_{L^2} \xrightarrow[\Lambda \to +\infty]{} 0.$$

Past this boundary layer, we can show that the nonlinearity remains negligible in the limit $\varepsilon \to 0$, essentially thanks to the operator J^ε, like in the case of the isotropic harmonic potential. The details can be found in [Carles and Miller (2004)].

To conclude this paragraph, and to prepare the next one, we compare the results of Proposition 8.13 with some others, concerning the case of a focusing nonlinearity:

$$\begin{cases} i\varepsilon\partial_t u^\varepsilon + \dfrac{\varepsilon^2}{2}\Delta u^\varepsilon = Vu^\varepsilon - \varepsilon^{n\sigma}|u^\varepsilon|^{2\sigma}u^\varepsilon, \\[2mm] u^\varepsilon(0,x) = \dfrac{1}{\varepsilon^{n/2}}Q\left(\dfrac{x-x_0}{\varepsilon}\right)e^{ix\cdot\xi_0/\varepsilon}. \end{cases} \tag{8.24}$$

Now, $V = V(x)$ is not supposed to be exactly a polynomial. Here, Q is the unique positive, radially symmetric ([Kwong (1989)]), solution of:

$$-\frac{1}{2}\Delta Q + Q = |Q|^{2\sigma}Q.$$

The problem (8.24) was introduced in [Bronski and Jerrard (2000)]. This first result was then refined in [Keraani (2002, 2006)]. The focusing nonlinearity is an obstruction to dispersive phenomena, which were measured by the presence of the scattering operator in the previous defocusing case. The solution u^ε is expected to keep the ground state Q as a leading order profile. Nevertheless, the point where it is centered in the phase space, initially (x_0, ξ_0), should evolve according to the Hamiltonian flow. In the absence of external potential, $V \equiv 0$, we have explicitly:

$$u^\varepsilon(t,x) = \frac{1}{\varepsilon^{n/2}}Q\left(\frac{x-x(t)}{\varepsilon}\right)e^{ix\cdot\xi(t)/\varepsilon+i\theta(t)/\varepsilon},$$

where $(x(t), \xi(t)) = (x_0 + t\xi_0, \xi_0)$ solves the Hamiltonian system with initial data (x_0, ξ_0), and $\theta(t) = t - t|\xi_0|^2/2$. When V is not trivial, seek u^ε of the form of a rescaled WKB expansion:

$$u^\varepsilon(t,x) \underset{\varepsilon\to 0}{\sim} \frac{1}{\varepsilon^{n/2}}\left(\sum_{j\geqslant 0}\varepsilon^j U_j\left(\frac{t}{\varepsilon}, \frac{x-x(t)}{\varepsilon}\right)\right)e^{i\phi(t,x)/\varepsilon}.$$

Plugging this expansion into Eq. (8.24) and canceling the $\mathcal{O}(\varepsilon^0)$ term, we get:

$$i\partial_t U_0 + \frac{1}{2}\Delta U_0 + U_0\left(-\partial_t\phi - \frac{1}{2}|\nabla\phi|^2 - V + |U_0|^{2\sigma}\right) - i\left(\dot{x}(t) - \nabla\phi\right)\cdot\nabla U_0 = 0.$$

Impose the leading order profile to be the standing wave given by

$$U_0(t,x) = e^{it}Q(x).$$

Then the above equation becomes:

$$U_0\left(-\partial_t\phi - \frac{1}{2}|\nabla\phi|^2 - V\right) - i\left(\dot{x}(t) - \nabla\phi\right)\cdot\nabla U_0 = 0.$$

Since $U_0 e^{-it}$ is real-valued, and since we seek a real-valued phase ϕ, this yields:

$$\partial_t \phi + \frac{1}{2}|\nabla \phi|^2 + V = 0 \quad ; \quad \phi(0,x) = x \cdot \xi_0.$$
$$\dot{x}(t) = \nabla \phi(t,x).$$

The first equation is the eikonal equation. We infer that we have exactly

$$\nabla \phi(t, x(t)) = \xi(t).$$

See Lemma 1.5. The form of U_0 and the exponential decay of Q show that we can formally assume that $x = x(t) + \mathcal{O}(\varepsilon)$. In this case,

$$\nabla \phi(t,x) = \nabla \phi(t, x(t)) + \mathcal{O}(\varepsilon) = \xi(t) + \mathcal{O}(\varepsilon) = \dot{x}(t) + \mathcal{O}(\varepsilon).$$

Thus, we have canceled the $\mathcal{O}(\varepsilon^0)$ term, up to adding extra terms of order ε, that would be considered in the next step of the analysis, which we stop here. Back to u^ε, this formal computation yields

$$u^\varepsilon(t,x) \sim \frac{1}{\varepsilon^{n/2}} Q\left(\frac{x-x(t)}{\varepsilon}\right) e^{i\phi(t,x)} \sim \frac{1}{\varepsilon^{n/2}} Q\left(\frac{x-x(t)}{\varepsilon}\right) e^{ix\cdot\xi(t)/\varepsilon + i\theta(t)/\varepsilon},$$

where $\theta(t) = t\left(1 - |\xi_0|^2/2 - V(x_0)\right) + \int_0^t x(s) \cdot \nabla V(x(s))ds$.

To give the above formal analysis a rigorous justification, the following assumptions are made in [Keraani (2006)]:

Assumption 8.18. The nonlinearity is L^2-subcritical: $\sigma < 2/n$. The potential $V = V(x)$ is real-valued, and can be written as $V = V_1 + V_2$, where

- $V_1 \in W^{3,\infty}(\mathbb{R}^n)$.
- $\partial^\alpha V_2 \in W^{2,\infty}(\mathbb{R}^n)$ for every multi-index α with $|\alpha| = 2$.

For instance, V can be a polynomial of degree at most two ($V_1 = 0$).

Theorem 8.19 ([Keraani (2006)]). *Let $x_0, \xi_0 \in \mathbb{R}^n$. Under Assumption 8.18, the solution u^ε to Eq. (8.24) can be approximated as follows:*

$$u^\varepsilon(t,x) = \frac{1}{\varepsilon^{n/2}} Q\left(\frac{x-x(t)}{\varepsilon}\right) e^{ix\cdot\xi(t)/\varepsilon + i\theta^\varepsilon(t)/\varepsilon} + \mathcal{O}(\varepsilon) \quad in \ L^\infty_{\text{loc}}(\mathbb{R}_t; H^1_\varepsilon),$$

where $(x(t), \xi(t))$ is given by the Hamiltonian flow, the real-valued function θ^ε depends on t only, and H^1_ε is defined in Assumption 7.27.

Remark 8.20. The assumption $\sigma < 2/n$ is crucial for the above result to hold. Indeed, if $\sigma = 2/n$, $V(x) = |x|^2/2$ is an isotropic harmonic potential, and $x_0 = \xi_0 = 0$, then (see [Carles (2002); Keraani (2006)]):

$$u^\varepsilon(t,x) = \frac{1}{(\cos t)^{n/2}} Q\left(\frac{x}{\varepsilon \cos t}\right) e^{i\frac{\tan t}{\varepsilon} - i\frac{|x|^2}{2\varepsilon}\tan t}, \quad 0 \leqslant t < \frac{\pi}{2}.$$

The proof of the above result heavily relies on the orbital stability of the ground state, which holds when $\sigma < 2/n$. For $v \in H^1(\mathbb{R}^n)$, denote

$$\mathcal{E}(v) = \frac{1}{2}\|\nabla v\|_{L^2}^2 - \frac{1}{\sigma+1}\|v\|_{L^{2\sigma+2}}^{2\sigma+2}.$$

The ground state Q is the unique solution, up to translation and rotation, to the minimization problem:

$$\mathcal{E}(Q) = \inf\{\mathcal{E}(v) \; ; \; v \in H^1(\mathbb{R}^n) \text{ and } \|v\|_{L^2} = \|Q\|_{L^2}\}.$$

The orbital stability is given by the following result:

Proposition 8.21 ([Weinstein (1985)]). *Let* $\sigma < 2/n$. *There exist* $C, h > 0$ *such that if* $\phi \in H^1(\mathbb{R}^n)$ *is such that* $\|\phi\|_{L^2} = \|Q\|_{L^2}$ *and* $\mathcal{E}(\phi) - \mathcal{E}(Q) < h$, *then:*

$$\inf_{y\in\mathbb{R}^n,\theta\in\mathbb{T}} \left\|\phi - e^{i\theta}Q(\cdot - y)\right\|_{H^1}^2 \leqslant C\left(\mathcal{E}(\phi) - \mathcal{E}(Q)\right).$$

The strategy in [Keraani (2006)] consists in applying the above result to the function

$$v^\varepsilon(t,x) = u^\varepsilon\left(t, \varepsilon x + x(t)\right) e^{-i(\varepsilon x + x(t))\cdot\xi(t)/\varepsilon}.$$

The proof eventually relies on Gronwall lemma and a continuity argument. In order to invoke these arguments, S. Keraani uses Proposition 8.21 and the scheme of the proof of [Bronski and Jerrard (2000)], based on duality arguments and estimates on measures. This yields the result on a time interval $[-T_0, T_0]$. Since this T_0 depends only on constants of the motion, the argument can be repeated, to get the L^∞_{loc} estimate of Theorem 7.33.

In the particular case where the external potential V is an harmonic potential (isotropic or anisotropic), the proof can be simplified. We invite the reader to pay attention to the short note [Keraani (2005)], where this simplification is available.

There are several differences between Proposition 8.13 and Theorem 8.19. First, the profile is very particular in Theorem 8.19: it has to be the ground state. Moreover, in order for the orbital stability property to hold, the assumption $\sigma < 2/n$ is necessary. The assumption is not inconsistent with the requirement $\sigma_0(n) \leqslant \sigma < 2/(n-2)$ in Proposition 8.13, since $\sigma_0(n) < 2/n$. On the other hand, the external potential V does not have to be exactly a polynomial in Theorem 8.19. In the next paragraph, we discuss the validity of Proposition 8.13 when V is a subquadratic potential.

8.3   About general subquadratic potentials

Consider the case of a more general external potential, with initial data concentrated at the origin in the phase space:

$$i\varepsilon\partial_t u^\varepsilon + \frac{\varepsilon^2}{2}\Delta u^\varepsilon = Vu^\varepsilon + \varepsilon^{n\sigma}|u^\varepsilon|^{2\sigma}u^\varepsilon \ ; \ u^\varepsilon(0,x) = \frac{1}{\varepsilon^{n/2}}\varphi\left(\frac{x}{\varepsilon}\right), \quad (8.25)$$

where $V = V(t,x)$ may depend on time. We have seen that for several issues, it is convenient to work with subquadratic potentials. Such a property was more than useful to solve the eikonal equation globally in space (see Proposition 1.9). Moreover, this assumption is fairly natural to study the nonlinear Cauchy problem (fix $\varepsilon > 0$ in Eq. (8.25)), to construct strong solutions (see Sec. 1.4.2), as well as to construct mild solutions thanks to local in time Strichartz estimates, in view of the results in [Fujiwara (1979)] (see Sec. 1.4.3). Suppose that V is smooth and subquadratic:

$$V \in C^\infty(\mathbb{R} \times \mathbb{R}^n), \quad \partial_x^\alpha V \in C(\mathbb{R}; L^\infty(\mathbb{R}^n)), \ \forall|\alpha| \geqslant 2.$$

The first point in Proposition 8.13 can be adapted to this more general framework. Up to considering $u^\varepsilon(t,x)e^{i\int_0^t V(\tau,0)d\tau/\varepsilon}$ instead of $u^\varepsilon(t,x)$, we may assume that $V(t,0) = 0$. Since for $|t| \leqslant \Lambda\varepsilon$, the curvature of rays of geometric optics is negligible, we can prove that for all $\Lambda > 0$,

$$\sum_{\mathcal{A}^\varepsilon \in \{\mathrm{Id}, x/\varepsilon + it\nabla, \varepsilon\nabla\}} \limsup_{\varepsilon \to 0} \sup_{0 \leqslant t \leqslant \Lambda\varepsilon} \left\|\mathcal{A}^\varepsilon(t)\left(u^\varepsilon(t) - u_{\mathrm{app}}^\varepsilon(t)\right)\right\|_{L^2} = 0,$$

for the same $u_{\mathrm{app}}^\varepsilon$ as in Proposition 8.13 (with $\Phi \equiv 0$). By assuming first that $\varphi \in \mathcal{S}(\mathbb{R}^n)$, we conclude as in Proposition 8.13. Moreover, the transition as $\Lambda \to \infty$ occurs in the same way.

Problems arise when trying to prove the analogue of the second régime in Proposition 8.13. Assume for simplicity that V is time-independent, $V = V(x)$. In all the previous cases, the proof that the nonlinearity is negligible for $\varepsilon \ll |t| \leqslant T$ relies on the Heisenberg derivative

$$J^\varepsilon(t) = U^\varepsilon(t)\frac{x}{\varepsilon}U^\varepsilon(-t), \quad \text{where } U^\varepsilon(t) = \exp\left(-i\frac{t}{\varepsilon}\left(-\frac{\varepsilon^2}{2}\Delta + V\right)\right).$$

We start with the good news:

Lemma 8.22. *Suppose that V is time-independent, and subquadratic. The operator J^ε defined above satisfies the following properties:*
(1) It commutes with the linear part of Eq. (8.25).
(2) Weighted Gagliardo–Nirenberg inequalities are available: there exists $\delta > 0$ independent of ε, such that the following holds. If $0 \leqslant \delta(r) < 1$, then there exists C_r such that for all $u \in \Sigma$,

$$\|u\|_{L^r} \leqslant \frac{C_r}{|t|^{\delta(r)}} \|u\|_{L^2}^{1-\delta(r)} \|J^\varepsilon(t)u\|_{L^2}^{\delta(r)}, \quad |t| \leqslant \delta.$$

Proof. The first point stems from the definition of J^ε. For the second point, we use the local dispersive estimate established in [Fujiwara (1979)]: there exist C and $\delta > 0$ independent of ε such that as soon as $|t| \leqslant \delta$,

$$\|U^\varepsilon(t)\|_{L^1 \to L^\infty} \leqslant \frac{C}{(\varepsilon|t|)^{n/2}}.$$

Since $U^\varepsilon(t)$ is unitary on L^2, interpolation yields, if $0 \leqslant \delta(r) < 1$:

$$\|U^\varepsilon(t)f\|_{L^r} \lesssim |\varepsilon t|^{-\delta(r)} \|f\|_{L^{r'}}, \quad |t| \leqslant \delta, \ \forall f \in L^{r'}.$$

Let $g^\varepsilon(t, x) = U^\varepsilon(-t)u(x)$. We have

$$\|U^\varepsilon(t)g^\varepsilon(t)\|_{L^r} \lesssim |\varepsilon t|^{-\delta(r)} \|g^\varepsilon(t)\|_{L^{r'}}.$$

Let $\lambda > 0$, and write

$$\|g^\varepsilon(t)\|_{L^{r'}}^{r'} = \int_{|x| \leqslant \lambda} |g^\varepsilon(t, x)|^{r'} dx + \int_{|x| > \lambda} |g^\varepsilon(t, x)|^{r'} dx.$$

Estimate the first term by Hölder's inequality, by writing $|g^\varepsilon|^{r'} = 1 \times |g^\varepsilon|^{r'}$,

$$\int_{|x| \leqslant \lambda} |g^\varepsilon(t, x)|^{r'} dx \lesssim \lambda^{n/p'} \left(\int_{|x| \leqslant \lambda} |g^\varepsilon(t, x)|^{r'p} dx \right)^{1/p},$$

and choose $p = 2/r' (\geqslant 1)$. Estimate the second term by the same Hölder's inequality, after inserting the factor x as follows,

$$\int_{|x| > \lambda} |g^\varepsilon(t, x)|^{r'} dx = \int_{|x| > \lambda} |x|^{-r'} |x|^{r'} |g^\varepsilon(t, x)|^{r'} dx$$

$$\leqslant \left(\int_{|x| > \lambda} |x|^{-r'p'} dx \right)^{1/p'} \left(\int_{|x| > \lambda} |xg^\varepsilon(t, x)|^2 dx \right)^{1/p}$$

$$\lesssim \lambda^{n/p'-r'} \|xg^\varepsilon(t, x)\|_{L^2}^{2/p}.$$

In summary, we have the following estimate, for any $\lambda > 0$:

$$\|g^\varepsilon(t)\|_{L^{r'}} \lesssim \lambda^{n/(p'r')}\|g^\varepsilon(t)\|_{L^2} + \lambda^{n/(p'r')-1}\|xg^\varepsilon(t,x)\|_{L^2}.$$

Notice that $n/(p'r') = \delta(r)$, and equalize both terms of the right hand side:

$$\lambda = \frac{\|xg^\varepsilon(t,x)\|_{L^2}}{\|g^\varepsilon(t)\|_{L^2}}.$$

This yields $\|g^\varepsilon(t)\|_{L^{r'}} \lesssim \|g^\varepsilon(t)\|_{L^2}^{1-\delta(r)}\|xg^\varepsilon(t,x)\|_{L^2}^{\delta(r)}$. Therefore,

$$\|U^\varepsilon(t)g^\varepsilon(t)\|_{L^r} \lesssim |\varepsilon t|^{-\delta(r)}\|g^\varepsilon(t)\|_{L^2}^{1-\delta(r)}\|xg^\varepsilon(t,x)\|_{L^2}^{\delta(r)}$$

$$\lesssim |t|^{-\delta(r)}\|g^\varepsilon(t)\|_{L^2}^{1-\delta(r)}\left\|\frac{x}{\varepsilon}g^\varepsilon(t,x)\right\|_{L^2}^{\delta(r)}.$$

Back to u, write $g^\varepsilon(t) = U^\varepsilon(-t)u$. This completes the proof of the lemma, since $U^\varepsilon(t)$ is unitary on L^2. $\qquad\square$

Recall that in applying Lemma 8.16, we have invoked the first point and the last two points only. The second point was used to infer the last two points. From the above lemma, the operator J^ε that we now consider also satisfies the first point and an analogue of the third point of Lemma 8.16. To be able to prove the analogue of Proposition 8.13 when the external potential is a general subquadratic potential, we would need to know how J^ε acts on gauge invariant nonlinearities. Unfortunately, this question seems to be open. Note that we may not need exactly the fourth point of Lemma 8.16: J^ε may act on gauge invariant nonlinearities "approximately" like a derivative. The goal would be to find what an acceptable definition of "approximately" could be, and to show that it is satisfied by J^ε. What can be proved, at least, is that the first two points in Lemma 8.16 hold *only* when the external potential is exactly a polynomial:

Proposition 8.23. *Let $V \in C^\infty(\mathbb{R}^n; \mathbb{R})$ be a smooth, subquadratic potential. Let $\phi \in C^4(]0,T] \times \mathbb{R}^n; \mathbb{R})$ and $f \in C^1(]0,T])$ for some $T > 0$. Assume that f has no zero on the interval $]0,T]$. Define the operator A^ε by:*

$$A^\varepsilon(t) = if(t)e^{i\phi(t,x)/\varepsilon}\nabla\left(e^{-i\phi(t,x)/\varepsilon}\cdot\right) = \frac{f(t)}{\varepsilon}\nabla\phi(t,x) + if(t)\nabla.$$

Then A^ε commutes with the linear part of Eq. (8.25) if and only if V is exactly a polynomial of degree at most two.

Remark 8.24. In Lemma 8.16, the phases ϕ_1 and ϕ_2 solve the eikonal equation

$$\partial_t\phi + \frac{1}{2}|\nabla\phi|^2 + V = 0. \tag{8.26}$$

In the above statement, we do not assume that ϕ solves the eikonal equation. However, we will see in the proof that it is essentially necessary (up to a phase shift which depends only on time, and thereby does not affect the definition of A^ε).

Proof. In view of Lemma 8.16, we only have to prove the "only if" part. Computations yield

$$
\begin{aligned}
\left[i\varepsilon\partial_t + \frac{\varepsilon^2}{2}\Delta - V, A_j^\varepsilon(t)\right] &= f'(t)\partial_j\phi + f(t)\partial_{jt}^2\phi + f(t)\partial_j V \\
&+ \varepsilon\Big(-f'(t)\partial_j + f(t)\nabla(\partial_j\phi)\cdot\nabla + \frac{1}{2}f(t)\Delta(\partial_j\phi)\Big).
\end{aligned}
\tag{8.27}
$$

This bracket is zero if and only if the terms in ε^0 and ε^1 are zero. The term in ε is the sum of an operator of order one and of an operator of order zero. It is zero if and only if both operators are zero. The operator of order one is zero if and only if

$$
f(t)\partial_{jj}^2\phi = f'(t), \quad \partial_{jk}^2\phi \equiv 0 \text{ if } j \neq k.
$$

In particular, $\partial_{jj}^2\phi$ is a function of time only, independent of x, and we have

$$
\frac{1}{2}f(t)\Delta(\partial_j\phi) \equiv 0.
$$

From the above computations, the first two terms in ε^0 also write

$$
f'(t)\partial_j\phi + f(t)\partial_{jt}^2\phi = \sum_{k=1}^n f(t)\partial_k\phi\partial_{jk}^2\phi + f(t)\partial_{jt}^2\phi = f(t)\partial_j\left(\partial_t\phi + \frac{1}{2}|\nabla\phi|^2\right).
$$

Canceling the term in ε^0 in (8.27) therefore yields, since f is never zero on $]0, T]$,

$$
\partial_j\left(\partial_t\phi + \frac{1}{2}|\nabla\phi|^2 + V\right) = 0.
\tag{8.28}
$$

Differentiating the above equation with respect to x_k and x_ℓ, all the terms with ϕ vanish, since we noticed that the derivatives of order at least three of ϕ are zero. We deduce that for any triplet (j, k, ℓ), $\partial_{jk\ell}^3 V \equiv 0$, that is, V is a polynomial of degree at most two.

Notice that since (8.28) holds for any $j \in \{1, \ldots, n\}$, there exists a function Ξ of time only such that

$$
\partial_t\phi + \frac{1}{2}|\nabla\phi|^2 + V = \Xi(t).
$$

This means that ϕ is almost a solution to the eikonal equation (8.26). Replacing ϕ by $\widetilde{\phi}(t, x) := \phi(t, x) - \int_0^t \Xi(\tau)d\tau$ does not affect the definition of A^ε, and $\widetilde{\phi}$ solves (8.26). $\qquad\square$

Remark 8.25. In view of Eq. (8.27), if not zero, the commutator could be at best of order $\mathcal{O}(\varepsilon)$. Unfortunately, a term of this size cannot be considered as a small source term, because of the factor ε in front of the time derivative of the semi-classical Schrödinger operator; see Lemma 1.2.

We leave out the discussion on general subquadratic potentials at this stage. Recall that for instance, we have no complete picture when V is as in Assumption 8.18 and Theorem 8.19. Finding tools to study this more general case would be of course the key to extend Proposition 8.13. It might also open new possibilities for similar problems.

In a similar spirit, we invite the reader to consult the very interesting reference [Sacchetti (2005)]. There, a global in time analysis is obtained for a problem which is slightly different from Eq. (8.25). The external potential V is particular (double well), and the initial data are concentrated on the two eigenfunctions associated to the semi-classical Hamiltonian (this implies that the concentration of the initial data is not at scale ε as in Eq. (8.25), but at scale $\sqrt{\varepsilon}$). Also, the order of magnitude of the coupling constant is much weaker than in Eq. (8.25) (it is exponentially decreasing as $\varepsilon \to 0$). However, it is critical as far as leading order nonlinear effects are concerned. A global in time analysis is then established (in the same fashion as in Theorem 8.2), which yields a precise description of the interaction between the linear effects due to the double well, and the nonlinear effects. The main typically nonlinear effect is the following: if the initial data are concentrated on a single well, and if the coupling factor is sufficiently large, then the solution remains localized on the same well, while the linear solution would oscillate between the two wells (beating motion). See also the more recent work [Bambusi and Sacchetti (2007)] on this subject.

Chapter 9

Some Ideas for Supercritical Cases

To conclude these notes, we present partial results for a focal point in the supercritical case $\alpha < n\sigma$, without external potential:

$$i\varepsilon\partial_t u^\varepsilon + \frac{\varepsilon^2}{2}\Delta u^\varepsilon = \varepsilon^\alpha |u^\varepsilon|^{2\sigma} u^\varepsilon \quad ; \quad u^\varepsilon(0,x) = a_0(x)e^{-i|x|^2/(2\varepsilon)}.$$

Using a semi-classical lens transform (which amounts to adapting Eq. (9.6) in view of the approach of [Carles (2002)], see Remark 9.5), we could also consider the case with an isotropic harmonic potential. To observe nonlinear effects which are due to focusing phenomenon, we assume $\alpha > 1$: the nonlinearity is negligible in a WKB régime. Also, we suppose that the nonlinearity is exactly cubic: $\sigma = 1$. All in all, we assume $1 < \alpha < n$. Note that in particular, this assumption excludes the one-dimensional case. This point appears several times in the proof. We therefore consider:

$$i\varepsilon\partial_t u^\varepsilon + \frac{\varepsilon^2}{2}\Delta u^\varepsilon = \varepsilon^\alpha |u^\varepsilon|^2 u^\varepsilon \quad ; \quad u^\varepsilon(0,x) = a_0(x)e^{-i|x|^2/(2\varepsilon)}. \tag{9.1}$$

In [Carles (2007a)], preliminary results are shown concerning the asymptotic behavior of u^ε. Besides these results, which we state and prove below, the computations in [Carles (2007a)] yield an interesting example. Indeed, a formal computation yields a function that solves Eq. (9.1), up to a source term which can be taken very small as $\varepsilon \to 0$. However, the usual stability analysis does not make it possible to conclude that it is close to the exact solution, since the exponential factor in the Gronwall lemma counterbalances the smallness of the source term. This problem is similar to the one encountered in the supercritical régime for WKB analysis (Chap. 4). Moreover, we can prove that this function is not a good approximation of the exact solution, when Gronwall lemma ceases to be interesting. The reason is more subtle than a spectral instability: it is related to the notion of good unknown functions, in order to observe the right oscillations.

First, qualitative arguments yield relevant scalings for Eq. (9.1). Recall that the mass and energy associated to Eq. (9.1) do not depend on time:

$$\|u^\varepsilon(t)\|_{L^2} = \|a_0\|_{L^2},$$
$$\|\varepsilon \nabla u^\varepsilon(t)\|_{L^2}^2 + \varepsilon^\alpha \|u^\varepsilon(t)\|_{L^4}^4 = \text{const.} = \mathcal{O}(1) \underset{\varepsilon \to 0}{\sim} \|xa_0\|_{L^2}^2.$$

Intuitively, the nonlinear term is relevant at a focal point if the potential energy is of order $\mathcal{O}(1)$ exactly, while the modulus of u^ε is described by a concentrating profile,

$$|u^\varepsilon(t,x)| \underset{\varepsilon \to 0}{\sim} \frac{1}{\varepsilon^{n\gamma/2}} \varphi\left(\frac{x}{\varepsilon^\gamma}\right), \tag{9.2}$$

for some $\gamma > 0$. The power of ε in front of φ is to ensure the L^2-norm conservation. We consider only the modulus of u^ε, because the forthcoming discussion will show that phenomena affecting the phase are crucial, and not completely understood. We then compute:

$$\varepsilon^\alpha \|u^\varepsilon(t)\|_{L^4}^4 \underset{\varepsilon \to 0}{\sim} \varepsilon^{\alpha - n\gamma} \|\varphi\|_{L^4}^4$$

We check that the "linear" value $\gamma = 1$ is forbidden, because in that case, the potential energy is unbounded ($\alpha < n$). Note also that the value $\gamma = 1$ was the one encountered in the critical case $\alpha = n\sigma$ (see Proposition 7.19). For the potential energy to be of order $\mathcal{O}(1)$ exactly, we have to choose:

$$\gamma = \frac{\alpha}{n} < 1. \tag{9.3}$$

We will not prove that the above argument is correct, but we will show that the scale ε^γ is an important feature of this problem. We show that there is a time at which u^ε behaves like in Eq. (9.2), but our analysis stops before $t = 1$. Notice also that the above argument suggests that the amplification of the solution u^ε as time goes to 1 is less important than in the linear case; supercritical phenomena may occur in the phase, and also affect the amplitude. We also point out that estimates agreeing with these heuristic arguments were established by S. Masaki [Masaki (2007)] in the analogous case, where the cubic nonlinearity is replaced by a Hartree type nonlinearity.

Definition 9.1. If $T > 0$, $(k_j)_{j \geqslant 1}$ is an increasing sequence of real numbers, $(\phi_j)_{j \geqslant 1}$ is a sequence in $H^\infty(\mathbb{R}^n)$, and $\phi \in C([0,T]; H^\infty)$, the asymptotic relation

$$\phi(t,x) \sim \sum_{j \geqslant 1} t^{k_j} \phi_j(x) \quad \text{as } t \to 0$$

means that for every integer $J \geqslant 1$ and every $s > 0$,

$$\left\| \phi(t, \cdot) - \sum_{j=1}^{J} t^{k_j} \phi_j \right\|_{H^s(\mathbb{R}^n)} = o\left(t^{k_J} \right) \quad \text{as } t \to 0.$$

Theorem 9.2. *Let* $a_0 \in \mathcal{S}(\mathbb{R}^n)$. *Assume* $n > \alpha > 1$. *Then there exist* $T > 0$ *independent of* $\varepsilon \in]0, 1]$, *a sequence* $(\phi_j)_{j \geqslant 1}$ *in* H^∞, *and* $\phi \in C([0, T]; H^\infty)$, *such that:*
1. $\phi(t, x) \sim \sum_{j \geqslant 1} t^{jn-1} \phi_j(x)$ *as* $t \to 0$.
2. *For* $1 - t \gg \varepsilon^\gamma$ $(\gamma = \alpha/n < 1)$, *the asymptotic behavior of* u^ε *is given by:*

$$\limsup_{\varepsilon \to 0} \sup_{0 \leqslant t \leqslant 1 - \Lambda \varepsilon^\gamma} \| u^\varepsilon(t) - v^\varepsilon(t) \|_{L^2(\mathbb{R}^n)} \xrightarrow[\Lambda \to +\infty]{} 0,$$

where $v^\varepsilon(t, x) = \dfrac{e^{i \frac{|x|^2}{2\varepsilon(t-1)}}}{(1-t)^{n/2}} a_0 \left(\dfrac{x}{1-t} \right) \exp \left(i \varepsilon^{\gamma-1} \phi \left(\dfrac{\varepsilon^\gamma}{1-t}, \dfrac{x}{1-t} \right) \right).$

Some comments are in order. In the linear case as well as in the critical case $\alpha = n > 1$, the above result holds with $\gamma = 1$ and $\phi \equiv 0$. The case $\alpha < n$ is supercritical as far as nonlinear effects near $t = 1$ are concerned. We emphasize two important features in the above result: the analysis stops sooner than $1 - t \gg \varepsilon$, and nonlinear effects cause the presence of the (non-trivial) phase ϕ. For $1 - t \gg \varepsilon^\gamma$, we have

$$\varepsilon^{\gamma-1} \phi \left(\frac{\varepsilon^\gamma}{1-t}, \frac{x}{1-t} \right) \sim \sum_{j \geqslant 1} \frac{\varepsilon^{j\alpha-1}}{(1-t)^{jn-1}} \phi_j \left(\frac{x}{1-t} \right).$$

The above phase shift starts being relevant for $1 - t \approx \varepsilon^{\frac{\alpha-1}{n-1}}$ (recall that $n > \alpha > 1$); this is the first boundary layer where nonlinear effects appear at leading order, measured by ϕ_1. We will check that this phase shift is relevant: ϕ_1 is not zero (unless $u^\varepsilon \equiv 0$, see (9.11) below). We then have a countable number of boundary layers in time, of size

$$1 - t \approx \varepsilon^{\frac{j\alpha-1}{jn-1}},$$

which reach the layer $1 - t \approx \varepsilon^\gamma$ in the limit $j \to +\infty$. At each new boundary layer, a new phase ϕ_j becomes relevant at leading order. In general, none of the ϕ_j's is zero: see e.g. (9.13) for ϕ_2. The result of a cascade of phases can be compared to the one discovered by C. Cheverry [Cheverry (2006)] in the case of fluid dynamics, although the phenomenon seems to be different. Yet, our result shares another property with [Cheverry (2006)], which does not appear in the above statement. Theorem 9.2 shows perturbations of the phase (the ϕ_j's), but not of the amplitude: the main profile is the same as in the linear case, that is, a rescaling of a_0. However, to compute the first N phase shifts, $(\phi_j)_{1 \leqslant j \leqslant N}$, one has to compute $N - 1$ corrector terms of the main profile a_0. This is due to the fact that the above result can be connected rather explicitly to the supercritical WKB analysis of Sec. 4.2.1, *via* a semi-classical conformal transform.

Each phase shift oscillates at a rate between $\mathcal{O}(1)$ (when it starts being relevant) and $\mathcal{O}(\varepsilon^{\gamma-1})$ (when it reaches the layer of size ε^γ). Since $\gamma > 0$, this means that each phase shift is rapidly oscillating at the scale of the amplitude, but oscillating strictly more slowly than the geometric phase $\frac{|x|^2}{2\varepsilon(t-1)}$, for $1 - t \gg \varepsilon^\gamma$. We will see that for $1 - t = \mathcal{O}(\varepsilon^\gamma)$, all the terms in ϕ, plus the geometric phase, have the same order: all these phases become comparable, see (9.16).

If we transpose the results of [Masaki (2007)] to the case of the cubic nonlinearity, we should have the uniform estimate:

$$\|J^\varepsilon(t) u^\varepsilon(t)\|_{L^2} \lesssim \varepsilon^{\alpha/n-1} = \varepsilon^{\gamma-1}.$$

Recalling that

$$J^\varepsilon(t) = i(t-1)e^{i|x|^2/(2\varepsilon(t-1))} \nabla \left(e^{-i|x|^2/(2\varepsilon(t-1))}. \right),$$

the above estimate suggests that besides the oscillations carried by the solution of the eikonal equation, u^ε oscillates at scale $\varepsilon^{1-\gamma}$, which is between ε and 1. This estimate is in perfect agreement with the previous discussions, and with Theorem 9.2. One of the aspects in [Masaki (2007)] is that the above estimate is proved to be valid for all time, and not only for $t < 1$. An interesting question would be to know if this estimate is sharp (as $\varepsilon \to 0$) for $t > 1$. In other words, do intermediary scales of oscillations have appeared near $t = 1$, persisting for $t > 1$?

9.1   Cascade of phase shifts

In this paragraph, we prove Theorem 9.2. We start by some formal computations, whose conclusion is in good agreement with the statement of Theorem 9.2. Then, we prove Theorem 9.2, and notice that the two approaches disagree, past the first boundary layer. This aspect is explained in Sec. 9.1.3.

9.1.1 *A formal computation*

For simplicity, we assume $a_0 \in \mathcal{S}(\mathbb{R}^n)$. Resume the approach *via* Lagrangian integral (see Sec. 7.1),

$$u^\varepsilon(t,x) = \frac{1}{(2\pi\varepsilon)^{n/2}} \int_{\mathbb{R}^n} e^{-i\frac{t-1}{2\varepsilon}|\xi|^2 + i\frac{x\cdot\xi}{\varepsilon}} A^\varepsilon(t,\xi) d\xi.$$

From Lemma 7.1, we know that the Lagrangian symbol A^ε converges at time $t = 0$, as $\varepsilon \to 0$. For $t \neq 1$, we apply stationary phase formula at a very formal level, as in Sec. 7.3 (with now $\sigma = 1$):

$$\mathcal{F}\left(|u^\varepsilon|^2 u^\varepsilon\right)\left(t, \frac{\xi}{\varepsilon}\right) \approx \frac{\varepsilon^{n/2}}{|t-1|^n} |A^\varepsilon|^2 A^\varepsilon(t,\xi) e^{-i\frac{t-1}{2\varepsilon}|\xi|^2}.$$

Using the second part of Lemma 7.1, we infer that the evolution of A^ε should be described by

$$i\partial_t A^\varepsilon(t,\xi) = \frac{\varepsilon^{\alpha-1}}{|t-1|^n} |A^\varepsilon|^2 A^\varepsilon(t,\xi).$$

To be consistent, we should say that the right hand side is negligible, since $\alpha > 1$. We keep it though. Like several times before, we notice that the modulus of A^ε is independent of time, and we compute explicitly:

$$A^\varepsilon(t,\xi) = A^\varepsilon(0,\xi) \exp\left(-i\varepsilon^{\alpha-1}|A^\varepsilon(0,\xi)|^2 \int_0^t \frac{d\tau}{(1-\tau)^n}\right)$$

$$\approx A_0(\xi) \exp\left(-i\varepsilon^{\alpha-1}|A_0(\xi)|^2 \int_0^t \frac{d\tau}{(1-\tau)^n}\right),$$

where $A_0(\xi) = e^{-in\pi/4} a_0(-\xi)$. Back to u^ε, this yields, for $t < 1$:

$$u^\varepsilon(t,x) \approx \frac{e^{in\pi/4}}{(1-t)^{n/2}} A^\varepsilon\left(t, \frac{x}{t-1}\right) e^{i\frac{|x|^2}{2\varepsilon(t-1)}}$$

$$\approx \frac{1}{(1-t)^{n/2}} a_0\left(\frac{x}{1-t}\right) e^{i\frac{|x|^2}{2\varepsilon(t-1)}} e^{-i\left|a_0\left(\frac{x}{1-t}\right)\right|^2 \varepsilon^{\alpha-1} \int_0^t \frac{d\tau}{(1-\tau)^n}}.$$

Since $n \geqslant 2$,

$$\int_0^t \frac{d\tau}{(1-\tau)^n} \underset{t\to 1}{\sim} \frac{1}{(n-1)(1-t)^{n-1}},$$

we see that the last phase in the approximation of u^ε becomes relevant for

$$\varepsilon^{\alpha-1} \approx (1-t)^{n-1} : 1 - t \approx \varepsilon^{\frac{\alpha-1}{n-1}}.$$

This is exactly the first boundary layer appearing in Theorem 9.2. To go further into this formal analysis, we adopt another point of view, which may appear as more explicit. Using a generalized WKB expansion, seek

$$u^\varepsilon(t,x) \underset{\varepsilon\to 0}{\sim} v_1^\varepsilon(t,x) = b^\varepsilon(t,x) e^{i\phi(t,x)/\varepsilon},$$

and change the usual hierarchy to force the contribution of the nonlinear term to appear in the transport equation:

$$\begin{cases} \partial_t \phi + \dfrac{1}{2}|\nabla \phi|^2 = 0 & ; \quad \phi(0,x) = -\dfrac{|x|^2}{2}. \\[2mm] \partial_t b^\varepsilon + \nabla \phi \cdot \nabla b^\varepsilon + \dfrac{1}{2}b^\varepsilon \Delta \phi = -i\varepsilon^{\alpha-1}|b^\varepsilon|^2 b^\varepsilon & ; \quad b^\varepsilon(0,x) = a_0(x). \end{cases}$$

The eikonal equation is the same as the linear case, as well as its solution. The transport equation is an ordinary differential equation along the rays of geometric optics $\frac{x}{1-t} = \text{const.}$, of the form

$$\dot{y} = -i\varepsilon^{\alpha-1}|y|^2 y.$$

The modulus of b^ε is constant along rays, and

$$b^\varepsilon(t,x) = \frac{1}{(1-t)^{n/2}} a_0\left(\frac{x}{1-t}\right) e^{-i\varepsilon^{\alpha-1}\left|a_0\left(\frac{x}{1-t}\right)\right|^2 \int_0^t \frac{d\tau}{(1-\tau)^n}}.$$

We retrieve the same approximation as with the Lagrangian integral approach. By construction,

$$i\varepsilon \partial_t v_1^\varepsilon + \frac{\varepsilon^2}{2}\Delta v_1^\varepsilon = \varepsilon^\alpha |v_1^\varepsilon|^2 v_1^\varepsilon + r_1^\varepsilon \quad ; \quad v_1^\varepsilon(0,x) = a_0(x)e^{-i|x|^2/(2\varepsilon)},$$

where

$$r_1^\varepsilon(t,x) = \frac{\varepsilon^2}{2} e^{i|x|^2/(2\varepsilon(t-1))} \Delta b^\varepsilon(t,x).$$

Denote $g_1^\varepsilon(t,x) = -\varepsilon^{\alpha-1}\left|a_0\left(\frac{x}{1-t}\right)\right|^2 \int_0^t \frac{d\tau}{(1-\tau)^n}$. We compute:

$$\Delta b^\varepsilon(t,x) = \frac{e^{ig_1^\varepsilon(t,x)}}{(1-t)^{2+n/2}} \Delta a_0\left(\frac{x}{1-t}\right)$$

$$- \frac{2ie^{ig_1^\varepsilon(t,x)}}{(1-t)^{2+n/2}} \nabla a_0\left(\frac{x}{1-t}\right) \cdot \nabla |a_0|^2\left(\frac{x}{1-t}\right) \varepsilon^{\alpha-1} \int_0^t \frac{d\tau}{(1-\tau)^n}$$

$$- \frac{ie^{ig_1^\varepsilon(t,x)}}{(1-t)^{2+n/2}} a_0\left(\frac{x}{1-t}\right) \Delta |a_0|^2\left(\frac{x}{1-t}\right) \varepsilon^{\alpha-1} \int_0^t \frac{d\tau}{(1-\tau)^n}$$

$$- \frac{e^{ig_1^\varepsilon(t,x)}}{(1-t)^{2+n/2}} a_0\left(\frac{x}{1-t}\right) \left|\nabla |a_0|^2\right|^2\left(\frac{x}{1-t}\right) \left(\varepsilon^{\alpha-1} \int_0^t \frac{d\tau}{(1-\tau)^n}\right)^2.$$

We infer

$$\|r_1^\varepsilon(t)\|_{L^2} \lesssim \frac{\varepsilon^2}{(1-t)^2} + \frac{\varepsilon^{\alpha+1}}{(1-t)^{n+1}} + \frac{\varepsilon^{\alpha+1}}{(1-t)^{n+1}} + \frac{\varepsilon^{2\alpha}}{(1-t)^{2n}}.$$

Therefore,

$$\frac{1}{\varepsilon}\int_0^t \|r_1^\varepsilon(\tau)\|_{L^2} d\tau \lesssim \frac{\varepsilon}{1-t} + \frac{\varepsilon^\alpha}{(1-t)^n} + \frac{\varepsilon^\alpha}{(1-t)^n} + \frac{\varepsilon^{2\alpha-1}}{(1-t)^{2n-1}}.$$

Following the energy estimates of Lemma 1.2, this quantity might be the one that dictates the size of the error $u^\varepsilon - v_1^\varepsilon$. The first term is the "linear" one: it is small for $1-t \gg \varepsilon$. The second and third ones (which are the same, but which we keep to keep track of the corresponding terms in Δb^ε) are small for $1 - t \gg \varepsilon^{\alpha/n}$, and the last one is small for $1 - t \gg \varepsilon^{(2\alpha-1)/(2n-1)}$. Since

$$\frac{2\alpha - 1}{2n - 1} < \frac{\alpha}{n} = \gamma < 1,$$

the last term in Δb^ε is the first to cease to be negligible. We note that in the equation for v_1^ε, the corresponding term has the same argument as v_1^ε. This suggests that the last term in r_1^ε might be eliminated by adding an extra phase term in the approximate solution, in view of the identity

$$i\varepsilon\partial_t\left(ae^{i\theta^\varepsilon}\right) = -\partial_t\theta^\varepsilon \times ae^{i\theta^\varepsilon} + e^{i\theta^\varepsilon}i\varepsilon\partial_t a.$$

Seek an approximate solution of the form:

$$v^\varepsilon(t,x) = \frac{1}{(1-t)^{n/2}}a_0\left(\frac{x}{1-t}\right)e^{i\phi^\varepsilon(t,x)}, \quad \phi^\varepsilon(t,x) = \frac{|x|^2}{2\varepsilon(t-1)} + g^\varepsilon(t,x).$$

We find

$$i\varepsilon\partial_t v^\varepsilon + \frac{\varepsilon^2}{2}\Delta v^\varepsilon = \left(i\frac{\varepsilon^2}{2}\Delta g^\varepsilon - \varepsilon\partial_t g^\varepsilon - \frac{\varepsilon^2}{2}|\nabla g^\varepsilon|^2 + \frac{\varepsilon}{1-t}x\cdot\nabla g^\varepsilon\right)v^\varepsilon$$

$$+ i\frac{\varepsilon^2}{(1-t)^{1+n/2}}\nabla g^\varepsilon\cdot\nabla a_0\left(\frac{x}{1-t}\right)e^{i\phi^\varepsilon}$$

$$+ \frac{1}{2}\left(\frac{\varepsilon}{1-t}\right)^2\frac{e^{i\phi^\varepsilon}}{(1-t)^{n/2}}\Delta a_0\left(\frac{x}{1-t}\right).$$

As suggested by the previous computations, write

$$g^\varepsilon(t,x) = \frac{1}{\varepsilon}\int_0^t h\left(\frac{\varepsilon^\alpha}{(1-\tau)^n},\frac{x}{1-t}\right)d\tau, \quad \text{with } h(z,\xi) \sim \sum_{j\geqslant 1}z^j g_j(\xi). \quad (9.4)$$

In the equation verified by v^ε, the last term is the "same" as in the linear case: it becomes relevant only in a boundary layer of size ε. Since our approach will lead us to the boundary layer of size ε^γ ($\gamma < 1$), we ignore that term. The remaining terms with a factor i are of order, in L^2,

$$\varepsilon^2\|\Delta g^\varepsilon(t)\|_{L^\infty} + \frac{\varepsilon^2}{1-t}\|\nabla g^\varepsilon(t)\|_{L^\infty} \lesssim \frac{\varepsilon}{(1-t)^2}\int_0^t\frac{\varepsilon^\alpha}{(1-\tau)^n}d\tau \lesssim \frac{\varepsilon^{\alpha+1}}{(1-t)^{n+1}},$$

and their contribution is also left out in this computation, since they become of order one only for $1 - t \approx \varepsilon^\gamma$.

Now we require that v^ε be an approximate solution to (9.1):

$$\left(\partial_t - \frac{x}{1-t} \cdot \nabla\right) g^\varepsilon + \frac{\varepsilon}{2}|\nabla g^\varepsilon|^2 = -\varepsilon^{\alpha-1}|v^\varepsilon(t,x)|^2$$

$$= -\frac{\varepsilon^{\alpha-1}}{(1-t)^n}\left|a_0\left(\frac{x}{1-t}\right)\right|^2.$$

Using (9.4), we get:

$$g_1(\xi) = -|a_0(\xi)|^2,$$

$$\text{for } j \geqslant 2, \quad g_j(\xi) = -\frac{1}{2}\sum_{p+q=j}\frac{1}{(pn-1)(qn-1)}\nabla g_p \cdot \nabla g_q, \qquad (9.5)$$

with the convention $g_0 \equiv 0$. This algorithm produces smooth solutions, since $a_0 \in \mathcal{S}(\mathbb{R}^n)$. Define

$$g_N^\varepsilon(t,x) = \frac{1}{\varepsilon}\sum_{j=1}^N \int_0^t \left(\frac{\varepsilon^\gamma}{1-\tau}\right)^{nj} d\tau \times g_j\left(\frac{x}{1-t}\right),$$

$$v_N^\varepsilon(t,x) = \frac{1}{(1-t)^{n/2}}a_0\left(\frac{x}{1-t}\right)e^{i\frac{|x|^2}{2\varepsilon(t-1)}+g_N^\varepsilon(t,x)}.$$

Proposition 9.3 (Formal approximation to (9.1)). *Let* $n > \alpha > 1$, $a_0 \in \mathcal{S}(\mathbb{R}^n)$, *and fix* $N \in \mathbb{N}^*$. *The function* v_N^ε *solves*

$$i\varepsilon\partial_t v_N^\varepsilon + \frac{\varepsilon^2}{2}\Delta v_N^\varepsilon = \varepsilon^\alpha|v_N^\varepsilon|^2 v_N^\varepsilon + r_N^\varepsilon \quad ; \quad v_N^\varepsilon(0,x) = a_0(x)e^{-i|x|^2/(2\varepsilon)}.$$

For $1 - t \geqslant \varepsilon^\gamma = \varepsilon^{\alpha/n}$, *the source term satisfies:*

$$\frac{1}{\varepsilon}\int_0^t \|r_N^\varepsilon(\tau)\|_{L^2}d\tau \lesssim \frac{\varepsilon^{(N+1)\alpha-1}}{(1-t)^{(N+1)n-1}} + \frac{\varepsilon^\alpha}{(1-t)^n}.$$

For $1 \leqslant j \leqslant N$, the jth term of the series defining g_N^ε becomes relevant in a boundary layer of size $\varepsilon^{\frac{j\alpha-1}{jn-1}}$; since

$$\frac{1}{\varepsilon}\int_0^t\left(\frac{\varepsilon^\gamma}{1-\tau}\right)^{nj}d\tau \underset{t\to 1}{\sim} \frac{1}{jn-1}\frac{\varepsilon^{j\alpha-1}}{(1-t)^{jn-1}}.$$

In the limit $N \to +\infty$, a countable family of boundary layers appear. Qualitatively, this agrees with Theorem 9.2.

Remark 9.4. In the critical case $\alpha = n > 1$, we have $\beta = \gamma = 1$: the above boundary layers "collapse" one on another. There are no such phase shifts as above.

The sole estimate of the source term does not prove much. In a standard (semi-linear) stability argument, the nonlinearity $|u^\varepsilon|^2 u^\varepsilon - |v^\varepsilon_N|^2 v^\varepsilon_N$ is usually treated by a Gronwall type argument. If the nonlinearity is "too strong", then the above estimate, which is completely relevant in the linear case, does not necessarily account for the size of the error. Since we are in a supercritical case, it is not surprising that Proposition 9.3 is only a formal result. To see what information could be hoped from Proposition 9.3, set $w^\varepsilon_N = u^\varepsilon - v^\varepsilon_N$. It solves

$$i\varepsilon\partial_t w^\varepsilon_N + \frac{\varepsilon^2}{2}\Delta w^\varepsilon_N = \varepsilon^\alpha \left(|u^\varepsilon|^2 u^\varepsilon - |v^\varepsilon_N|^2 v^\varepsilon_N\right) - r^\varepsilon_N \quad ; \quad w^\varepsilon_{N|t=0} = 0.$$

Applying Lemma 1.2, we find, for $t > 0$:

$$\|w^\varepsilon_N(t)\|_{L^2} \lesssim \varepsilon^{\alpha-1} \int_0^t \left\| |u^\varepsilon|^2 u^\varepsilon - |v^\varepsilon_N|^2 v^\varepsilon_N \right\|_{L^2} d\tau + \frac{1}{\varepsilon} \int_0^t \|r^\varepsilon_N(\tau)\|_{L^2} d\tau$$

$$\lesssim \varepsilon^{\alpha-1} \int_0^t \left(\|u^\varepsilon(\tau)\|_{L^\infty}^2 + \|v^\varepsilon_N(\tau)\|_{L^\infty}^2 \right) \|w^\varepsilon_N(\tau)\|_{L^2} d\tau$$

$$+ \frac{\varepsilon^{(N+1)\alpha-1}}{(1-t)^{(N+1)n-1}}.$$

Assume that for $1 - t \gg \varepsilon^\gamma$, u^ε is of the same order of magnitude as v^ε_N, that is

$$\|u^\varepsilon(t)\|_{L^\infty} \lesssim \frac{1}{(1-t)^{n/2}}.$$

(This estimate will actually be proved in Sec. 9.1.2, see Remark 9.8.)
Gronwall lemma then yields, for $t < 1$:

$$\|w^\varepsilon_N(t)\|_{L^2} \lesssim \frac{\varepsilon^{(N+1)\alpha-1}}{(1-t)^{(N+1)n-1}} e^{C\varepsilon^{\alpha-1}/(1-t)^{n-1}}.$$

Formally, take $N = \infty$: in L^2, the error w^ε is controlled by

$$\frac{\varepsilon^\alpha}{(1-t)^n} e^{C\varepsilon^{\alpha-1}/(1-t)^{n-1}}.$$

For $1 - t \gtrsim \varepsilon^{(\alpha-1)/(n-1)}$, the exponential is bounded. Moreover, if

$$\frac{\varepsilon^{\alpha-1}}{(1-t)^{n-1}} = c\log\frac{1}{\varepsilon},$$

we still infer

$$\|w^\varepsilon_N(t)\|_{L^2} \ll 1,$$

provided that c (independent of ε) is sufficiently small. This shows that the first phase shift, also derived with the Lagrangian integral approach, is the right one, since the approximation is good past the first boundary layer. On the other hand, the second boundary layer corresponds to

$$1 - t \approx \varepsilon^{\frac{2\alpha-1}{2n-1}}.$$

We check that

$$\frac{\varepsilon^\alpha}{(1-t)^n} e^{C\varepsilon^{\alpha-1}/(1-t)^{n-1}} \Big|_{t=1-\varepsilon^{(2\alpha-1)/(2n-1)}} \gg 1.$$

Therefore, we cannot conclude anything by this approach, as soon as the second boundary layer is reached (and even before). We will see in Sec. 9.1.3 that this approximation ceases indeed to be good between the first two boundary layers. We will also explain why.

9.1.2 *A rigorous computation*

Introduce the new unknown function ψ given by:

$$u^\varepsilon(t,x) = \frac{1}{(1-t)^{n/2}} \psi^\varepsilon \left(\frac{\varepsilon^\gamma}{1-t}, \frac{x}{1-t} \right) e^{i\frac{|x|^2}{2\varepsilon(t-1)}}. \tag{9.6}$$

Recalling that $\gamma = \alpha/n < 1$, denote

$$h = \varepsilon^{1-\gamma} \xrightarrow[\varepsilon\to 0]{} 0. \tag{9.7}$$

Changing the notation $\psi^\varepsilon(\tau, \xi)$ into $\psi^h(t,x)$, we check that for $t < 1$, Eq. (9.1) is equivalent to:

$$ih\partial_t \psi^h + \frac{h^2}{2}\Delta\psi^h = t^{n-2}|\psi^h|^2\psi^h \quad ; \quad \psi^h\left(h^{\frac{\gamma}{1-\gamma}}, x\right) = a_0(x). \tag{9.8}$$

Except for two aspects, this equation is the same as in Sec. 4.2.1:

- There is a factor t^{n-2} in front of the nonlinearity.
- The data are prescribed at $t = h^{\frac{\gamma}{1-\gamma}}$, instead of $t = 0$.

Note that the factor t^{n-2} is not present if $n = 2$, and would be singular at $t = 0$ if we wanted to treat the case $n = 1$.

Remark 9.5. It is easy to adapt this analysis to the case of a supercritical focal point, in an isotropic harmonic potential:

$$i\varepsilon\partial_t u^\varepsilon + \frac{\varepsilon^2}{2}\Delta u^\varepsilon = \frac{|x|^2}{2}u^\varepsilon + \varepsilon^\alpha |u^\varepsilon|^2 u^\varepsilon \quad ; \quad u^\varepsilon(0,x) = a_0(x).$$

Indeed, resuming the reduction of the end of Sec. 5.3, we can set

$$\psi^h(t,x) = U^\varepsilon\left(\frac{t}{\varepsilon^\gamma} - 1, x\right),$$

where U^ε is given by

$$U^\varepsilon(t,x) = \frac{1}{(1+t^2)^{n/4}} e^{i\frac{t}{1+t^2}\frac{|x|^2}{2\varepsilon}} u^\varepsilon\left(\arctan t, \frac{x}{\sqrt{1+t^2}}\right).$$

We find:

$$\begin{cases} ih\partial_t\psi^h + \dfrac{h^2}{2}\Delta\psi^h = \left((t_0^h)^2 + (t-t_0^h)^2\right)^{n/2-1} |\psi^h|^2\,\psi^h, \\ \psi^h(t_0^h, x) = a_0(x), \end{cases}$$

with $t_0^h = h^{\gamma/(1-\gamma)}$. This reduced problem is closely akin to Eq. (9.8). In particular, it is easy to adapt Theorem 9.2 to this case.

To study Eq. (9.8), we naturally adapt the approach of [Grenier (1998)], presented in Sec. 4.2.1. We want to write

$$\psi^h(t,x) = a^h(t,x)e^{i\phi^h(t,x)/h},$$

where

$$\begin{cases} \partial_t\phi^h + \dfrac{1}{2}|\nabla\phi^h|^2 + t^{n-2}|a^h|^2 = 0 & ; \quad \phi^h_{|t=t_0^h} = 0, \\ \partial_t a^h + \nabla\phi^h \cdot \nabla a^h + \dfrac{1}{2}a^h\Delta\phi^h = i\dfrac{h}{2}\Delta a^h & ; \quad a^h_{|t=t_0^h} = a_0, \end{cases} \qquad (9.9)$$

and $t_0^h = h^{\gamma/(1-\gamma)}$. We introduce the velocity $v^h = \nabla\phi^h$, and force the initial time to be zero by a shift in time:

$$\widetilde{v}^h(t,x) = v^h\left(t + t_0^h, x\right) \quad ; \quad \widetilde{a}^h(t,x) = a^h\left(t + t_0^h, x\right).$$

We now have to study a quasi-linear equation:

$$\partial_t\mathbf{u}^h + \sum_{j=1}^n A_j(\mathbf{u}^h)\partial_j\mathbf{u}^h = \frac{h}{2}L\mathbf{u}^h,$$

with $\quad \mathbf{u}^h = \begin{pmatrix} \operatorname{Re}\widetilde{a}^h \\ \operatorname{Im}\widetilde{a}^h \\ \widetilde{v}_1^h \\ \vdots \\ \widetilde{v}_n^h \end{pmatrix} = \begin{pmatrix} \widetilde{a}_1^h \\ \widetilde{a}_2^h \\ \widetilde{v}_1^h \\ \vdots \\ \widetilde{v}_n^h \end{pmatrix}, \quad L = \begin{pmatrix} 0 & -\Delta & 0 & \dots & 0 \\ \Delta & 0 & 0 & \dots & 0 \\ 0 & 0 & & 0_{n\times n} & \end{pmatrix}, \quad$ and

$$A(\mathbf{u},\xi) = \sum_{j=1}^n A_j(\mathbf{u})\xi_j = \begin{pmatrix} \widetilde{v}\cdot\xi & 0 & \frac{\widetilde{a}_1}{2}{}^t\xi \\ 0 & \widetilde{v}\cdot\xi & \frac{\widetilde{a}_2}{2}{}^t\xi \\ 2\left(t+t_0^h\right)^{n-2}\widetilde{a}_1\,\xi & 2\left(t+t_0^h\right)^{n-2}\widetilde{a}_2\,\xi & \widetilde{v}\cdot\xi I_n \end{pmatrix}.$$

The matrix $A(\mathbf{u}, \xi)$ can be symmetrized by

$$S^h = \begin{pmatrix} I_2 & 0 \\ 0 & \dfrac{1}{4(t+t_0^h)^{n-2}} I_n \end{pmatrix},$$

which depends only on t and h, since the nonlinearity that we consider is exactly cubic. In estimating $\partial_t S$, the assumption $n \geqslant 2$ is again helpful. Indeed, we can mimic the computations of Sec. 4.2.1. For $s > n/2 + 2$, we bound

$$\langle S^h \Lambda^s \mathbf{u}^h, \Lambda^s \mathbf{u}^h \rangle,$$

(scalar product in $L^2(\mathbb{R}^{n+2})$) by computing its time derivative:

$$\frac{d}{dt} \langle S^h \Lambda^s \mathbf{u}^h, \Lambda^s \mathbf{u}^h \rangle = \langle \partial_t S^h \Lambda^s \mathbf{u}^h, \Lambda^s \mathbf{u}^h \rangle + 2 \langle S^h \partial_t \Lambda^s \mathbf{u}^h, \Lambda^s \mathbf{u}^h \rangle,$$

since S^h is symmetric. Because $n \geqslant 2$,

$$\langle \partial_t S^h \Lambda^s \mathbf{u}^h, \Lambda^s \mathbf{u}^h \rangle \leqslant 0.$$

Therefore, we can easily infer the analogue of Theorem 4.1:

Proposition 9.6. *Let* $n > \alpha > 1$ *and* $a_0 \in \mathcal{S}(\mathbb{R}^n)$. *There exist* $T^* > 0$ *independent of* $h \in]0, 1]$ *and a unique pair* $(\phi^h, a^h) \in C([t_0^h, T^* + t_0^h]; H^\infty)^2$, *solution to* (9.9). *Moreover,* a^h *and* ϕ^h *are bounded in* $L^\infty([t_0^h, T^* + t_0^h]; H^s)$, *uniformly in* $h \in]0, 1]$, *for all* s.

Remark 9.7. In [Carles (2007a)], the homogeneous nonlinearity $\varepsilon^\alpha |u^\varepsilon|^2 u^\varepsilon$ is replaced with $f\left(\varepsilon^\alpha |u^\varepsilon|^2\right) u^\varepsilon$, where $f' > 0$. Computations are not more difficult, just a little lengthier to write. We point out that in estimating the time derivative of the symmetrizer S, the "new" term (compared to Sec. 4.2.1) is non-positive, as above, and therefore can be left out in the energy estimates leading to the analogue of the above proposition.

Remark 9.8. We infer that ψ^h is bounded in $L^\infty([t_0^h, T^* + t_0^h] \times \mathbb{R}^n)$. In view of Eq. (9.6), this shows that for $1 - t \geqslant \varepsilon^\gamma / T^*$,

$$\|u^\varepsilon(t)\|_{L^\infty} \lesssim \frac{1}{(1-t)^{n/2}}.$$

We have a similar result for the expected limit of (ϕ^h, a^h):

Proposition 9.9. *Let* $n \geqslant 2$ *and* $a_0 \in \mathcal{S}(\mathbb{R}^n)$. *There exists* $T > 0$ *such that the system*

$$\begin{cases} \partial_t \phi + \dfrac{1}{2} |\nabla \phi|^2 + t^{n-2} |a|^2 = 0 & ; \ \phi_{|t=0} = 0, \\[2mm] \partial_t a + \nabla \phi \cdot \nabla a + \dfrac{1}{2} a \Delta \phi = 0 & ; \ a_{|t=0} = a_0 \end{cases} \tag{9.10}$$

has a unique the solution $(a, \phi) \in C([0,T]; H^\infty)$. *In addition, there exist sequences* $(\phi_j)_{j \geqslant 1}$ *and* $(a_j)_{j \geqslant 1}$ *in* $H^\infty(\mathbb{R}^n)$, *such that*

$$\phi(t,x) \sim \sum_{j \geqslant 1} t^{nj-1} \phi_j(x), \quad \text{and} \quad a(t,x) \sim \sum_{j \geqslant 0} t^{nj} a_j(x) \quad \text{as } t \to 0.$$

The last part of the proposition is easily verified, by induction: plugging such asymptotic series into Eq. (9.10), a formal computation yields a source term which is $\mathcal{O}(t^\infty)$ as $t \to 0$. We can now measure the error:

Proposition 9.10. *Let* $s \in \mathbb{N}$. *We have* $T^* \geqslant T$, *and there exists* C_s *independent of* h *such that for every* $t_0^h \leqslant t \leqslant T$,

$$\|a^h(t) - a(t)\|_{H^s} + \|\phi^h(t) - \phi(t)\|_{H^s} \leqslant C_s \left(ht + h^{\frac{\gamma(n-1)}{1-\gamma}} \right).$$

Proof. We keep the same notations as above. Define $(\tilde{a}, \tilde{v})$ from (a, v) by the same shift in time. Denote by $\mathbf{u}$ the analogue of $\mathbf{u}^h$ corresponding to $(\tilde{a}, \tilde{v})$. We have

$$\partial_t \left(\mathbf{u}^h - \mathbf{u} \right) + \sum_{j=1}^n A_j(\mathbf{u}^h) \partial_j \left(\mathbf{u}^h - \mathbf{u} \right) + \sum_{j=1}^n \left(A_j(\mathbf{u}^h) - A_j(\mathbf{u}) \right) \partial_j \mathbf{u} = \frac{h}{2} L \mathbf{u}^h.$$

We know that $\mathbf{u}^h$ and $\mathbf{u}$ are bounded in $L^\infty([0, \min(T^*,T) - t_0^h]; H^s)$. Denoting $\mathbf{w}^h = \mathbf{u}^h - \mathbf{u}$, we get, for $s > 2 + n/2$:

$$\frac{d}{dt} \left(S^h \Lambda^s \mathbf{w}^h, \Lambda^s \mathbf{w}^h \right) \leqslant C \left(\|\mathbf{u}\|_{H^{s+2}}, \|\mathbf{u}^h\|_{H^s} \right) \left(S^h \Lambda^s \mathbf{w}^h, \Lambda^s \mathbf{w}^h \right)$$
$$+ Ch \|\mathbf{u}\|_{H^{s+2}} \|\mathbf{w}^h\|_{H^s}$$
$$\lesssim \left(S^h \Lambda^s \mathbf{w}^h, \Lambda^s \mathbf{w}^h \right) + h^2,$$

where S^h is the previous symmetrizer, which depends only on t and h. Using Gronwall lemma, we infer:

$$\|\mathbf{w}^h(t)\|_{H^s} \lesssim ht + \left\| (a,v)|_{t=t_0^h} - (a,v)|_{t=0} \right\|_{H^s}$$
$$\lesssim ht + \left(t_0^h \right)^{n-1} \|(\partial_t a, \partial_t v)\|_{L^\infty([0,T];H^s)} \quad \lesssim ht + \left(t_0^h \right)^{n-1},$$

where we have used Proposition 9.9. Along with a continuity argument, this proves that we also have $T^* \geqslant T$, since T^* does not depend on h. This completes the proof of Proposition 9.10. $\qquad \square$

Note that since $\alpha > 1$, $\frac{\gamma(n-1)}{1-\gamma} > 1$. Back to ψ^h, we infer:

$$
\begin{aligned}
\left\| \psi^h - a_0 e^{i\phi/h} \right\|_{L^\infty([t_0^h,\tau];L^2)} &\lesssim \left\| a^h e^{i\phi^h/h} - a e^{i\phi/h} \right\|_{L^\infty([t_0^h,\tau];L^2)} \\
&\quad + \left\| a e^{i\phi/h} - a_0 e^{i\phi/h} \right\|_{L^\infty([t_0^h,\tau];L^2)} \\
&\lesssim \left\| a^h - a \right\|_{L^\infty([t_0^h,T];L^2)} + \frac{1}{h} \left\| \phi^h - \phi \right\|_{L^\infty([t_0^h,\tau]\times\mathbb{R}^n)} \\
&\quad + \left\| a - a_0 \right\|_{L^\infty([t_0^h,\tau];L^2)} \\
&\lesssim h + \tau + h^{\frac{\alpha-1}{1-\gamma}} + \tau^n.
\end{aligned}
$$

Letting $\tau \to 0$, Theorem 9.2 follows, by using Eq. (9.6), and the last point of Proposition 9.9. Note that the cascade of phase shifts proceeds along the same spirit as in Sec. 5.3, since it stems from the Taylor expansion, as time goes to zero, of the rapidly oscillatory phase.

Remark 9.11. The cascade of phase shifts can be understood as the creation of a new phase, appearing discretely in time. With the transform (9.6) in mind, the asymptotic expansion of the phase shift ϕ stems from the last part of Proposition 9.9. The coupling in (9.10) shows that even if $\phi_{|t=0} = 0$, $\phi(\delta, x)$ is not identically zero, for $\delta > 0$ arbitrarily small. The phase of ψ^h is given asymptotically by

$$
\frac{\phi(t,x)}{h} \sim \sum_{j \geqslant 1} \frac{t^{jn-1}}{h} \phi_j(x).
$$

With the same line of reasoning as above, a phase shift appears for t of order $h^{1/(n-1)}$, then a second for t of order $h^{1/(2n-1)}$, and so on. The superposition of these phase shifts, which are oscillating faster and faster, finally leads to a continuous phase, corresponding to an oscillation associated to the wavelength h.

9.1.3   *Why do the results disagree?*

The construction of Section 9.1.1 and the results of the previous paragraph do not agree. To see this, we come back to Proposition 9.9. Plugging the

Taylor expansions in time for ϕ and a into Eq. (9.10), we find:

$$\mathcal{O}\left(t^{n-2}\right): \quad (n-1)\phi_1 + |a_0|^2 = 0, \tag{9.11}$$

$$\mathcal{O}\left(t^{n-1}\right): \quad na_1 + \nabla\phi_1 \cdot \nabla a_0 + \frac{1}{2}a_0\Delta\phi_1 = 0, \tag{9.12}$$

$$\mathcal{O}\left(t^{2n-2}\right): \quad (2n-1)\phi_2 + \frac{1}{2}|\nabla\phi_1|^2 + 2\,\mathrm{Re}(\overline{a_0}a_1) = 0. \tag{9.13}$$

The function ϕ_1 is the same as the one obtained by the approach of Sec. 9.1.1: the two approximate solutions are close to each other up to the first boundary layer, when the first phase shift appears. This also stems from the computations at the end of Sec. 9.1.1, in view of Remark 9.8. On the other hand, we see that to get ϕ_2, the modulation of the amplitude (a_1) must be taken into account (note that we consider approximations as $t \to 0$ here, not as $\varepsilon \to 0$: a_1 does not denote the same quantity as in Chap. 4); in Eq. (9.5), g_2 is computed without evaluating Δa_0, unlike ϕ_2, so $g_2 \neq \phi_2$ in general. This means in particular that the two approximate solutions diverge when reaching the second boundary layer: the approach of Sec. 9.1.1 is only formal, and does not lead to a good approximation. And yet, the source term in Proposition 9.3 is small. We will see below that this divergence is not due to a spectral instability, but to the fact that the approach followed to construct the formal approximation was too crude.

We apply the transform (9.6) to the intermediary approximate solution v_N^ε. We show that the formal approximation stops being a good approximation between the first and the second boundary layer exactly, as suspected at the end of Sec. 9.1.1. Like for the exact solution, write

$$v_N^\varepsilon(t,x) = \frac{1}{(1-t)^{n/2}}\psi_N^\varepsilon\left(\frac{\varepsilon^\gamma}{1-t}, \frac{x}{1-t}\right)e^{i\frac{|x|^2}{2\varepsilon(t-1)}}.$$

We check that ψ_N^h is given by

$$\psi_N^h(\tau,\xi) = a_0(\xi)e^{i\varphi_N^h(\tau,\xi)}, \quad \text{where } g_N^\varepsilon(t,x) = \varphi_N^h\left(\frac{\varepsilon^\gamma}{1-t}, \frac{x}{1-t}\right).$$

We compute

$$\varphi_N^h(t,x) = \frac{1}{ht}\sum_{j=1}^{N}\frac{1}{nj-1}\left(t^{nj} - \left(t_0^h\right)^{nj}\right)g_j(x).$$

Therefore, ψ_N^h satisfies

$$ih\partial_t\psi_N^h + \frac{h^2}{2}\Delta\psi_N^h = t^{n-2}|\psi_N^h|^2\psi_N^h + \theta_N^h(t,x) \quad ; \quad \psi_N^h\left(t_0^h, x\right) = a_0(x),$$

where:

$$\theta_N^h(t,x) = \left(t^{(N+1)n-2}K_0(x) + ihK_1(t,x) + \frac{\left(t_0^h\right)^n}{t^2}\Xi_0^h(x) \right) \psi_N^h(t,x)$$

$$+ ihK_2(t,x) + h^2K_3(t,x) + ih\frac{\left(t_0^h\right)^n}{t^2}\Xi_1^h(x),$$

where the functions K_j are smooth and independent of h, and Ξ_j^h are bounded in all Sobolev spaces, uniformly in h. Note that the factor in front of Ξ_j^h is not singular, since we assume $t \geqslant t_0^h$, and $n \geqslant 2$. Now write $\psi_N^h(t,x) = a_N^h(t,x)e^{i\phi_N^h(t,x)/h}$, with

$$\partial_t \mathbf{v}^h + \sum_{j=1}^{n} A_j(\mathbf{v}^h)\partial_j \mathbf{v}^h = \frac{h}{2}L\mathbf{v}^h + \mathsf{S}^h(t,x), \text{ with } \mathbf{v}^h(t,x) = \begin{pmatrix} \operatorname{Re} a_N^h \\ \operatorname{Im} a_N^h \\ \partial_1 \phi_N^h \\ \vdots \\ \partial_n \phi_N^h \end{pmatrix},$$

$$\mathsf{S}^h(t,x) = \begin{pmatrix} \operatorname{Re}\left(K_1 a_N^h\right) + \operatorname{Re}\left(\left(K_2 - ihK_3 + \frac{\left(t_0^h\right)^n}{(t+t_0^h)^2}\Xi_1^h\right)e^{-i\phi_N^h/h}\right) \\ \operatorname{Im}\left(K_1 a_N^h\right) + \operatorname{Im}\left(\left(K_2 - ihK_3 + \frac{\left(t_0^h\right)^n}{(t+t_0^h)^2}\Xi_1^h\right)e^{-i\phi_N^h/h}\right) \\ -(t+t_0^h)^{(N+1)n-2}\partial_1 K_0 - \frac{\left(t_0^h\right)^n}{(t+t_0^h)^2}\partial_1\Xi_0^h \\ \vdots \\ -(t+t_0^h)^{(N+1)n-2}\partial_n K_0 - \frac{\left(t_0^h\right)^n}{(t+t_0^h)^2}\partial_n\Xi_0^h \end{pmatrix},$$

where the matrices A_j are the same as in Sec. 9.1.2 and the functions in the definitions of $\mathbf{v}^h$ and S^h are evaluated at $(t+t_0^h, x)$. We can proceed like in Sec. 9.1.2: the new term is the source S^h. Unlike for the exact solution, the oscillatory aspect of the problem has not disappeared: the first two components of S^h contain a highly oscillatory factor. Therefore, we cannot expect h-independent energy estimates here. To measure the effect of this oscillatory term, forget the shift in time, and take $t_0^h = 0$. Then assuming that for small times, $\partial_x^\beta \phi_N^h(t,x) = \mathcal{O}(t^{n-1})$ for any multi-index β (like for the exact solution), the H^s norms of the first two components of S^h are controlled by

$$\mathcal{O}\left(\frac{t^{1+s(n-1)}}{h^s}\right).$$

Back to the initial variables, this yields a control by

$$\left(\frac{\varepsilon^\gamma}{1-t}\right)^{1+s(n-1)} \varepsilon^{-s(1-\gamma)} = \frac{\varepsilon^{\gamma+s\alpha-s}}{(1-t)^{1+s(n-1)}}.$$

This is small for $1 - t \gg \varepsilon^\omega$, with
$$\omega = \frac{\gamma + s\alpha - s}{1 + s(n-1)}.$$
We check that for $n > \alpha > 1$, we have
$$\frac{\alpha - 1}{n - 1} < \omega = \frac{\gamma + s(\alpha - 1)}{1 + s(n - 1)} < \frac{2\alpha - 1}{2n - 1}, \text{ for all } s \geqslant 0.$$
The first inequality means that we can expect the formal approximation to be a good approximation of the exact solution beyond the first boundary layer (which holds true). The second one explains why the approximation ceases to be relevant before the second boundary layer.

A possible way to understand the above computation is that the choice of the variables is crucial: working with the "usual" unknown v^ε (as in Sec. 9.1.1) is not very efficient. On the other hand, with the variables introduced by E. Grenier for his generalized WKB methods, a precise and rigorous analysis is possible, *via* the transform (9.6).

Remark 9.12. Even though there is stability in a reasonable sense for the limiting hyperbolic system (9.10), small perturbations of the initial amplitude a_0 may drastically alter the asymptotic behavior of u^ε; see Sec. 5.3.

9.2   And beyond?

Using the analysis of Sec. 4.2.1, we could not only describe u^ε for $t \leqslant 1 - \Lambda \varepsilon^\gamma$ in the limit $\Lambda \to +\infty$, but also for $t \leqslant 1 - \varepsilon^\gamma / T$, where T is given by Proposition 9.9. The main differences with the approximate solution of Theorem 9.2 is that the amplitude a can no longer be approximated by its initial value a_0, and a phase modulation must be inserted (which was denoted $\Phi^{(1)}$ in Sec. 4.2.1). This shows that for $t = 1 - \varepsilon^\gamma / T$, the amplitude of u^ε is of order $\varepsilon^{-n\gamma/2} = \varepsilon^{-\alpha/2}$ in L^∞. Moreover, the potential term in the nonlinear energy is of order $\mathcal{O}(1)$ exactly:
$$\varepsilon^\alpha \left\| u^\varepsilon(1 - \varepsilon^\gamma/T) \right\|_{L^4}^4 = \varepsilon^\alpha \int_{\mathbb{R}^n} \left(\frac{T}{\varepsilon^\gamma} \right)^{2n} \left| \psi^h \left(T, \frac{Tx}{\varepsilon^\gamma} \right) \right|^4 dx$$
$$= \varepsilon^\alpha \left(\frac{T}{\varepsilon^\gamma} \right)^n \left\| \psi^h(T) \right\|_{L^4}^4 = T^n \left\| \psi^h(T) \right\|_{L^4}^4 \approx 1.$$
Since we know that the potential energy is bounded for all time, from the conservation of the energy, this suggests that u^ε might have reached its maximal order of magnitude at time $t = 1 - \varepsilon^\gamma / T$, as guessed in the introduction of this chapter.

To know how u^ε evolves past this time, it seems reasonable to introduce the (L^2 unitary) scaling transform,

$$u^\varepsilon(t,x) = \frac{1}{\varepsilon^{n\gamma/2}} \varphi^\varepsilon\left(\frac{t-1}{\varepsilon^\gamma}, \frac{x}{\varepsilon^\gamma}\right) = \frac{1}{\varepsilon^{\alpha/2}} \varphi^\varepsilon\left(\frac{t-1}{\varepsilon^\gamma}, \frac{x}{\varepsilon^\gamma}\right). \tag{9.14}$$

With the same change of notation as for ψ, we have

$$ih\partial_t\varphi^h + \frac{h^2}{2}\Delta\varphi^h = \left|\varphi^h\right|^2 \varphi^h. \tag{9.15}$$

For $-1/t_0^h \leqslant t \leqslant -1/(T + t_0^h)$, we have:

$$\varphi^h(t,x) = \left(\frac{-1}{t}\right)^{n/2} \psi^h\left(\frac{-1}{t}, \frac{-x}{t}\right) e^{i|x|^2/(2ht)}$$

$$= \left(\frac{-1}{t}\right)^{n/2} a^h\left(\frac{-1}{t}, \frac{-x}{t}\right) \exp\left(\frac{i}{h}\left(\frac{|x|^2}{2t} + \phi^h\left(\frac{-1}{t}, \frac{-x}{t}\right)\right)\right)$$

$$=: \mathbf{a}^h(t,x) e^{i\Phi^h(t,x)/h}. \tag{9.16}$$

The phase Φ^h is no longer in Sobolev spaces, because of the quadratic term. However, it enters the class studied in Sec. 4.2.2. So to go further into the analysis, we meet again a question which was natural in Chap. 4: what happens when singularities appear in the limiting Euler equation? What does it mean for u^ε?

We have seen that because we study a focal point in a supercritical régime, new frequencies have appeared. There are potentially other possible effects. For instance, is there a (different) notion of caustic for φ^h, that is, in a supercritical WKB régime? Indeed, we know that we have to expect the solution of the limiting Euler system to develop singularities in finite time ([Chemin (1990); Makino et al. (1986); Xin (1998)]), but this tells us nothing about φ^h. Typically, we do not expect the L^∞ norm of φ^h to be unbounded, since both its L^2 norm and its L^4 norm are bounded for all time and all h. Of course, this does not imply that the L^∞ norm of φ^h is bounded, but this is a rather appealing property. As suggested in Chap. 6, this might mean that two notions of caustic could be distinguished in supercritical régimes: a geometrical notion (the rays along which the amplitude is carried cease to form a diffeomorphism of the whole space), and an analytical notion (the L^∞ norm of the wave function becomes unbounded as the semi-classical parameter goes to zero). These two notions might be disconnected in supercritical cases, and other analytical mechanisms may be involved.

In the linear, subcritical and critical cases, we have seen that past a single focal point, one phase suffices to describe the rapid oscillations of the wave function, unlike what happens for a cusped caustic in the linear case, for instance. In the present supercritical case, it is not clear even how many phases are necessary to describe the wave function for $t > 1$. As a matter of fact, the geometry of singularities is not understood either: it is not clear that for $t > 1$, u^ε is of order $\mathcal{O}(1)$ again. It might for instance keep the form it has reached at time $t = 1 - \varepsilon^\gamma/T$.

This informal discussion shows that many questions remain open, both in a WKB régime and in a caustic régime. Moreover, these questions are more connected than it may seem at first glance.

Bibliography

Alazard, T. (2006). Low Mach number limit of the full Navier-Stokes equations, *Arch. Ration. Mech. Anal.* **180**, 1, pp. 1–73.

Alazard, T. and Carles, R. (2007a). Loss of regularity for super-critical nonlinear Schrödinger equations, archived as `math.AP/0701857`.

Alazard, T. and Carles, R. (2007b). Supercritical geometric optics for nonlinear Schrödinger equations, archived as `arXiv:0704.2488`.

Alinhac, S. (1995a). *Blowup for nonlinear hyperbolic equations* (Birkhäuser Boston Inc., Boston, MA).

Alinhac, S. (1995b). Explosion géométrique pour des systèmes quasi-linéaires, *Amer. J. Math.* **117**, 4, pp. 987–1017.

Alinhac, S. (2002). A minicourse on global existence and blowup of classical solutions to multidimensional quasilinear wave equations, in *Journées "Équations aux Dérivées Partielles" (Forges-les-Eaux, 2002)* (Univ. Nantes, Nantes), pp. Exp. No. I, 33.

Alinhac, S. and Gérard, P. (1991). *Opérateurs pseudo-différentiels et théorème de Nash-Moser*, Savoirs Actuels (InterEditions, Paris).

Bahouri, H. and Gérard, P. (1999). High frequency approximation of solutions to critical nonlinear wave equations, *Amer. J. Math.* **121**, 1, pp. 131–175.

Bambusi, D. and Sacchetti, A. (2007). Exponential times in the one-dimensional Gross–Pitaevskii equation with multiple well potential, *Comm. Math. Phys.* **275**, 1, pp. 1–36.

Barab, J. E. (1984). Nonexistence of asymptotically free solutions for nonlinear Schrödinger equation, *J. Math. Phys.* **25**, pp. 3270–3273.

Bégout, P. and Vargas, A. (2007). Mass concentration phenomena for the L^2-critical nonlinear Schrödinger equation, *Trans. Amer. Math. Soc.* **359**.

Bensoussan, A., Lions, J.-L. and Papanicolaou, G. (1978). *Asymptotic analysis for periodic structures*, Vol. 5 (North-Holland Publishing Co., Amsterdam).

Bourgain, J. (1995). Some new estimates on oscillatory integrals, in *Essays on Fourier analysis in honor of Elias M. Stein (Princeton, NJ, 1991)*, Princeton Math. Ser., Vol. 42 (Princeton Univ. Press, Princeton, NJ), pp. 83–112.

Bourgain, J. (1998). Refinements of Strichartz' inequality and applications to 2D-NLS with critical nonlinearity, *Internat. Math. Res. Notices*, 5, pp. 253–283.

233

Boyd, R. W. (1992). *Nonlinear Optics* (Academic Press, New York).

Brenier, Y. (2000). Convergence of the Vlasov–Poisson system to the incompressible Euler equations, *Comm. Partial Differential Equations* **25**, 3-4, pp. 737–754.

Bronski, J. C. and Jerrard, R. L. (2000). Soliton dynamics in a potential, *Math. Res. Lett.* **7**, 2-3, pp. 329–342.

Burq, N. (1997). Mesures semi-classiques et mesures de défaut, *Astérisque*, 245, Exp. No. 826, 4, pp. 167–195, séminaire Bourbaki, Vol. 1996/97.

Burq, N., Gérard, P. and Tzvetkov, N. (2004). Strichartz inequalities and the nonlinear Schrödinger equation on compact manifolds, *Amer. J. Math.* **126**, 3, pp. 569–605.

Burq, N., Gérard, P. and Tzvetkov, N. (2005). Multilinear eigenfunction estimates and global existence for the three dimensional nonlinear Schrödinger equations, *Ann. Sci. École Norm. Sup. (4)* **38**, 2, pp. 255–301.

Burq, N. and Zworski, M. (2005). Instability for the semiclassical non-linear Schrödinger equation, *Comm. Math. Phys.* **260**, 1, pp. 45–58.

Carles, R. (2000a). Focusing on a line for nonlinear Schrödinger equations in $\mathbb{R}^2$, *Asymptot. Anal.* **24**, 3-4, pp. 255–276.

Carles, R. (2000b). Geometric optics with caustic crossing for some nonlinear Schrödinger equations, *Indiana Univ. Math. J.* **49**, 2, pp. 475–551.

Carles, R. (2001a). Geometric optics and long range scattering for one-dimensional nonlinear Schrödinger equations, *Comm. Math. Phys.* **220**, 1, pp. 41–67.

Carles, R. (2001b). Remarques sur les mesures de Wigner, *C. R. Acad. Sci. Paris, t. 332, Série I* **332**, 11, pp. 981–984.

Carles, R. (2002). Critical nonlinear Schrödinger equations with and without harmonic potential, *Math. Models Methods Appl. Sci.* **12**, 10, pp. 1513–1523.

Carles, R. (2003a). Nonlinear Schrödinger equations with repulsive harmonic potential and applications, *SIAM J. Math. Anal.* **35**, 4, pp. 823–843.

Carles, R. (2003b). Semi-classical Schrödinger equations with harmonic potential and nonlinear perturbation, *Ann. Inst. H. Poincaré Anal. Non Linéaire* **20**, 3, pp. 501–542.

Carles, R. (2007a). Cascade of phase shifts for nonlinear Schrödinger equations, *J. Hyperbolic Differ. Equ.* **4**, 2, pp. 207–231.

Carles, R. (2007b). Geometric optics and instability for semi-classical Schrödinger equations, *Arch. Ration. Mech. Anal.* **183**, 3, pp. 525–553.

Carles, R. (2007c). WKB analysis for nonlinear Schrödinger equations with potential, *Comm. Math. Phys.* **269**, 1, pp. 195–221.

Carles, R. (2008). On the Cauchy problem in Sobolev spaces for nonlinear Schrödinger equations with potential, *Port. Math. (N. S.)* **65**, to appear.

Carles, R., Fermanian, C. and Gallagher, I. (2003). On the role of quadratic oscillations in nonlinear Schrödinger equations, *J. Funct. Anal.* **203**, 2, pp. 453–493.

Carles, R. and Keraani, S. (2007). On the role of quadratic oscillations in nonlinear Schrödinger equations II. The L^2-critical case, *Trans. Amer. Math. Soc.*

359, 1, pp. 33–62.

Carles, R., Markowich, P. A. and Sparber, C. (2004). Semiclassical asymptotics for weakly nonlinear Bloch waves, *J. Stat. Phys.* **117**, 1-2, pp. 343–375.

Carles, R. and Miller, L. (2004). Semiclassical nonlinear Schrödinger equations with potential and focusing initial data, *Osaka J. Math.* **41**, 3, pp. 693–725.

Carles, R. and Nakamura, Y. (2004). Nonlinear Schrödinger equations with Stark potential, *Hokkaido Math. J.* **33**, 3, pp. 719–729.

Carles, R. and Rauch, J. (2002). Focusing of spherical nonlinear pulses in $\mathbb{R}^{1+3}$, *Proc. Amer. Math. Soc.* **130**, 3, pp. 791–804.

Carles, R. and Rauch, J. (2004a). Focusing of Spherical Nonlinear Pulses in $\mathbb{R}^{1+3}$ II. Nonlinear Caustic, *Rev. Mat. Iberoamericana* **20**, 3, pp. 815–864.

Carles, R. and Rauch, J. (2004b). Focusing of Spherical Nonlinear Pulses in $\mathbb{R}^{1+3}$ III. Sub and Supercritical cases, *Tohoku Math. J.* **56**, 3, pp. 393–410.

Cazenave, T. (2003). *Semilinear Schrödinger equations, Courant Lecture Notes in Mathematics*, Vol. 10 (New York University Courant Institute of Mathematical Sciences, New York).

Cazenave, T. and Haraux, A. (1998). *An introduction to semilinear evolution equations, Oxford Lecture Series in Mathematics and its Applications*, Vol. 13 (The Clarendon Press Oxford University Press, New York), translated from the 1990 French original by Yvan Martel and revised by the authors.

Cazenave, T. and Weissler, F. (1989). Some remarks on the nonlinear Schrödinger equation in the critical case, in *Lect. Notes in Math.*, Vol. 1394 (Springer-Verlag, Berlin), pp. 18–29.

Cazenave, T. and Weissler, F. (1990). The Cauchy problem for the critical nonlinear Schrödinger equation in H^s, *Nonlinear Anal. TMA* **14**, pp. 807–836.

Cazenave, T. and Weissler, F. (1992). Rapidly decaying solutions of the nonlinear Schrödinger equation, *Comm. Math. Phys.* **147**, pp. 75–100.

Chemin, J.-Y. (1990). Dynamique des gaz à masse totale finie, *Asymptotic Anal.* **3**, 3, pp. 215–220.

Chemin, J.-Y. (1998). *Perfect incompressible fluids, Oxford Lecture Series in Mathematics and its Applications*, Vol. 14 (The Clarendon Press Oxford University Press, New York), translated from the 1995 French original by I. Gallagher and D. Iftimie.

Cheverry, C. (2004). Propagation of oscillations in real vanishing viscosity limit, *Comm. Math. Phys.* **247**, 3, pp. 655–695.

Cheverry, C. (2005). Sur la propagation de quasi-singularités, in *Séminaire: Équations aux Dérivées Partielles. 2004-2005* (École Polytech., Palaiseau), pp. Exp. No. VIII, 20.

Cheverry, C. (2006). Cascade of phases in turbulent flows, *Bull. Soc. Math. France* **134**, 1, pp. 33–82.

Cheverry, C. and Guès, O. (2007). Counter-examples to concentration-cancellation and supercritical nonlinear geometric optics for the incompressible Euler equations, *Arch. Ration. Mech. Anal.* To appear.

Christ, M., Colliander, J. and Tao, T. Ill-posedness for nonlinear Schrödinger and wave equations, archived as `arXiv:math.AP/0311048`.

Cicognani, M. and Colombini, F. (2006a). Loss of derivatives in evolution Cauchy problems, *Ann. Univ. Ferrara Sez. VII Sci. Mat.* **52**, 2, pp. 271–280.

Cicognani, M. and Colombini, F. (2006b). Modulus of continuity of the coefficients and loss of derivatives in the strictly hyperbolic Cauchy problem, *J. Differential Equations* **221**, 1, pp. 143–157.

Dalfovo, F., Giorgini, S., Pitaevskii, L. P. and Stringari, S. (1999). Theory of Bose-Einstein condensation in trapped gases, *Rev. Mod. Phys.* **71**, 3, pp. 463–512.

Dereziński, J. and Gérard, C. (1997). *Scattering theory of quantum and classical N-particle systems* (Texts and Monographs in Physics, Springer Verlag, Berlin Heidelberg).

Duistermaat, J. J. (1974). Oscillatory integrals, Lagrange immersions and unfolding of singularities, *Comm. Pure Appl. Math.* **27**, pp. 207–281.

Dunford, N. and Schwartz, J. T. (1963). *Linear operators. Part II: Spectral theory. Self adjoint operators in Hilbert space*, With the assistance of William G. Bade and Robert G. Bartle (Interscience Publishers John Wiley & Sons New York-London).

Evans, L. C. (1998). *Partial differential equations, Graduate Studies in Mathematics*, Vol. 19 (American Mathematical Society, Providence, RI).

Feynman, R. P. and Hibbs, A. R. (1965). *Quantum mechanics and path integrals (International Series in Pure and Applied Physics)* (Maidenhead, Berksh.: McGraw-Hill Publishing Company, Ltd., 365 p.).

Foschi, D. (2005). Inhomogeneous Strichartz estimates, *J. Hyperbolic Differ. Equ.* **2**, 1, pp. 1–24.

Fujiwara, D. (1979). A construction of the fundamental solution for the Schrödinger equation, *J. Analyse Math.* **35**, pp. 41–96.

Gallagher, I. and Gérard, P. (2001). Profile decomposition for the wave equation outside a convex obstacle, *J. Math. Pures Appl. (9)* **80**, 1, pp. 1–49.

Gérard, P. (1993). Remarques sur l'analyse semi-classique de l'équation de Schrödinger non linéaire, in *Séminaire sur les Équations aux Dérivées Partielles, 1992–1993* (École Polytech., Palaiseau), pp. Exp. No. XIII, 13.

Gérard, P. (1996). Oscillations and concentration effects in semilinear dispersive wave equations, *J. Funct. Anal.* **141**, 1, pp. 60–98.

Gérard, P. (1998). Description du défaut de compacité de l'injection de Sobolev, *ESAIM Control Optim. Calc. Var.* **3**, pp. 213–233 (electronic).

Gérard, P., Markowich, P. A., Mauser, N. J. and Poupaud, F. (1997). Homogenization limits and Wigner transforms, *Comm. Pure Appl. Math.* **50**, 4, pp. 323–379.

Ginibre, J. (1995). Introduction aux équations de Schrödinger non linéaires, Cours de DEA, Paris Onze Édition.

Ginibre, J. (1997). An introduction to nonlinear Schrödinger equations, in R. Agemi, Y. Giga and T. Ozawa (eds.), *Nonlinear waves (Sapporo, 1995)*, GAKUTO International Series, Math. Sciences and Appl. (Gakkōtosho, Tokyo), pp. 85–133.

Ginibre, J. and Ozawa, T. (1993). Long range scattering for nonlinear Schrödinger and Hartree equations in space dimension $n \geq 2$, *Comm. Math. Phys.* **151**, 3, pp. 619–645.

Ginibre, J., Ozawa, T. and Velo, G. (1994). On the existence of the wave operators for a class of nonlinear Schrödinger equations, *Ann. IHP (Physique Théorique)* **60**, pp. 211–239.

Ginibre, J. and Velo, G. (1979). On a class of nonlinear Schrödinger equations. II Scattering theory, general case, *J. Funct. Anal.* **32**, pp. 33–71.

Ginibre, J. and Velo, G. (1985a). The global Cauchy problem for the nonlinear Schrödinger equation revisited, *Ann. Inst. H. Poincaré Anal. Non Linéaire* **2**, pp. 309–327.

Ginibre, J. and Velo, G. (1985b). Scattering theory in the energy space for a class of nonlinear Schrödinger equations, *J. Math. Pures Appl. (9)* **64**, 4, pp. 363–401.

Ginibre, J. and Velo, G. (1992). Smoothing properties and retarded estimates for some dispersive evolution equations, *Comm. Math. Phys.* **144**, 1, pp. 163–188.

Grenier, E. (1998). Semiclassical limit of the nonlinear Schrödinger equation in small time, *Proc. Amer. Math. Soc.* **126**, 2, pp. 523–530.

Grigis, A. and Sjöstrand, J. (1994). *Microlocal analysis for differential operators, London Mathematical Society Lecture Note Series*, Vol. 196 (Cambridge University Press, Cambridge), an introduction.

Hayashi, N. and Naumkin, P. (1998). Asymptotics for large time of solutions to the nonlinear Schrödinger and Hartree equations, *Amer. J. Math.* **120**, 2, pp. 369–389.

Hayashi, N. and Naumkin, P. (2006). Domain and range of the modified wave operator for Schrödinger equations with a critical nonlinearity, *Comm. Math. Phys.* **267**, 2, pp. 477–492.

Hayashi, N. and Tsutsumi, Y. (1987). Remarks on the scattering problem for nonlinear Schrödinger equations, in *Differential equations and mathematical physics (Birmingham, Ala., 1986), Lectures Notes in Math.*, Vol. 1285 (Springer, Berlin), pp. 162–168.

Hörmander, L. (1994). *The analysis of linear partial differential operators* (Springer-Verlag, Berlin).

Hörmander, L. (1995). Symplectic classification of quadratic forms, and general Mehler formulas, *Math. Z.* **219**, 3, pp. 413–449.

Hunter, J. and Keller, J. (1987). Caustics of nonlinear waves, *Wave motion* **9**, pp. 429–443.

Ibrahim, S. (2004). Geometric Optics for Nonlinear Concentrating Waves in a Focusing and non Focusing two geometries, *Commun. Contemp. Math.* **6**, 1, pp. 1–23.

Joly, J.-L., Métivier, G. and Rauch, J. (1995). Focusing at a point and absorption of nonlinear oscillations, *Trans. Amer. Math. Soc.* **347**, 10, pp. 3921–3969.

Joly, J.-L., Métivier, G. and Rauch, J. (1996a). Nonlinear oscillations beyond caustics, *Comm. Pure Appl. Math.* **49**, 5, pp. 443–527.

Joly, J.-L., Métivier, G. and Rauch, J. (1996b). Several recent results in nonlinear geometric optics, in *Partial differential equations and mathematical physics (Copenhagen, 1995; Lund, 1995)* (Birkhäuser Boston, Boston, MA), pp. 181–206.

Joly, J.-L., Métivier, G. and Rauch, J. (1997a). Caustics for dissipative semilinear oscillations, in F. Colombini and N. Lerner (eds.), *Geometrical Optics and Related Topics* (Birkäuser), pp. 245–266.

Joly, J.-L., Métivier, G. and Rauch, J. (1997b). Estimations L^p d'intégrales oscillantes, in *Séminaire Équations aux Dérivées Partielles, 1996–1997* (École Polytech., Palaiseau), pp. Exp. No. VII, 17.

Joly, J.-L., Métivier, G. and Rauch, J. (2000). Caustics for dissipative semilinear oscillations, *Mem. Amer. Math. Soc.* **144**, 685, pp. viii+72.

Kato, T. (1987). On nonlinear Schrödinger equations, *Ann. IHP (Phys. Théor.)* **46**, 1, pp. 113–129.

Kato, T. (1989). Nonlinear Schrödinger equations, in *Schrödinger operators (Sønderborg, 1988), Lecture Notes in Phys.*, Vol. 345 (Springer, Berlin), pp. 218–263.

Keel, M. and Tao, T. (1998). Endpoint Strichartz estimates, *Amer. J. Math.* **120**, 5, pp. 955–980.

Keraani, S. (2001). On the defect of compactness for the Strichartz estimates of the Schrödinger equations, *J. Differential Equations* **175**, 2, pp. 353–392.

Keraani, S. (2002). Semiclassical limit for a class of nonlinear Schrödinger equations with potential, *Comm. Partial Differential Equations* **27**, 3-4, pp. 693–704.

Keraani, S. (2005). Limite semi-classique pour l'équation de Schrödinger non-linéaire avec potentiel harmonique, *C. R. Math. Acad. Sci. Paris* **340**, 11, pp. 809–814.

Keraani, S. (2006). Semiclassical limit for nonlinear Schrödinger equation with potential. II, *Asymptot. Anal.* **47**, 3-4, pp. 171–186.

Klainerman, S. (1985). Uniform decay estimates and the Lorentz invariance of the classical wave equation, *Comm. Pure Appl. Math.* **38**, 3, pp. 321–332.

Kolomeisky, E. B., Newman, T. J., Straley, J. P. and Qi, X. (2000). Low-dimensional Bose liquids: Beyond the Gross-Pitaevskii approximation, *Phys. Rev. Lett.* **85**, 6, pp. 1146–1149.

Kossioris, G. T. (1993). Formation of singularities for viscosity solutions of Hamilton-Jacobi equations in higher dimensions, *Comm. Partial Differential Equations* **18**, 7-8, pp. 1085–1108.

Kuksin, S. B. (1995). On squeezing and flow of energy for nonlinear wave equations, *Geom. Funct. Anal.* **5**, 4, pp. 668–701.

Kwong, M. K. (1989). Uniqueness of positive solutions of $\Delta u - u + u^p = 0$ in $\mathbb{R}^n$, *Arch. Rational Mech. Anal.* **105**, 3, pp. 243–266.

Landau, L. and Lifschitz, E. (1967). *Physique théorique ("Landau-Lifchitz"). Tome III: Mécanique quantique. Théorie non relativiste* (Éditions Mir, Moscow), deuxième édition, Traduit du russe par Édouard Gloukhian.

Lax, P. D. (1957). Asymptotic solutions of oscillatory initial value problems, *Duke Math. J.* **24**, pp. 627–646.

Lebeau, G. (1992). Contrôle de l'équation de Schrödinger, *J. Math. Pures Appl.* (9) **71**, 3, pp. 267–291.

Lebeau, G. (2001). Non linear optic and supercritical wave equation, *Bull. Soc. Roy. Sci. Liège* **70**, 4-6, pp. 267–306 (2002), hommage à Pascal Laubin.

Lebeau, G. (2005). Perte de régularité pour les équations d'ondes sur-critiques, *Bull. Soc. Math. France* **133**, pp. 145–157.

Lin, F. and Zhang, P. (2005). Semiclassical limit of the Gross–Pitaevskii equation in an exterior domain, *Arch. Rational Mech. Anal.* **179**, 1, pp. 79–107.

Lions, P.-L. (1996). *Mathematical topics in fluid mechanics. Vol. 1, Oxford Lecture Series in Mathematics and its Applications*, Vol. 3 (The Clarendon Press Oxford University Press, New York), incompressible models, Oxford Science Publications.

Lions, P.-L. and Paul, T. (1993). Sur les mesures de Wigner, *Rev. Mat. Iberoamericana* **9**, 3, pp. 553–618.

Ludwig, D. (1966). Uniform asymptotic expansions at a caustic, *Comm. Pure Appl. Math.* **19**, pp. 215–250.

Majda, A. (1984). *Compressible fluid flow and systems of conservation laws in several space variables, Applied Mathematical Sciences*, Vol. 53 (Springer-Verlag, New York).

Makino, T., Ukai, S. and Kawashima, S. (1986). Sur la solution à support compact de l'équation d'Euler compressible, *Japan J. Appl. Math.* **3**, 2, pp. 249–257.

Masaki, S. (2007). Semi-classical analysis for Hartree equations in some super-critical cases, *Ann. Henri Poincaré* **8**, 6, pp. 1037–1069.

Maslov, V. P. and Fedoriuk, M. V. (1981). *Semiclassical approximation in quantum mechanics, Mathematical Physics and Applied Mathematics*, Vol. 7 (D. Reidel Publishing Co., Dordrecht), translated from the Russian by J. Niederle and J. Tolar, Contemporary Mathematics, 5.

Merle, F. and Vega, L. (1998). Compactness at blow-up time for L^2 solutions of the critical nonlinear Schrödinger equation in 2D, *Internat. Math. Res. Notices*, 8, pp. 399–425.

Métivier, G. (2004a). Exemples d'instabilités pour des équations d'ondes non linéaires (d'après G. Lebeau), *Astérisque*, 294, pp. vii, 63–75.

Métivier, G. (2004b). *Small viscosity and boundary layer methods*, Modeling and Simulation in Science, Engineering and Technology (Birkhäuser Boston Inc., Boston, MA), theory, stability analysis, and applications.

Métivier, G. (2005). Remarks on the well-posedness of the nonlinear Cauchy problem, in *Geometric analysis of PDE and several complex variables, Contemp. Math.*, Vol. 368 (Amer. Math. Soc., Providence, RI), pp. 337–356.

Métivier, G. and Schochet, S. (1998). Trilinear resonant interactions of semilinear hyperbolic waves, *Duke Math. J.* **95**, 2, pp. 241–304.

Moyua, A., Vargas, A. and Vega, L. (1999). Restriction theorems and maximal operators related to oscillatory integrals in $\mathbb{R}^3$, *Duke Math. J.* **96**, 3, pp. 547–574.

Nakanishi, K. and Ozawa, T. (2002). Remarks on scattering for nonlinear Schrödinger equations, *NoDEA Nonlinear Differential Equations Appl.* **9**, 1, pp. 45–68.

Nier, F. (1996). A semi-classical picture of quantum scattering, *Ann. Sci. École Norm. Sup. (4)* **29**, 2, pp. 149–183.

Ozawa, T. (1991). Long range scattering for nonlinear Schrödinger equations in one space dimension, *Comm. Math. Phys.* **139**, pp. 479–493.

Pitaevskii, L. and Stringari, S. (2003). *Bose-Einstein condensation, International Series of Monographs on Physics*, Vol. 116 (The Clarendon Press Oxford University Press, Oxford).

Rauch, J. and Keel, M. (1999). Lectures on geometric optics, in *Hyperbolic equations and frequency interactions (Park City, UT, 1995)* (Amer. Math. Soc., Providence, RI), pp. 383–466. See also *Lectures on nonlinear geometric optics,* available at http://www.math.lsa.umich.edu/~rauch/courses.html

Reed, M. and Simon, B. (1975). *Methods of modern mathematical physics. II. Fourier analysis, self-adjointness* (Academic Press [Harcourt Brace Jovanovich Publishers], New York).

Robert, D. (1987). *Autour de l'approximation semi-classique, Progress in Mathematics*, Vol. 68 (Birkhäuser Boston Inc., Boston, MA).

Robert, D. (1998). Semi-classical approximation in quantum mechanics. A survey of old and recent mathematical results, *Helv. Phys. Acta* **71**, 1, pp. 44–116.

Sacchetti, A. (2005). Nonlinear double well Schrödinger equations in the semiclassical limit, *J. Stat. Phys.* **119**, 5-6, pp. 1347–1382.

Schwartz, J. T. (1969). *Nonlinear functional analysis* (Gordon and Breach Science Publishers, New York), notes by H. Fattorini, R. Nirenberg and H. Porta, with an additional chapter by Hermann Karcher, Notes on Mathematics and its Applications.

Serre, D. (1997). Solutions classiques globales des équations d'Euler pour un fluide parfait compressible, *Ann. Inst. Fourier* **47**, pp. 139–153.

Sjöstrand, J. (1982). Singularités analytiques microlocales, in *Astérisque*, Vol. 95 (Soc. Math. France, Paris), pp. 1–166.

Sone, Y., Aoki, K., Takata, S., Sugimoto, H. and Bobylev, A. V. (1996). Inappropriateness of the heat-conduction equation for description of a temperature field of a stationary gas in the continuum limit: examination by asymptotic analysis and numerical computation of the Boltzmann equation, *Phys. Fluids* **8**, 2, pp. 628–638.

Sparber, C., Markowich, P. A. and Mauser, N. J. (2003). Wigner functions versus WKB-methods in multivalued geometrical optics, *Asymptot. Anal.* **33**, 2, pp. 153–187.

Stein, E. M. (1993). *Harmonic analysis: real-variable methods, orthogonality, and oscillatory integrals, Princeton Mathematical Series*, Vol. 43 (Princeton University Press, Princeton, NJ), with the assistance of Timothy S. Murphy, Monographs in Harmonic Analysis, III.

Strauss, W. A. (1974). Nonlinear scattering theory, in J. Lavita and J. P. Marchand (eds.), *Scattering theory in mathematical physics* (Reidel).

Strauss, W. A. (1981). Nonlinear scattering theory at low energy, *J. Funct. Anal.* **41**, pp. 110–133.

Sulem, C. and Sulem, P.-L. (1999). *The nonlinear Schrödinger equation, Self-focusing and wave collapse* (Springer-Verlag, New York).

Szeftel, J. (2005). Propagation et réflexion des singularités pour l'équation de Schrödinger non linéaire, *Ann. Inst. Fourier (Grenoble)* **55**, 2, pp. 573–671.

Tao, T. (2006). *Nonlinear dispersive equations, CBMS Regional Conference Series in Mathematics,* Vol. 106 (Published for the Conference Board of the

Mathematical Sciences, Washington, DC), local and global analysis.

Taylor, M. (1981). *Pseudodifferential operators, Princeton Mathematical Series*, Vol. 34 (Princeton University Press, Princeton, N.J.).

Taylor, M. (1997). *Partial differential equations. III, Applied Mathematical Sciences*, Vol. 117 (Springer-Verlag, New York), nonlinear equations.

Teufel, S. (2003). *Adiabatic perturbation theory in quantum dynamics, Lecture Notes in Mathematics*, Vol. 1821 (Springer).

Thirring, W. (1981). *A course in mathematical physics. Vol. 3* (Springer-Verlag, New York), quantum mechanics of atoms and molecules, Translated from the German by Evans M. Harrell, Lecture Notes in Physics, 141.

Thomann, L. (2007). Instabilities for supercritical Schrödinger equations in analytic manifolds, archived as arXiv:0707.1785.

Tsutsumi, Y. (1987). L^2–solutions for nonlinear Schrödinger equations and nonlinear groups, *Funkcial. Ekvac.* **30**, 1, pp. 115–125.

Tsutsumi, Y. and Yajima, K. (1984). The asymptotic behavior of nonlinear Schrödinger equations, *Bull. Amer. Math. Soc. (N.S.)* **11**, 1, pp. 186–188.

Weinstein, M. I. (1985). Modulational stability of ground states of nonlinear Schrödinger equations, *SIAM J. Math. Anal.* **16**, 3, pp. 472–491.

Whitham, G. B. (1999). *Linear and nonlinear waves, Pure and Applied Mathematics* (John Wiley & Sons Inc., New York).

Xin, Z. (1998). Blowup of smooth solutions of the compressible Navier-Stokes equation with compact density, *Comm. Pure Appl. Math.* **51**, pp. 229–240.

Yajima, K. (1979). The quasiclassical limit of quantum scattering theory, *Comm. Math. Phys.* **69**, 2, pp. 101–129.

Yajima, K. (1987). Existence of solutions for Schrödinger evolution equations, *Comm. Math. Phys.* **110**, pp. 415–426.

Yajima, K. (1996). Smoothness and non-smoothness of the fundamental solution of time dependent Schrödinger equations, *Comm. Math. Phys.* **181**, 3, pp. 605–629.

Zakharov, V. E. and Shabat, A. B. (1971). Exact theory of two-dimensional self-focusing and one-dimensional self-modulation of waves in nonlinear media, *Ž. Èksper. Teoret. Fiz.* **61**, 1, pp. 118–134.

Index